GO: ON THE GEOGRAPHIES OF GUNNAR OLSSON

GO: On the Geographies of Gunnar Olsson

Edited by

CHRISTIAN ABRAHAMSSON
Lund University, Sweden

MARTIN GREN
Linnaeus University, Sweden

ASHGATE

Published by
Ashgate Publishing Limited
Wey Court East
Union Road
Farnham
Surrey, GU9 7PT
England

Ashgate Publishing Company
Suite 420
101 Cherry Street
Burlington
VT 05401-4405
USA

www.ashgate.com

British Library Cataloguing in Publication Data
GO : on the geographies of Gunnar Olsson.
1. Olsson, Gunnar, 1935– 2. Geography. I. Abrahamsson, Christian. II. Gren, Martin. III. Olsson, Gunnar, 1935–
910–dc22

Library of Congress Cataloging-in-Publication Data
Go : on the geographies of Gunnar Olsson / by Christian Abrahamsson and Martin Gren.
    p. cm.
Includes bibliographical references and index.
    ISBN 978-1-4094-1237-3 (hardback : alk. paper) – ISBN 978-1-4094-1238-0 (ebook : alk. paper) 1. Geography – Philosophy. 2. Human geography – Philosophy. 3. Cartography – Philosophy. 4. Olsson, Gunnar, 1935– I. Abrahamsson, Christian. II. Gren, Martin.
    G70.G53 2012
    910.92–dc23

                                                                                    2011040155

ISBN 978-1-4094-1237-3 (hbk)
ISBN 978-1-4094-1238-0 (ebk)

MIX
Paper from
responsible sources
FSC
www.fsc.org
FSC® C018575

Printed and bound in Great Britain by the MPG Books Group, UK.

Contents

List of Figures

Notes on Contributors

Christian Abrahamsson is a Lecturer at the Department of Geography, Lund University (Sweden).

Trevor J. Barnes is Professor and Distinguished University Scholar, Department of Geography, at University of British Columbia (Canada).

Alessandra Bonazzi is Associate Professor, Dipartimento di Discipline della Comunicazione, at University of Bologna (Italy).

Michael Dear is Professor of City and Regional Planning, College of Environmental Design, at University of California, Berkeley (USA).

Marcus A. Doel is Professor of Human Geography, Department of Geography, at Swansea University (Wales).

Franco Farinelli is Professor, Dipartimento di Discipline della Comunicazione, at University of Bologna (Italy).

Reginald G. Golledge (1937–2009) was Professor of Geography, Department of Geography, at University of California, Santa Barbara (USA).

Martin Gren is Associate Professor, School of Business and Economics, Linnaeus University (Sweden).

Jette Hansen-Møller is Associate Professor, Forest and Landscape, at University of Copenhagen (Denmark).

David Jansson is Associate Professor, Department of Social and Economic Geography, at Uppsala University (Sweden).

Gunnael Jennson is the alias of Gunnar Olsson and Ole Michael Jensen.

Ole Michael Jensen is Associate Professor, Statens Byggeforskningsinstitut, at University of Aalborg (Denmark).

Tom Mels is Associate Professor, Department of Human Geography, at Gotland University College (Sweden).

Gunnar Olsson is Professor Emeritus, Department of Social and Economic Geography, at Uppsala University (Sweden).

Chris Philo is Professor of Geography, Geographical and Earth Sciences, at University of Glasgow (Scotland).

Michael J. Watts is Class of 1963 Professor of Geography and Development Studies, at University of California, Berkeley (USA).

Acknowledgements

Editing a book is one of the most pleasurable and rewarding things one can ever do in academia, but it can also at times be a quite frustrating endeavour. As we are now, at last, completing this project we are truly sorry and apologetic for the delays. Yet, we are also extremely happy for the newly born. From the bottom of our hearts we would like to thank all the authors for their contributions. We must also here take the opportunity to express our gratitude to *Valerie Rose* at Ashgate, who made it all possible, and the two anonymous referees involved in the early stage. *Pion* must be acknowledged for granting us copyrights, and then there were a number of people who helped us along the way in various ways: *Sara Westin, Birgitta Olsson, Brett Christophers, Jamie Doucette, David Jansson*, and *Richard Peet*. The final acknowledgement must however go to *Gunnar Olsson*, and it has to be in Swedish: *Tack Gunnar för all oumbärlig inspiration, stöd och uppmuntran under åren!*

Copyrights

BILL VIOLA
Surrender, 2001
Color video diptych on two plasma displays mounted vertically on wall
80 2/5 × 24 x 3 1/2 in (204.2 cm × 61 cm × 8.9 cm)
Performers, John Fleck, Weba Garretson
Photo: Kira Perov

"Geography 1984," Department of Geography, University of Bristol, Seminar Paper, Ser. A, No. 7, 1967, pp. 1–28. (Copyright: Gunnar Olsson)
"Inference Problems in Locational Analysis," in K.R. Cox and R.G. Golledge (eds): *Behavioral Models in Geography: A Symposium*. Evanston, IL: Northwestern University Press, 1969, pp. 14–34. (Copyright: Gunnar Olsson)
"Servitude and Inequality in Spatial Planning: Methodology and Ideology in Conflict," *Antipode*, Vol. 6, No. 1, 1974. (Copyright: Richard Peet)
"–/–", in Peter Gould and Gunnar Olsson (eds): *A Search for Common Ground*. London: Pion, 1982, pp. 223–231. (Copyright: Pion Limited)
"The Social Space of Silence," *Environment and Planning D: Society and Space*, Vol. 5, 1987, pp. 249–262. (Copyright: Pion Limited)
"Chiasm of Thought-and-Action," *Environment and Planning D: Society and Space*, Vol. 11, 1993, pp. 279–294. (Copyright: Pion Limited)

"Projection of Desire/Desire of Projection," English original of "Die Projektion des Begehrens/Das Begehren der Projektion," in Dagmar Reichert (Hrsg): *Räumliches Denken*. Zürich: vdf Hochschulverlag AG an der ETH Zürich, 1996, pp. 225–249. (Copyright: Gunnar Olsson)

"Mapping the Forbidden," *Fennia*, Vol. 188, 2010, pp. 3–10. (Copyright: Gunnar Olsson)

Gunnael Jensson: "*Mappa Mundi Universalis*: A Commentary on the Power of Cartographical Reason," Uppsala: Uppsala International Contemporary Art Biennal, 2000. (Copyright: Gunnar Olsson and Ole Michael Jensen)

Part A
INTRODUCTION

Chapter 1
Preamble

Christian Abrahamsson and Martin Gren

MG: Now we are almost done, except for this introduction that we have saved for last. My suggestion is that we do it as a short "preamble", and in the form of a dialogue between the two of us. What do you think?

CA: In more ways than one I think that beginning with a dialogue is appropriate, particularly as the dialogic form comes very close to the pedagogy that Gunnar has been espousing throughout his career. It is almost a form of maieutics – if this has to do with the fact that his mother was a midwife I cannot say. But there is, I think, something in the dialogue that comes very close to defining his geography. In one way I guess one could define it as a form of communication, of a drawing together or of an existence in the limit *between* Buber's I and thou. This sensibility is, one could argue, present throughout his work. It is there in its embryonic form in the early work on human interaction, easier to detect in the deconstruction of the equal sign and reaching its maturity in the dialectics of individual and society. Whether or not the past two decades work on the cartographic(al) reason can be described as an attempt to synthesize the minimalistic figures of "/" and "–" is another question.

MG: Yes, that is another question. But a desire to grasp the world of understanding in minimalistic ways, to boil it down and compress it to a minimum, that is certainly a trade mark of Gunnar's geographies. In terms of dialogue one could describe much of his work as a set of literary, personal reports on the conversations he has had with the works of other persons in a quest not only for understanding but for understanding the understanding. That is also a form of communication, although in his work "to make common" certainly appears as a profound problem rather than as a solution. When we plunge into the "I and thou" we are also bound to meet those forces of the collective which, in Gunnar's country, are opposed to the uniqueness of individual originals. In fact, I suppose, that this feeling is not unlike what we are experiencing now when we are trying to turn his geographies and this book into common property. Anyway, since you were the one who once started the project, and then later asked me to join as co-editor, maybe you should say something about the background of the book?

CA: My first thought was quite simply to give Gunnar a gift. To gather together a group of people who have benefited from gifts he has given to them. In other words, to take seriously the fact that intellectual work is first and foremost a gift economy à la Marcel Mauss; an economy in which a gift has to be reciprocated. I think that Gunnar would say that this reciprocation has already taken place. Though that might well be true, I nevertheless thought that a book was needed.

MG: I suppose a "gift" is an example as good as any of the world of human intangibles which Gunnar's invisible geography is concerned, if not obsessed, with. Thus it is pertinent that the term seems to stem from a Scandinavian root, where it means "marriage" as well as "poison"; and you and I have both been "gifted" by his work, though some critics may perhaps say "poisoned" instead. Maybe we should mention something about our respective relations with him, even though there are contributions in the book that give personal accounts. What is your first memory of Gunnar?

CA: My first memory of him is that he was late, which is indeed very uncharacteristic of him. He was supposed to give a lecture but never showed up. Eventually the departmental secretary phoned him and soon after he arrived. What then followed was a lecture that literally changed the course of my studies or more precisely my formation: "lure[d] … into the forbidden grounds … the romantic comedy [has] proceed[ed]" ever since. (Olsson 1982: 264). I had previously planned to become a political theorist, intellectual historian or philosopher but after Gunnar's lecture I realized that I could stay in geography (though staying in Swedish geography has proven more difficult than I could foresee). It hadn't occurred to me before that geography could be philosophy or intellectual history. Then a number of years went by without much contact before we happened to be in northern California at the same time, he at Stanford and I at Berkeley. This was the beginning of a closer relationship and soon after returning to Sweden Gunnar became my supervisor. How about your first encounter with Gunnar?

MG: It was when I visited him in Uppsala, as a newly enrolled PhD student in the late 1980s. My supervisor had mentioned at some point that there was this "maverick" Swedish professor whose work might be of interest and beneficial for a guy like me. Eventually, I sent Gunnar a letter and received a postcard in return, in which he invited me to his home for a talk since "we seem to have common interests". After a train ride from Göteborg, where I desperately tried to re-read as much as possible of *Birds in Egg/Eggs in Bird* in order to prepare myself for the meeting, I arrived at around 11.00 in the morning, possibly in 1989, to his house. Gunnar opened the door, invited me in, and we sat down in two leather chairs for coffee and biscuits. Small talk began, but it did not take long before Gunnar half-closed his eyes and then approached the somewhat nervous student with his question: "Do you know what self-reference is?" Since I had read Gregory Bateson and some systems-theory, I knew a little about the problems involved when Cretans truthfully claim that they are liars. What came out of my mouth was at least sufficient enough for being accepted as a regular visitor to the Olssonian retreat. Gunnar did not become my supervisor; we just met on a regular basis and talked. After I finished my PhD, we have continued to do so. And I have continued to be grateful, "less for the sake of understanding, more for the sake of living" (Olsson 2007: 437), if these two endeavors could ever be separated. In Gunnar's house, and with Birgitta (his wife), they can't.

But enough of personal confessions, let's bring in the others. What one often does in an introduction is to present the authors and their contributions, but I feel that it would be better to just let them speak for themselves in and by their own contributions. What do you think?

CA: We both should and should not. Often editors see it as their role to explain, interpret or just plainly describe what the various authors of a collection *actually* have written. There is not necessarily anything wrong with that, but I think that doing so in this particular book would not only be foolish but would in some ways also constitute a type of violence. All forms of translation are, of course, potentially violent, but I think we should leave it to the readers themselves to commit that kind of transgression.

MG: What you are also saying is that we should abstain from introducing the authors. But could you briefly describe how they were selected?

CA: In a way they chose themselves, in another they were of course selected by us. I think that early on we both agreed that we wanted to include former students, friends and critics. To our good fortune almost everyone we asked was willing to participate. One could say that the authors, as the geographical beings they are, neatly capture some of the primary coordinates in Gunnar's own geographies: Uppsala, Ann Arbor, Bologna, Berkeley, Copenhagen, with outliers in Glasgow, Vancouver, Visby, Swansea, Santa Barbara and Iceland.

MG: Do you think that the contributions give a reasonably representative mapping of Gunnar's geographies? Is something missing perhaps?

CA: Not only some things, but also some people. I have in mind friends and comrades like Allan Pred, Dagmar Reichert, José Ramirez and Peter Gould. I am also certain that the book can be critiqued for not engaging more fully with Gunnar's work during the quantitative period.

MG: And maybe the book could be considered not critical enough of Gunnar's geographies?

CA: I am sure it can be critiqued for that reason, though I am not sure how to respond. Perhaps one answer could be that the book certainly contains a critique but not a criticism of Gunnar's work. Even though the difference between these two attitudes is not always obvious, I think that it is an important one to make, especially as we wanted to put together a book that contained careful readings of his work– readings that in a way build on, or extend, a certain tone, a sensibility which can easily evaporate. This is, once again, a question of the dialogic form, expressed in the form of trust. I am talking about the form of trust that can emerge if one doesn't presuppose faulty reasoning, bad faith or, worse, that the author is trying to manipulate the world.

MG: Alright. But let us not go into details about the politics of reading here. Since the two of us are both Swedish and aligned with Anglo-Saxon variants of geography, should we try and place Gunnar's work in that context?

CA: Well, you are, of course, familiar with the joke that Sweden's contribution to the Enlightenment was that Descartes died here in 1650, and that we topped this stellar reputation by not granting Michel Foucault a doctorate at Uppsala in the 1950s. Although these stories are more complex and although there are some good examples of speculative thought in Sweden, the angelic Emanuel Swedenborg perhaps foremost among them, I still think that Sweden is, for good or bad, one of the most utilitarian countries in the world. I think this is very interesting, particularly if one considers the fact that Sweden has never had a great philosophical tradition – as has Denmark, Finland and Norway. Little wonder then that Gunnar probably has had fewer students in Sweden than in Denmark, Finland and Norway. You can say much more than I on this, as you have done works also on Torsten Hägerstrand and time-geography, but I sometimes think of Gunnar and Torsten as representing Swedenborg and Linnaeus. One lived his life in exile, while the other was ennobled.

MG: What the geographies of Gunnar and Torsten have in common, as I understand them, is that both have been guided by a desire to excavate the bare onto-epistemological skeleton of geo and graphein. Exile and nobility might just come as parts of such a project, and also Gunnar, in search of the intangible skeleton of invisible truth, may well be considered a hybrid of Linnaeus and Swedenborg. But let us not turn ourselves into preachers, therapists or missionaries and bait more traps. Shouldn't we, finally, say something about how Gunnar's texts were selected?

CA: Well, we wanted to include texts from every decade of his career and give the reader a sense of both continuities and discontinuities in his work. We also wanted to make available for the first time texts that are either previously unpublished or published in another language. This book also offered an opportunity to gather texts, although they may not be difficult to find, that have never been previously anthologised. Apart from that there really isn't much left to say other than that I hope the forbidden grounds will be tempting not only to the two of us but to others as well, thereby guaranteeing that the romantic comedy will go on.

Chapter 2
Gunnar Olsson and Humans as Geo-Graphical Beings

Martin Gren

This chapter is a mapping of those ingredients in the geographies of Gunnar Olsson which I myself take to be major contributions in, of, at, and to the understanding of humans as geo-graphical beings. In addition, I hope to provide a general introduction to his geographies. After a throw of dice, the fix-point presented itself: "What does it mean to be human?" (Olsson 2007: 4).

Like a restless point this question has been spinning around on the plane of Gunnar Olsson's geographies, weaving and leaving behind a set of lines which may be interpreted as visible traces *of* invisible relations *in* his own imagination. Yet, Olsson's inability to suppress his desires to speak and write from any other *topos* than his own, most often explicitly using his own I as a de-materialized fix-point for mapping expeditions in and of the human territory of thought, should certainly not be mistaken for a solipsistic geography of private affairs.

When the chips are down, also you and I must find our ways through the unknown of whatever has not yet happened. As we traverse the shallow surface of a dematerialized human territory we too have to rely on navigational tools which ensure that we do not get lost when we encounter what we never encountered before. We too will have to depend on devices and procedures that enable us to sail across the sea also when its waves are made not of water but of promises, trust, power, love, reasoning, affects, fears, hopes and all other kinds of invisibles that constitute human interaction. As human beings negotiating various trails on life's way, within our own internal, untouchable, invisible subsistence, we must all be identifiable, predictable, obedient, and, yes, also creative, inventive, imaginative.

Human, all thoroughly human. And when reflecting on how our doings and territories are braided together, we are in actuality in the company of geographer Olsson. Like anyone else, we are then engaging with the strangely familiar combination of something "too evasive to capture" and "too forbidden to leave alone", that "godly question as impossible to answer as not to pose" (Olsson 2007: 4).

In the meantime, we live forwards and understand backwards, inevitably moving from the here and now to the then and there.

From the Earth to space

In 1957 Olsson boarded the train that took him to Uppsala University, where since 2000 he is professor emeritus of geography (Olsson 2002: 245). In-between, and up to this day, he has been on what he himself is most likely to describe as a lifelong journey of self-conscious reevaluations. Marked by a refusal to limit himself to studies that seek referential anchorage solely in visible things, a widespread tendency in modern human geography, Olsson has continued throughout the years to pursue his own theme of *human interaction in search of its geographical essences.*

To simply state, or to accept at face value the taken-for-granted relations between proper names and definite descriptions, or to staunchly claim that a *is* b, has perhaps never been a real option for Olsson. Trained in geography at the time when spatial science and quantitative geography was marketed as the new road to salvation, with a promise to turn the discipline into a hard science proper, Olsson early insisted that the connections between the variables, measurement data, and parameter values are more important than what these entities stand for. To him the relation has always mattered more than the related: Is human interaction to be conceptualized as a function of spatial behavior or as an outcome of spatial distributions? What is the dependent and what is the independent variable? How can one distinguish effect from cause, cause from effect? What came first, the birds or the eggs?

Olsson early knew that he wanted to use a quantitative approach, one of the formative moments being his reading of a licentiate thesis by Esse Lövgren, "a brilliant man obsessed with the idea of translating the vagaries of human behavior into the precise language of mathematics" (Olsson 2002: 245). Eventually, Olsson himself was to become a "space cadet", a young man deeply involved in the search for the hidden laws of spatial interaction. The theoretical issue was focused on how concrete earthly human behavior, such as migration, can be translated into the precise language of mathematics and the abstract language of space, the latter most often operationalized as distance. In the making was not only geography as a social science, but a social science of a particular kind:

> for what is the glue of the social if it is not human interaction, and what is the
> fundamental variable of geography if it is not distance? (Olsson 2002, 246.)

The obvious objective of the interaction models was that they should lead to more knowledge about human spatial behavior. However, experiments showed that the parameter values, which were designed to yield an understanding of such behavior, might have more to do with spatial origins and destinations than with human action itself. Also Olsson's own studies, which were to form the basis of what was then a double doctorate in Sweden (his licentiate and doctoral theses in 1965 and 1968), suggested that the spatial interaction models which were intended to catch migration over distance in fact revealed more about the spatial distribution of opportunities than about human interaction *per se*. As he later concluded:

> In short, it seems (a) that travel reflects the spatial distribution of opportunities; (b) that geographic space is a prison; (c) that the models predict well. The implication is that neither gravity nor entropy probes beneath the surface features of revealed preference. They are blind to the deep structure of institutional facts. (Olsson 1980: 197b.)

Although the spatial interaction models were supposed to tell stories about human interaction, they seemed instead to reflect the surface features of the spatial distribution of opportunities. In being "blind to the deep structure of institutional facts" they were measuring map rather than behavior, the external geometry rather than the internal logics of the social. In other words, at stake was the classic geographic inference problem of form and process, a problem which has continued to be fundamental for Olsson. For good reasons too, situated as it is at the very heart of geography. In his own words, geographers have since time immemorial:

> marched together in a kind of methodological two-step: first the investigated phenomenon is translated into the graven image of a map; then that idol is interpreted in terms of generating processes. Picture turns to story, matter to meaning, here to thus. (Olsson 2002: 248.)

In spatial science the geographic inference problem was slightly reformulated into issues about the relations between mapped *spatial* forms and the interpretation of their generating processes. However, serious problems were soon to bubble up in the muddy waters that the translation of the Earth and human interaction into the language of space had stirred up. According to Olsson, both his empirical studies and spatial science more generally had led to many analytical difficulties, especially the fact that the same spatial form can be generated through drastically different processes. Leaving here all technicalities aside, the suspicion grew that not even the most perfect description of a spatial form can be used as the basis for conclusions about how and why that form has come about. Although it is sometimes possible to reason from process to form, the opposite, so Olsson argued, is never appropriate. Based on evidence and findings the conclusion was as inevitable for him as it was devastating for human geography as a spatial science. One cannot make valid inferences about human behavior from mapped spatial distributions, or, in other words: one cannot reason from form to process.

The story could have ended, as a frozen idol in remembrance of the deity of geography as a quantitative spatial science, or of a group of worshippers caught in the net of their own creations. Yet, an important part of the story is the relationship which existed in those days between spatial science and applied social science. The idea that was abstract scientific theory should merge with concrete human practice, through social engineering, and all for the betterment of the citizens of the modern welfare state. That ambition, however, came with a problem because the relations between knowing the world through scientific models and the intentional politics of changing it on the basis of such models are far from clear cut.

A key to planning the future was supposed to be embedded in the specification of invariant relations, not only between theory and practice, but more precisely between cause and effect. If one could specify causal relations between a spatial distribution and its future social consequences, then one ought to be on safe grounds when one leapt into planning. As noted, though, Olsson's own work had suggested that even from a perfect description of a spatial form it is impossible to say anything definite about the processes which have generated it. It follows that the goal of foreseeing the future is also severely hampered. In addition, according to the principles of causal theory the consequences of actions based on observed autocorrelations cannot be foreseen, which adds to the difficulties in predicting the social consequences of a deliberate manipulation of spatial form. A delicate problem indeed, not the least because according to the principles of utilitarian ethics purposeful action should be judged not in terms of its good intentions, but precisely in terms of its consequences.

Based on evidence from human geography as spatial science Olsson concluded that to argue for planning based on spatial interaction models was scientifically, ethically and politically questionable. This conclusion was however quite at odds with the practice of social engineering in the 1960s and 1970s, particularly as it was carried out in Sweden, where a basic idea of the prevailing ideology was that a better and more just society should be constructed on the basis of exact scientific knowledge. Expressed in the ethos of spatial science:

> laws of social gravity should enable us to construct a human world of justice and efficiency. (Olsson 1991a: 186.)

The modern message was that applied geography could and should contribute to the planning of society with scientific models of human spatial behavior. Of crucial importance for Olsson was that there were instances in Sweden where planning was actually based on practical applications of the gravity model, one being a reform aimed at the revamping of administrative and political districts that was carried out during the early 1970s. An overarching purpose of that reform was to create spatial units large enough to carry the considerable burdens which were inevitable if the utopian ideal of modern welfare was to be realized. That every citizen should have the same opportunities and the same access to welfare services (Olsson 1977, 1980: 201b–205b) sounded all very reasonable in the realm of well-intended planning and politics. However, when thought was transformed into action, quite another picture emerged. According to Olsson's analyses this was directly related to the methodology through which the new degree of equality was to be implemented.

Most importantly, the new and larger spatial areas were designed to cause as little disturbance to the people as possible. The delimitations were therefore made to reflect observed spatial interaction patterns as they revealed themselves through a variant of the gravity model. Data were then fitted onto what was a particular version of the Pareto function, and what eventually evolved was a set of efficient boundaries to which the administrative and political areas were to be adjusted. The problem with Pareto's optimality principle, however, is that it defines an efficient equilibrium solution as a state in which no person can be made better off, unless someone else at the same time is made worse off. To put it crudely, a social distribution in which all resources are owned by one single individual is then an efficient equilibrium solution. If this principle is adopted in planning then the risks are obvious, because what *is* an efficient solution for the one easily slips into something to which the many *ought* to be adjusted.

The problem with the practiced methodology, in Olsson's view, was that it reflected the very opposite of the ideal of human equality because it preserved the status quo of the inequality inherent in the mapped distributions. Consequently, the spatial reform seemed to be "far more geared towards the growth and maintenance of its own bureaucracy than towards the interest of those sick and disadvantaged which it is supposed to serve" (Olsson 1977: 357–358). This, he claimed, was a serious case of "faulty reasoning", because when operationalizing and applying a variant of the gravity model the planners were in effect equating the pursuance of logical scientific reasoning and empirical truth with the advocacy of an elitist theory of society. It was this "profound misunderstanding of scientific methodology" which "led to the further institutionalization of those inequalities which the plans were meant to lessen" (Olsson 1980: 203b). A somewhat paradoxical state of affairs, since several of the people involved in Sweden were themselves geographers. It seems reasonable to assume that at least they should have known better and realized that embedded in the language of the Pareto function, in the optimality principle, and in the theory of the elites, was exactly the system of inequality and servitude which the plans were meant to alleviate.

When applied, the gravity model served to conserve rather than diminish existing inequalities, an outcome that was clearly contrary to political intention. For Olsson, this came to be a prime example of planning not only as a special case of human "thought-and-action", but as an instance of "Greek tragedy." Although the reform was supposed to pave the way to a better and more just society, in the course of implementation when thought was translated into action, the good intentions turned into tragic results. No one was to blame, though, because during the journey the actors had done nothing particularly wrong, merely remained faithful to their own internal rules of reasoning.

So it is that in the formal language of the gravity model, in its application in planning, and, most importantly, in the (mis)translations in-between, a host of ethical and power-filled issues were embedded. For Olsson, both continuing and extending his original studies of human interaction, these insights all pointed towards the relations between the individual and the collective. Hence his oft-repeated question: "By which forces are we made so obedient and so predictable?" (Olsson 1991b: 85).

From space to language

In 1966 Olsson accepted a job offer from the University of Michigan, where he was soon promoted to full professor, a sign both of the times and how his work in spatial science was being valued (he completed his Swedish doctorate in 1968). In what was to be eleven years at the Department of Geography at Ann Arbor, he continued to investigate human interaction in relation to theory, model, and reality. Through in-depth engagements with the philosophy and theory of science, Olsson became occupied with alternative logics, dialectics, epistemology, ontology, ethics, aesthetics, power and the like. What had initially started as an attempt to project earthly human behavior onto the reference plane of space was now being transformed into a concern with *language* as a general medium for human interaction.

As a runaway spatial scientist Olsson was to make contributions to a renewed humanistic geography, as he was one of the very first geographers to engage systematically with language. Thinkers like Jacques Derrida and Jacques Lacan, whom he engaged with early, were later to appear in the hands of a younger generation of (post)modern geographers who were busy exploring issues of language and representation. Yet, Olsson, on his way towards the conclusion that there is no point of understanding outside of the internal limits of language, refused to side either with modern certainty or with postmodern ambiguity. Gradually adopting the view that the human condition is a constant struggle between opposites, he eventually became a practicing dialectician seeking in particular to place himself – like his favorite God *Janus* – in the forbidden space of the excluded middle. Once there he began a long wrestling with paradoxes. At times he also performed experiments with, on, and at the limits of language.

As I hope to have shown, it was the work in spatial science, and specifically his experiments with the gravity model, that led Olsson to investigate the internal premises and reasoning rules which structure the models themselves. Reevaluating his practical engagements, and notably his own understanding, Olsson set out to unveil the seemingly objective methodologies of spatial science, including correlation and regression techniques, noting in particular how these formulations are based on a host of philosophical assumptions. This was eventually to lead him, quite famously, to claim that allegedly objective analyses, including those he himself had formerly been involved in, usually reveal more about the language they are written *in* than about the phenomena they are supposed to be speaking *about*.

In his book *Birds in Egg/Eggs in Bird*, Olsson explicated and summoned his evolving understanding of human interaction in thought, action, and language. Having moved beyond his previous attempts to translate human interaction, such as migration, into a particular spatial science language, for example variants of the gravity model, he now began to investigate the role of language as a medium for human interaction. As he formulated it in the beginning of the book: "Since language is the medium in which the mind operates, the issue is not the collection of facts but the communication of how these facts are ordered in the mind" (Olsson 1980: 6b).

A central premise of this work is that human language can be separated neither from thought nor from action, hence Olsson's frequent usage of the term "thought-and-action." Understanding language also means that the objects of observation necessarily also become the instrument of observation. Having thus opened the door to the circular chamber of self-reference, he realized that the prospects of conventional reasoning may be dim, because to achieve "that deep knowledge requires a metalanguage of a type we do not have and perhaps never will" (Olsson 1980: 7b). Nevertheless, Olsson would not be Olsson if he could resist the temptation to catch and to explicate "the reasoning rules which govern that thinking and speaking" (Olsson 1980: 7b). What went along was still the inference problem of form and process, but now slightly transformed and more stripped from an explicit inferential bond to the geographic and the spatial. In my own rephrasing: what is the *form* of thought and reasoning rules in language that governs human *processes* of thinking and speaking?

This challenge has remained and the desire has persisted. Launched as a prospect of "a series of investigations dealing with the psychology of space", in the synopsis accompanying the six studies comprising Olsson's doctoral dissertation (Olsson 1968: 5), the drive is evidenced also in his most recent book *Abysmal* (Olsson 2007). As for *Birds in Egg/Eggs in Bird*, it started out as an analysis of the logics, languages, and reasoning rules of spatial science and social science more generally. But part of the plot was already from the beginning in the idea that the internal domain of formal science is entangled in the external world of social affairs. This close relationship between world and language was to be repeatedly emphasized in *Birds in Egg/Eggs in Bird*, often in ways that reflected the author's engagements with Marxism and dialectics:

> The employed reasoning rules are internal relations which themselves reflect not only the phenomena we are dealing with but also our own lives and ways of thinking. (Olsson 1980: 18b.)

To Olsson, language was not an independent realm which exists on its own. Like the (late) Wittgenstein, he came to adopt and advocate the view that there is no objective independent point of support for thinking outside of language. Speaking and thinking are linguistic activities which have been acquired during millennia of socialization. How humans think and speak, the languages they use and the ideas they express, can therefore not be separated from the forms of life to which they belong, practice and express. Summarizing a broader trajectory of thought in the social sciences at that time, Olsson thus declared it to be:

> a central attitude that the world and our ideas are so entangled in each other that they cannot be separated. Another tenet is that social relations can only occur through the communication of ideas. It is in this manner that social

relations can be treated as internal relations, which in turn makes social relations between people like logical relations between propositions and logical relations between propositions like social relations between people. It follows that the subject matter of the social sciences is neither [hu]man[s] nor society but [hu]man[s] and society conceived relationally. This is crucial to understand, for it is by applying some particular rules of reasoning – rules which themselves are created, sustained, and changed as part of social relations between men [humans] – that we establish the logical relations between propositions. But those rules of reasoning are neither ethically nor aesthetically neutral, for "grammar tells us what kind of object anything is". (Olsson 1980: 7b–8b.)

Humans and society entangled, like the birds and the eggs. The objects and rules of reasoning that grammar permits will eventually have consequences also for the society that humans build and the social relations through which they live their lives. The science that they construct will inevitably mirror the language they use, and the language they practice will be reflected in whatever systems of representations they bring forth. For Olsson there was but a thin, and not neutral, permeable abyss that separated the logical relations between propositions and the social relations between people. When speaking about the negative exponential he could thus claim that:

Inherent in the truth of the negative exponential is consequently the authoritarian belief that society is always right, an observation which makes it easier to understand those forces which try to whip us all into the same thoughts-and-actions. But who are the masters of the formal axioms? And who are the slaves of the theorems? Does the cognitive order of thoughts become the order of things because the axioms are condensations of conventional truths? Is a language of extreme precision by necessity a language of extreme suppression because conventions without domination are as impossible as obligations without sanction? But is it the few over the many, or the many over the few? You or I, I or you? (Olsson 1980: 38e.)

So which translation is to be used, and whose order is to be privileged, in the ontological transformations between things and thoughts? If mean-values are internally privileged in the equations that translate human interaction, which would imply that the significance of small deviations of the few are simultaneously minimized, then what happens when one on that basis leaps over to the other side of action in society? Will one find a mirror-world of echoing reasoning rules that speak in the name of the collective but thereby also reduces the significance of the individual?

As could be expected, *Birds in Egg/Eggs in Bird* also drifted into questions of ontology. What emerged in particular was a kind of Cartesian understanding of human thought-and-action as embedded in two different ontological realms, by Olsson labeled "the physical and the mental." In turn, these realms appeared to be aligned with two different epistemologies, although his own warning should be kept in mind: "As any epistemologist knows, binary opposition is the first step towards dialectical transcendence" (Olsson 1980: 8e). Most importantly, the two realms may be understood as a mapping both of the human territory of thought-and-action *and* of its limits, thus reflecting what is a necessary and inevitable braiding of ontology and epistemology.

The physical realm is a world in which phenomena with physical extensionality *exist*. In principle, humans are able to taste, touch, smell, hear and see them, which means that these phenomena are relatively open for public inspection. Disputes about knowledge-claims can be both triggered and settled by pointing at these physical existences, or by projecting them onto a coordinate net in which time-space is used as a plane of reference. Thus the concept of representation can here be slowed down to the point where it starts to stutter, re-pre-sent, and thereby to show what it actually means: to once again present that which was given to the senses.

The other ontological realm that Olsson mapped in *Birds in Egg/Eggs in Bird* is the mental, which he by that time had become particularly interested in. Indeed, issues of this realm will continue to be central also in his later works, including his attempts to develop a cartography of thought and initiate a critique of cartographic(al) reason. In contrast to the physical realm, the invisible human mental realm is one not of existence but of *subsistence*. As a consequence we do not here find phenomena that exist as they do in the physical realm, but objects of thought that subsist in the human mind. Although these objects of thought, what Immanuel Kant referred to as "noumena", do not have physical extensionality, they nevertheless ground and sustain human life. As internally created and sustained by languaging humans, objects of thought do not exist apart from the acts that bring them forth. It follows that human subsistence can, as in *Birds in Egg/Eggs in Bird*, be understood in terms of verbal acts:

> And so it is that empires, prisons, and brothels are like all other human creations. Verbal acts. As a consequence, they stay alive only as long as the obligations inherent in those verbal acts are kept alive. When we cease to believe in a word, it no longer has the power. And when words lose their power, so do the institutions that are built upon them. (Olsson 1980: 12e.)

A distinguishing feature of subsistence in the invisible human realm is that objects of thought are difficult to empirically verify or falsify. If empirical knowledge entails the grasping of phenomena with a physical extensionality accessible by the five senses, then "empires, prisons, and brothels" are, strictly speaking, beyond reach. Existing as they are in the physical realm only as mere "bricks and mortar", they become part of human subsistence through naming and thus "verbal acts" are required. In turn, these acts will need to find their ways through social institutions like law, politics and religion. More formally expressed by Olsson:

the truth-value of propositions about existing objects are in the objects themselves, while the validity of propositions about subsisting objects are in the mind of the speaker. (Olsson 1980: 68b.)

Most importantly, for Olsson the ontological transformations that language enable and brings about were to become a key characteristic of human thought-and-action. And ontological transformations are always inseparable from power. One reason being that there is, in principle, almost an infinite possibility to name and to define, and therefore some kind of delimitation is sorely needed. So what is it that makes humans translate and name things-and-relations in certain ways? What is it that makes them so obedient and so predictable? As now could be expected, engaging in-depth with such questions was defined by Olsson as a "tremendous challenge, too intriguing to ignore." And that challenge was nothing less than to:

understand where it comes from, the power of those commanding thoughts and words we obey without hearing them. (Olsson 1980: 12e.)

In Olsson's mind, "power" is for humans inseparable from their language. It follows that relationships of power are also reflected in how internal relations in language are simultaneously held together and kept apart. Consequently, under the umbrella of language as a domain of internal relations and a medium for ontological transformations, Olsson analyzed and distilled in *Birds in Egg/Eggs in Bird* the internal structures of two languages; the "language of social science" and the "language of human action." Both idioms were crucially important in the context of his concerns, because each one revealed a particular *translation function of human interaction*. Taken together they illustrated essential aspects of the power-filled traffic of ontological transformations between theory and practice, description and prescription, thought and action.

The formal language of social science, which Olsson associated with the Galilean tradition, emanated primarily from the representation of phenomena which exist in the external, visible, touchable, and physical realm. Commanding in the background is therefore also the hidden, or taken-for-granted, assumption that identical words refer to, and should refer to, identical things. Yet, this also means that reasoning rules and truth statements depend on frozen, or stable, categories and definite descriptions. In accordance with the ideal of Liebniz's *salva veritate*, this is to ensure that meanings do not change when they travel to other contexts, or for that matter when they move between different individuals.

This circumstance provides the language of social science with certainty, and it does so by establishing stable descriptions and consensual communication about what is taken to be identical things. This further suggests that the language is clearly designed to privilege truth preservation, another way of saying that well-formed propositions are either true or false. In its methodology this type of language aims at causal inference, explanation and prediction. Yet, its pragmatics are in the goal of empirical description, which means that theory in practice tends to be a matter of formalizing law statements on the basis of an interpretative description of empirically observed things. For these reasons, explanation is in fact restricted to "the surface features of the external" (Olsson 1980: 47e).

The language of social science, as portrayed by the Olsson of that time, is a language in which time has been excluded or at least come to a full stop. Including time would mean that the presence of qualitative change would have to be acknowledged, if so with the unsettling consequence that the relationship between word and object may possibly change from time 1 to time 2, from context X to Y, or between me and you. Entrance forbidden, certainty at risk, and provided as security is the timeless logic of "if P then Q", regardless of the social context at hand. Of particular importance for Olsson, now orbiting both fast and slow in his dialectical spiral to the beat of the circulating drums of self-reference, was the insight that in the internal structure of conventional logic the middle is necessarily excluded. According to this timeless logic, if something is required to be *either* a *or* b, then it cannot at the same time be *both* a *and* b. Hence paradox, as well as tautology, are excluded from this language as threatening enemies to be expelled at all cost.

Matters are however quite different in the other language that Olsson portrayed and which he associated with the Aristotelian tradition. This language, of human action, is propelled by practical reasoning and is well equipped for addressing "particular issues as they arise in particular contexts" (Olsson 1980: 81b), a situation in which meaning must be allowed to change with context. Time, as qualitative change, is thus included in the language of human action, and the investigator is able to recognize that things may sometimes be *both* this *and* that. Being a language of and for practical action, its mode of reasoning is not restricted to the preservation of truth, in fact quite the opposite. Contrary to what the premises of truth-preservation in conventional logic stipulate, where the procedure is to travel faithfully and obediently from the given premises all the way to the inevitable conclusion, its internal structure allows for a reasoning that can make false whatever now happens to be true, thus reflecting what humans so often do in the course of life.

Of further importance for Olsson is that the language of human action and its practical reason are to be found in the domains of planning and politics. Most often the point of those practices is not to preserve what is true, but rather the opposite, that is, to make false what now is true in order to change and create something else that, eventually, will appear as true in the future, be it a new railroad or an alternative system of gender. The essential purpose of such practical reasoning is consequently not to infer the truth of what was, is or will be, but instead "to indicate to the actor what [she or] he should do in order to attain an intended result" (Olsson 1980: 81b). The goal of attaining a future state of affairs illustrates that the language of human action has teleological reasoning and understanding as its fundamentals, and also that it is deeply embedded in the domain of subsistence. In other words, this is a language which acknowledges that human interaction is constituted also through objects and relations of thought such as intentions, hopes, fears, and plans for the future.

In his analysis of the two languages, Olsson arrived at the conclusion that their respective purposes and subject matters are radically different. Just as the language of social science is characterized by causal reasoning, so the language of human action is characterized by practical reason. Different tools suitable for different purposes, and precisely therefore danger lies ahead:

> The languages of causal and practical reasoning are radically different. For this reason, it is dangerous to extend the language of descriptive social science into the realm of prescriptive social engineering. The point is that if we construct a society on the basis of current thought and practice, then we confine ourselves to a world which reflects the particular conception of [hu]man[s] inherent in the language of causal explanation. We will then impose on praxis a sterile strictness which it neither has nor ought to have. The result is a society that mirrors the language in which we describe it. (Olsson 1980: 61b.)

The story about the two languages illustrates some crucially important ingredients in Olsson's well known critical attitude towards planning and politics. If human thought-and-action is far more ambiguous than the certainty of conventional logic admits, then any transformation into action based on such a logic must be as problematic as reasoning from spatial form to process. To this could be added that just as potentially dangerous is, in Olsson's mind, the implementation of any ideology, regardless of brand and color, which reasons from aggregates to individuals or from the generalities of the collective to the specifics of you and me. As already noted, for him it all boils down to the relationship between society and the individual, the classical theme of social science which has reappeared in his work in many guises. For example:

> all political and social scientific theories at bottom concern one issue and one issue only. This is the relation of society and individual, which in other contexts is called the relation between the desire of the self as the same and the desire of the self as other. (Olsson 1980: 33e.)

At the end of *Birds in Egg/Eggs in Bird* Olsson folded back to the discipline of geography, and his verdict was summed up in quite harsh words:

> With its traditional stress on space, measurability and visual landscape, geography has committed itself to the surface features of the external. Since the external is in things rather than relations, we have produced studies of reifications in which man, woman, and child inevitably are treated as things and not as the sensitive, constantly evolving human beings we are. Our professional terms are seemingly well defined and the identities are as stable as any authoritarian ruler could demand. When we subsequently base our plans and actions on such knowledge,

we become bound to produce thingified people obedient to any social authority. (Olsson 1980: 47e.)

Yet, the bumpy road towards the secrets of human interaction in language had been only partly paved. In the first half of *Birds in Egg/Eggs in Bird* Olsson had arrived at the conclusion that crucial for language is categorization. And to categorize is to fix the meaning of words which requires some operation of distinction so that identity and difference can be established. If one wants to understand *a, b,* or both, then the place to travel is to *b, a,* or perhaps to both destinations. But if the aim is to understand how something could become and be transformed into something else, particularly in the realm of subsistence, then one must search elsewhere. This elsewhere, for Olsson, was not to be found in the related as such, but in the relation through which "things" are kept together and held apart.

In *Birds in Egg/Eggs in Bird* there was one invisible glue that came to occupy a particularly important place in Olsson's analysis of the drama of internal ontological relations and transformations: the equal sign. As the "ultimate symbol of the word *is*" (Olsson 2002: 258), the equal sign captured knowledge in its purest and most abstract form. Rephrased, the equal sign became to him the primary relational translation function that ontologically ties down human thought-and-action. Returning to the two languages outlined above, his analysis had also shown that even though both idioms pivot around the equal sign, they use it in quite different ways. In his own words:

> Thought and action are reflected in language and every language is defined by its equality [equal] sign. In the drama of reasoning, the equality [equal] sign plays two roles. One is to denote identity, the other to denote existence. In both cases, the purpose is to tie the reasoning down to an ultimate. In this sense, the equality sign serves the same function as the anchors of a ship. It grounds our thoughts and actions and keeps them from drifting into obscurity. (Olsson 1980: 64b.)

What finally hatched into the bird of social scientific reasoning was the ontological egg of Olsson's understanding of the human condition, a situation in which you and me are involved in a perpetual struggle between opposing forces. In an act of dialectical synthesis, he also made an attempt to reconcile the two different languages of social science and human action in an eye-like shaped depiction (Olsson 1980: 20e). Placed at its center was the kernel of "thought-and-action" which in turn was surrounded, traversed and held in place by lines of opposing forces: existence-subsistence, external-internal, logic-dialectics, practical reasoning-scientific reasoning, future-past, beliefs-facts. This "mandala of thought-and-action" was actually the first prototype of a set of similar models that were later to be developed further and appear in various guises (Olsson 1991a, 1991b, 1993, 2007). Crucially important for understanding his later work is the circumstance that the mandala was an explicit attempt to create something truly original. And so it was that Olsson, finally, set out to represent human thought-and-action:

> not from the outside of what others say it is, but from the inside of what I believe it is myself. (Olsson 1980: 20e.)

From language to a cartography of thought

In 1977 Olsson moved from Ann Arbor to Sweden, where he was appointed to the chair of economic geography and planning at the Nordic Institute in Urban and Regional Planning (Nordplan) in Stockholm. During the 20 years at the institute he carried out the responsibility for an ambitious doctoral program, with students from the Nordic countries. As the dedicated teacher he is known to be, it comes as no surprise that he has described his own major contribution as "twenty years of advanced transdisciplinary and transcultural teaching" (Olsson 2002: 259–260). The adventure at Nordplan was presumably guided by his dream of teaching as a:

> chimaera, a phantasm, a combination of that which must not be combined. Yet another reality impossible to leave alone, at the same time forbidden and extremely beautiful. Neither a scientific laboratory nor a government-sponsored Center for Brainwashing, rather an open environment in which the grafting proves to be so successful that everyone can develop one's own personality. The point is not to produce preordered copies but to allow some genuine originals the freedom they need to be what they are. (Olsson 2002: 264.)

As before teaching went hand in hand with research, one may here note that much of what eventually found its way into *Birds in Egg* (Olsson 1975) came from a college wide graduate course entitled "Thought and Action" that Olsson gave at the University of Michigan. Given both the orientation of Nordplan and his previous interests it is unsurprising that he continued to pursue questions about planning, ideology, socialization, power and politics. In his book *Antipasti* (Olsson 1990a), a collection of essays in Swedish written between 1983 and 1989, the focus was once again on the empirical case of Sweden and its politics of social engineering. Although Sweden has often been presented as a prototype for the modern welfare state, an ideal to follow, Olsson upon his return from the US found several cracks and paradoxes in its congregation. Among them, even an "institutionalized hatred of everything different" (Olsson 2002: 264).

Appearing almost as a companion to *Antipasti* was another book, *Lines of Power/Limits of Language* (Olsson 1991a), in which Olsson addressed theoretical problems of representation through "an exhibition of invisible lines" (Olsson 1991a, 3). The seat of power, if there ever were to be such a place, was one of lines within limits of language. And the most taboo-ridden, taken-for-granted and power-filled of all lines are those that mark the difference and identity between you and me, us and them, the individual and the collective. Again his oft-repeated question pointed out his specific concerns: "How do I know the difference between you and me and how do we share our beliefs in the same?" The overarching idea of the studies that went into *Antipasti* and *Lines of Power/Limits of Language* were firmly lodged in his conceptualization of power as:

> a game of ontological transformations, a magical performance in which visible things are turned to invisible relations, untouchable mind to touchable matter. (Olsson 2002: 260.)

What was also to emerge in Olsson's work during the latter part of the 1980s and much of the 1990s were explicit attempts to foreshadow *a cartography of thought*. The mapping expeditions that followed should however not to be taken as a radical break in his oeuvre, but must instead be understood as extensions of a trajectory already laid down in *Eggs in Bird/Birds in Egg*. For example, the two chapters "Squaring" and "Malevich torpedoed" in *Lines of Power/Limits of Language* were further elaborations of what had begun as the "mandala of thought-and-action" in an attempt to represent human thought-and-action from the inside of what he himself believed it was (Olsson 1980: 20e). As recalled, the major issue in *Birds in Egg/Eggs in Bird* hovered around the observation that even though humans automatically know how to speak and think, they do not quite know how to explicate the reasoning rules that govern these activities. Another egg given by the bird could be found in one of the book's last sentences where Olsson cited James Joyce: "the map of the soul's gruopography rose in relief within their quarterlings" (Olsson 1980: 48e). At least in his own quarter it eventually did.

What in the 1990s was to supplement Olsson's withstanding attempts to unveil the speaking and thinking part of human interaction was an explicit cartographic orientation. As before, the issue was not about how humans traverse their touchable physical realm, but how they navigate in and through the untouchable, internal and subsisting realm of thought. And since the ontology of thought, for Olsson, is one of invisible relations it follows that "invisible geography" and "geography of the invisible" are also quite fitting descriptions of his cartographic project. And as he once noted:

> Any artist worthy of the name tries to render not the visible but the non-visible, not what catches the eye but what hides in the taken-for-granted. (Olsson 1994: 221.)

Consequently, when Olsson set his agenda the challenge to be abstract was no longer deemed sufficient, as every mode of understanding is abstract almost by definition. Since his cartography of thought had to include also an understanding of the understanding, the real challenge was to become "abstract enough." Inspired by modern abstract art Olsson's mind was set towards a work in which "the expression of things gives way to the pure expression of relation" and a writing which was "not *about* something, but *is* that something itself" (Olsson

1993: 280). In this ambition he echoed earlier attempts to heed Samuel Beckett's call for unifying form and content, especially to write "so that the text is what it expresses and expresses what it is" (Olsson 1980: 48e). The specific task, then, was to foreshadow a cartography of thought in such a way that the resulting maps were not about the invisible human territory but *one with this territory itself*. For Olsson, the mission was clear:

> to map the taboo-ridden territory of the taken-for-granted, to enter the lands not yet discovered. (Olsson 1998: 145.)

Olsson's cartography of thought is thus a continuation of his long-standing interest in understanding the untouchable forces that make humans so obedient and predictable, but now reformulated and addressed as a problem of way-finding. As a self-licensed cartographer of the taken-for-granted he wants to investigate "how the navigational tools of invisible maps and social compasses are constructed" (Olsson 1998: 145). The key question presents itself: *How do humans find their ways in the invisible realm of thought and the hitherto unknown?* Olsson's answer to that "forbidding question" is both "embarrassingly simple" and "extremely complex", that is:

> for knowing where we are, we consult invisible maps of the invisible; for finding out where to go, we rely on compasses whose directional needle points not to the magnetic north pole, but to the ontological transformations of the taken-for-granted. (Olsson, 1998: 145.)

So it is that the map becomes crucially important for Olsson, even to the extent that it appears as a solution to the riddles of human way-finding in their taken-for-granted. There, in the power-filled invisible realm of thought and imagination, humans must "rely on maps of the invisible, themselves invisible" (Olsson 1994: 215). Yet, although the map can be conceived as an invisible coordinate net that provides form for thought by furnishing it with fix-points and orientations, it also comes with a problematic history as well as a problematic geography. To be precise, the map itself is a power-filled tool that privileges certain kinds of understandings at the cost of excluding others. It follows that Olsson's attempts to foreshadow a cartography of thought is intrinsically and thoroughly braided with his efforts to initiate a critique of cartographic(al) reason (Olsson 1998, 2007).

An intrinsic part of Olsson's cartographic orientation is his semiotic understanding of language that evolved after *Birds in Egg/Eggs in Bird*, and where the semiotics of Ferdinand de Saussure and Jacques Lacan has been particularly important. Consequently, a premise that guided his mapping expeditions is the conceptualization of humans as semiotic animals, a species that simultaneously defines both itself and its territory through the use of signs. To Olsson, humans appear almost as ontological replicas of the sign itself, equipped as they are with an ability to unite and separate oppositions like visible and invisible, untouchable and touchable, relations and things, that is, the type of ontological categories which in *Birds in Egg/Eggs in Bird* had been identified as inherent to the human condition. The concept of the sign, as Derrida has put it:

> cannot itself surpass this opposition between the sensible and the intelligible. The concept of the sign, in each of its aspects, has been determined by this opposition throughout the totality of its history. (Derrida 2002: 355.)

In accordance with a semiotic understanding, and corresponding to his previous characterization of human thought-and-action as a game of ontological transformations, Olsson thus came to conceptualize humans as ontological transformators whose very subsistence is located in the fraction line that separates and unites the two aspects of the sign known as the signified and the Signifier. It was also there, at the limiting penumbra through which signifier and signified are kept together and apart, that he came to spend a considerable amount of time. In his own words:

> It is this hidden Bar of Categorial Meetings that I go to search for fractions of taboo-ridden phenomena and taboo-ridden insights. It is exactly in the abyss of this power-filled void that visible becomes invisible, untouchable touchable. It is here that presence meets absence, absence meets presence. It is here that I meets the Other, full of dreams and full of realities. It is here that Descartes can be witnessed staring into his own mirror. (Olsson 1991a: 60.)

It was also in that abyss, where humans simultaneously are held together and apart by their use of signs, where the dematerialized essence of Olsson's old theme of human interaction was now to be found. It was there, at the Bar de Saussure, that human thought-and-action could be stripped bare as a never-ending imaginative probing of whatever cannot be fully combined. In essence, it was not the Signifier or signified but the fraction line itself that enabled humans to straddle the ontological divide. Thus the sign was:

> On the surface two worlds, one sensible, one intelligible, deeper down a kaleidoscope of socially constructed appearances. (Olsson 2007: 80.)

Driving forces of Olsson's attempts to foreshadow his cartography of thought have been an urge to move beneath the external surface features of language and a desire to be abstract enough. "Intelligible" and "sensible" pertains to the surface features of the external, but deeper down lies a domain of internal relations, objects of thought that reach us as "socially constructed appearances." Although the naming of things and relations would continue to be central for him, it is of crucial importance that the concept of the sign does not only bring forth external routes of signification for thought. In addition, it points also inwards to the non-visible internal relations, between signified and Signifier, between sound patterns and words, that are nurtured and given form in the cradle of social conventions. Somewhere in that internal domain of relations, even before they are expressed in words, hide the secrets that Olsson's cartography of thought ultimately aims to uncover.

And more abstract than words are "lines." For him lines had been material correlates of internal invisible relations for quite some time, in fact almost from the beginning of his career. As he recalls, one of the earliest beautiful catches was "when the world was pictured as a straight regression line" (Olsson 1991a, 187). Having previously spent over a decade with the two lines of the equal sign, which transformed into the dialectical slash, and along the way morphed into the hyphens that denote identity and difference, the minimalistic lines-man could conclude:

> The story of the straight line contains everything I know and everything I have
> not yet understood. (Olsson 1990b: 107.)

In accordance with a non-satisfiable desire to represent the non-representable - to be abstract enough – lines had become Olsson's way "to push the relation between expression and impression to its limits" (Olsson 2002: 261). More generally, his own studies may well be characterized as "experiments performed on straight lines, themselves serving as objective correlates of abstract relations" (Olsson 1991a: 24). Eventually, the Saussurean Bar was to become the line *par excellence*, the minimalist rendering of the power-filled void of the internal invisible relations of human thought-and-action, the stage of a never ending play of ontological transformations. Most importantly, the theme of human interaction, as well as the inference problem of form and process, continued to be pursued also in his cartography of thought, albeit now dressed in the garb of semiotics. His various sketches thus had two assumptions in common:

> One is that man is a semiotic animal, a species whose individuals are kept
> together and apart by their use of signs; the other that every sign within itself
> combines elements of drastically different ontologies. (Olsson 1993: 280.)

When practicing his art, in and of a cartography of thought, Olsson often treated the concept of the sign as if it were a mathematical function, an approach quite similar to his earlier work in spatial science. Different ontological settings and values could then be calculated by experimenting with the internal relations of nominator and denominator, the signified and the Signifier. When mapping the invisible territory of human thought-and-action, he could thus demarcate its dematerialized extension from inside the limits set by the pure signified and the pure Signifier, these two extremes being strictly speaking beyond signification (Olsson 1991a, 1991b, 1993, 1994). The land of human thought-and-action is thus located in-between the two realms of "mindscape" and "rockscape", the boundaries of the human territory set by the extreme end-values of signified (as pure meaning) and Signifier (as pure matter). In this dematerialized land humans bring forth their own thought-and-actions through significations woven into an invisible net of various mixtures and combinations of Signifiers and signifieds (Olsson 1993).

Consequently, Olsson is able to claim that humans also in their role as semiotic creatures have no choice but to live their lives "confined to the prison-house of language" (Olsson 1998: 150), thereby retaining a linguistic understanding of human thought-and-action. Yet, conceptualizing thought in terms of naming and categorization in language is not the only way to understand how humans become so obedient and predictable. Another route goes via form; as Olsson's depiction and desire of imagination suggest:

> Such is the desire of my imagination: a minimalist rendering of how we define ourselves, an acknowledgement that linguistic signs are forms of art shaped by forms in art. (Olsson 1993: 280.)

Along the way Olsson's mappings somehow shifted in emphasis from thought to form. If "[l]ife is form, form the modality of life" (Olsson 1993: 279), then one may eventually find also in the invisible domain of subsistence some kind of form of, and for, thought before signification. What he came to explore further was in particular how the visible content of human thought was cast in an invisible geometric form. When scrutinized, it seemed as if thought was embedded in an underlying coordinate net of geometric relations, not the least because humans are embodied creatures. As such they are beings who come to their world of subsistence as geometric ready-mades, and it would be strange if their invisible thoughts and reasoning patterns did not reflect that circumstance. For Olsson, geometry appeared as a formalization of the intuition of the tactile, as "the taken-for-granted constructed, the constructed taken-for-granted" (Olsson 1998: 148). Boiled down to its essentials, it was simply "impossible to think-and-act-and-experience without it" (Olsson 1998: 147).

It follows that geometry is also involved in the translation function of knowledge which Olsson, in a minimalist rendering characteristic of him, often refers to as "a = b." Knowledge is always indirect and always involves a translation of something ("a") into something else ("b"). The translation function stipulates that there is an inevitable and absolute friction of distance between "a" and "b" that humans need to overcome in order to get their knowledge claims approved. In Olsson's own phrasing:

> Knowledge can be defined as the ability and the opportunity of saying that a = b and be believed when one does it. (Olsson 1998: 148.)

As in the gravity model, distance minimization is once again at stake. Since word and object can never be the same, "the only way of knowing the world is to treat it as if it were other than it actually is" (Olsson 1998: 149). Following Immanuel Kant, Olsson distinguishes human knowledge and understanding as an exercise in translation in which the epistemological operator "as if" serves as the rhetorical wand. This is commensurate with his general conceptualization of human thought-and-action as a game of ontological transformations, especially so because the epistemological operator enables humans to treat their untouchable world *as if* it were touchable. In principle there is presumably nothing to be done about this operator, for Olsson it is an intrinsic part of what it means to be human. However, an important part of his story is that humans, in their translations from "a to b", are prone to accept and privilege reasoning and knowledge claims which conform to a *geometric mode of understanding*. They are therefore likely to put more trust in statements about things than about relations, and to place more confidence in the rhetoric of the visible than in that of the invisible.

This means that what Olsson previously in *Birds in Egg/Eggs in Bird* had identified as a reasoning mode that results in the thingification of human thought-and-action, he now understands as intrinsically tied to geometric (re)presentation. As an illustration he often invokes the two alternative expressions with which Euclid blessed the successful end of a proof: Q.E.D. and Q.E.F. Simplified, Q.E.D. – which was to be demonstrated - means that others are persuaded by a translation that adheres to the axioms and inference rules of geometric reasoning. In the case of Q.E.F. – which was to be shown - trust is generated through the body itself (Olsson 1991a, 193, 198). One may here note that Olsson signed his own "Torpedo of human action" with the letters Q.E.I. – which was to be discovered or invented – an approval stamp that "is so rare that it is almost incredible" (Olsson 1991a: 198–199). Yet, however original his torpedo of human action may be as an invention, the form in which it is presented fulfills the requirements of being stamped Q.E.D.

Given that there is a geometric mode of and for thought, and since Olsson is but one of many in stressing the importance of geometry, it was "natural" for him to depict human thought-and-action through the fundamental concepts of geometry itself, especially *points*, *lines*, and *planes*. In fact, Olsson sometimes even goes so far that he treats the geometric triad as "graven images of intimate communication" (Olsson 1991: 198). And when planes, lines, and points are combined, then one is very close to the definition of a *map*. Consequently, in Olsson's cartography human thought is an invisible map of the invisible that consists of points, lines and planes. It is by naming the lines, points and planes that the geometric endeavor is effectively anchored in geography. According to his own definition: "Geography is a Geometry with names" (Olsson 2007: 111).

It follows that the technique by which humans are socialized into their own invisible maps of the invisible is essentially one of teaching them how "to name the points, lines and planes" (Olsson 1991b: 87). And so it is that to Olsson the dematerialized land of thought-and-action is ruled by a Centre for Cartographic Socialization, a propaganda institute which is founded on the rhetoric of geometry and whose overall objective is to make the inhabitants geographically obedient and predictable. Consequently, nobody is admitted into this territory who does not know her geometry. And no one is let out who does not know his geography (Olsson 1991b: 86–87).

And so it is that by another round of self-conscious reevaluations Olsson found that the map holds the key to understanding understanding. As a self-appointed defender he could then answer the question of whether his own approach should be regarded as geography or not:

> Of course it is! For what is geography, if not the drawing and interpretation of lines. The only quality that makes my geography unusual is that it does not limit itself to the study of visible things. Instead it tries to foreshadow a cartography of thought. To practice this art, however, is incredibly difficult, for any attempt must face the challenge of being abstract enough. (Olsson 1991a: 181.)

Back to the Earth

In 2000 Olsson was promoted to the rank of professor emeritus. After having felt an urge to stage some kind of retrospective of scattered works from the previous decade, he gradually turned to a "forward-looking *pro*spective" (Olsson 2007: ix). Seven years later *Abysmal: A Critique of Cartographic Reason* was published, a *magnum opus* of more than 500 pages. The border-man himself suggested that perhaps the best way to approach the tome was "to read it as a minimalist guide to the landscape of western culture" (Olsson 2007: xi); a proposal likely to upset those who have grown wary and critical of big stories from the West, especially when they are written with their own location and particularity erased. But Olsson, as usual, did not hide from the fact that the world to be presented was *his* world. As he noted, many well-equipped and distinguished scholars have traveled through that same landscape before, but none:

> dressed in my shoes and adorned with my eyes, none with a map like mine, none with a compass of the same declination. No painter with my brush, palette and canvas, no poet of my rhyme and reason. (Olsson 2007: xii.)

Abysmal, as specified by its subtitle, is a critique of the particular kind of reason which Olsson calls "cartographic(al) reason", an adopted term which was originally coined by his friend the Italian geographer Franco Farinelli. What that term refers to is a mode of sharing and understanding the world that is underpinned by a certain tacit cartographical architecture, logic and form of and for reason. Oversimplified, and leaving all technicalities and historical details and complexities aside, this type of reason is intrinsically related to what I understand as a double cartographic articulation.

The first articulation entails the gradual translation and transformation of the Earth and human earthly existence into maps, a development in which the techniques for cartographic representation of certain features of the Earth have been particularly important. In principle, one eventually learned to transpose the terrestrial globe onto the flat surface of the map, where it appears as a cartographic object (not the Earth in its totality but as a set of topographic areas). The second cartographic articulation extends the cartographical translation function, the tools and procedures that enabled certain geometric features of the Earth to be transformed and projected onto maps, to human earthly existence. In other words, the cartographical derivatives were implicitly or explicitly turned into tools for describing and prescribing human thought-and-action on the Earth. In the context of western culture Franco Farinelli has indeed claimed that:

> All Greek history, from Anaximander's times onwards, is nothing but the concrete realization on the earth, in the city, of the order of the map, or *pinax*. (Farinelli 1998: 141.)

This suggests that from the very beginning the map was not only distinct from the territory, but also, and more importantly, that it *preceded* the territory. In that manner, the order of the map can be ascribed a crucial and significant constitutive role for thought. Indeed, reason itself can by analogy be brought forth by the map as a kind of "cartographical object" with its own location and boundaries. However, by the same operation the cartographic and architectural foundation of reason, the Farinellian "order of the map", seems also to descend into the taken-for-granted of western culture. Reason will exist and be treated as reason on its own as it were, and not as in essence being cartographically founded or intrinsically related to geographic representation.

Consequently, the "mappification" of the Earth and human earthly existence is a task which is usually equated with the work of land-surveying geographers, a technical enterprise that has little to do with reason *per se*, the latter a domain reserved for the philosophers. Accordingly, geography has most often been characterized as a descriptive science, well in line with the notion of description and mapping as mere recording. However, when cartographic(al) reason is being invoked, the cards are about to be turned. Thus, compressed into one sentence by Farinelli:

> Western thought (reason) is nothing else than a protocol of geographical representation. (Farinelli 1998: 135.)

Hence a basic idea of cartographic(al) reason is that the drawing and writing on and through maps ("carto-graphy"), where the transposition of the round solid Earth onto a flat medium has been particularly formative, is constitutive of (at least western) reason itself. Since this means that cartographic(al) reason underpins also the domain of philosophy, it follows that Olsson is able to claim that key philosophers like Plato, Immanuel Kant and Ludwig Wittgenstein in their way of thinking were "practicing geographers" (Olsson 2007: 214). Camouflaged as philosophers in theory, they were in practice:

secret agents stationed in the frontier regions of sensibility and intelligibility, courageous explorers of the taboo-ridden realm of the taken-for-granted, self-appointed land-surveyors commissioned to establish the Mason-Dixon line between the oikumene of the humans, on the one hand, and the anoikumene of gods and brutes, on the other. (Olsson 2007: 214.)

While carrying their identity cards as theoretical philosophers, these secret agents were enrolled in the common cartographic mission of mapping what Olsson refers to as the "*territory* of the humans" or the "*land* of thought-and-action." As explorers of what it means to be human, these practicing geographers were all involved in marking boundaries and making demarcations, especially between the intelligible and the sensible, the real and the imaginary, or that which belongs to the humans and that utterly different which lies outside. However, as they primarily have been mappers of separate worlds, the abyss in-between has been very much excluded from their investigations. While acknowledging the difficulties in mapping their blind spots of observation, Olsson whispers that although "the abysmal may never be said, it can sometimes be shown" (Olsson 2007: 235).

It is in *Abysmal* that Olsson stages that show. Since his project is motivated by a quest for understanding understanding, or rather understanding how he understands, it is for him not enough to describe the outcomes of how the territory of the humans has been mapped. In addition, and in accordance with his own cartography of thought, Olsson has to include also explications of how that mapping has been performed, and of how and why the results have been accepted. In a mission impossible that borders on contra espionage he is therefore obliged to investigate also the surveying instruments, the fix-points, the base-lines and whatever else of cartographic tools that both he and his comrades among the "secret agents" have been using in the mapping of what it means to be human.

In other words, what is at stake is the hidden workings of the (western) taken-for-granted vis-à-vis cartographic(al) reason. As illustrated by the "cryptic orders" Olsson gave to himself:

- lay bare the familiar of the unknown;
- find the principles of imagination and specify the rules of ontological transformation;
- draw a map of the Territory of Humans, (re)trace its fluctuating boundaries and find its stable center;
- produce an atlas of what it means to be human;
- initiate a critique of cartographical reason (Olsson 2007: 9).

Abysmal is the report of what geographer Olsson discovered during his clandestine operations into the landscape of western culture. Starting with the key geographical coordinates of "above, below, now, then", he eventually constructs an atlas of how the boundaries of the territory of humans have fluctuated over time between the outer limits of "Mindscape" and "Rockscape." As a crucial part of this operation he presents *Uruk, Peniel, Thebes*, and *Nicaea*, as a series of local habitations provided with proper names and definite descriptions. All meant to illustrate:

> how the boundaries of the Human Territory are determined through a continuous struggle with the gods, on the one hand, and the animals, on the other. (Olsson 2007: 273.)

As noted, it is the key coordinates of "above, below, now, then", that serve as the fundamental fix-points for Olsson's expeditions. Together these coordinates form the geometry of the cross, an image of "*Homo erectus* caught in its own net" (Olsson 2007: 12). Yet, the most important part of the cross is actually a fifth point placed at its center where the vertical and horizontal lines are crossing each other in the origo of the "and." It is this "preposition-turned-conjunctive" that functions as a "grammatical bridge" and enables the ontological translational and transformational traffic between above – below, now – then. The braiding of geometry and geography revisited, because "[d]isguised as a geometry with names, this cross begins to live, in the process transforming the system of spatial coordinates into a Vitruvian Man" (Olsson 2007: 12). And:

> thereby of how the abyss between the worlds of being and understanding is negotiated. And in the investigation of that abyss /.../ lies the hidden but well-established seat of power. (Olsson 2007: 5.)

To "lay bare the familiar of the unknown" was one of the orders Olsson gave himself for *Abysmal*. For him this means investigating the hidden abysmal relations between what it means to be human and the taken-for-granted, where the latter is approached as being steeped in cartographic(al) reason; that "dominating mode of thought-and-action which is constantly in the process of being resurrected" (Olsson 2007: 11). It may here be added that although the term may not be used as such, cartographic(al) reason is often inherent in the kinds of claims many (or most according to Olsson) scholars make and take for granted. An example which he himself often refers to is Wittgenstein who stated that a philosophical problem has the form "I don't know my way about." In one of Olsson's reformulations:

> show me your map and your compass and I shall show you who and where you are, whence you came and where you are heading. (Olsson 2007: x.)

A pillar of cartographic(al) reason is that (western) thinking and philosophy have been founded on a kind of hidden architecture rooted in the graphies of carto and geo. In Olsson's understanding this means that a taken for granted geometry, a coordinate-net of lines, points, and planes, or in short the map, has enabled philosophers and thinkers to capture and weave their stories of what it means to be human in particular ways. This habit points once again to the intimate relation between the rules of geometry and the rules of thought. Olsson's own message is that:

> the rules of geometry and the rules of thought are one and the same, the implication that whoever holds the keys to the former automatically knows the way also to the latter. (Olsson 2007: 99.)

For those who are interested in a critique of cartographic(al) reason it becomes important to lay bare its structure. By doing so one would possibly then be in a better position to explicate the mode of presentation which philosophers, like Immanuel Kant, lived and therefore could never fully understand. It is in that context that Olsson presents his own attempts to initiate a critique of cartographic(al) reason as a fourth critique, a modest continuation of Kant's critiques of pure, practical and judgmental reason. Since the map is the most important constituent of cartographic(al) reason, he can thus claim that:

> **The map** itself is the answer to Kant's question about the title by which we possess the solid ground of pure understanding. (Olsson 2007: 214.)

There is, then, no doubt that the map is of utmost importance in and for cartographic(al) reason, hence also in Olsson's investigations in *Abysmal*. Indeed, for him the map becomes the only place to settle upon in the world of understanding. When he presents what is needed in order to make a map, the relations to geometry are of course clearly pronounced. All that is needed is three concepts:

- a point which remains at rest;
- a line better known as a scale or a coordinate-net; and
- a plane onto which the merging picture / story is projected (Olsson 2007: 126).

For Olsson, all maps are structured around these cartographical primitives; a point which at rest becomes a fix-point, a line which can evolve into a scale or a coordinate-net, and a plane which can serve as a projection screen. The socializing power that makes humans so obedient and predictable thus rests in "the combined principles of geometry and naming, i.e. in the interface between the theories of picture-making and story-telling, on the one hand, and the practices of pointing and baptizing, on the other" (Olsson 1998: 147). It follows that the socializing cartographic machine of power is most strongly at work in the invisible abyss where the five senses of the body meet the sixth sense of culture. As it is presented in *Abysmal*, humans find their way:

> in the Terra Incognitae by unconsciously adhering to a rhetorical contract in which the cardinal directions of up and down, front and back, left and right serve as fix-points in a collection of invisible maps. On this view the map becomes the sign *par excellence*, a text to be read and interpreted, a tapestry in which

> Signifiers and signifieds are interwoven into mysterious patterns of meaning,
> power and submission. (Olsson 2007: 110.)

The map thus appears as the sign *par excellence* in the transparent abode of
cartographic(al) reason, where it anchors thought-and-action into the taken-for-
granted. If it were not for the fix-points and lines of direction humans would have
to enter the unknown as if every time it were a *tabula rasa*, a free-floating world
without bearings. Deprived of a coordinate-net to hang on to, or a plane sturdy
enough to capture and sustain their objects of thought, they would be altogether
lost. As Olsson puts it:

> As my taken-for-granted are unhinged, my fix-points begin to float and my
> compass to spin. In the resulting confusion I no longer know my way about.
> (Olsson 2007: 189.)

Most importantly, in this perspective the sign itself is also a map. It is "a
weaving together of picture and narrative, a power-filled statement which tells
me both where I am and where I should go, indicative and imperative in the same
breath. Squeezed into its own minimum the map is a double fold, verb turned to
noun, noun to verb" (Olsson 2007: 115). Rephrased, Olsson here suggests that the
most important tool he has used in his own mappings, the concept of the sign, can
itself be understood as a condensed form of a map, and thus a ready-made in and
of cartographic(al) reason. The map called "sign" glues together what is said with
what is shown, a material image with meaning, and establishes points of location
and direction for thought.

The sign and the map thus serve at the same time as both creators and curators
of the abyss; an indication as good as any other that cartographic(al) reason cannot
be separated from its own instruments, among them the cartographical primitives.
And for Olsson the most significant of them is neither the point nor the line, but
the plane. This is a consequence of a particular problematic that the art of his
cartography necessarily has to confront: "How do I project a dematerialized point
onto a transparent plane?" (Olsson 1994: 215).

As long as the cartographic task is to transform geographical objects like
rivers, mountains, cities, coastlines and provinces from the surface of the Earth
into a map, there is often little need to problematize the plane onto which these
phenomena are being projected. When it comes to the invisible domain of thought
and reason, however, the situation is quite different. Subsisting points and lines in
thought and geographical object on the Earth are too different to be captured or
constructed in the same categorical net. A way of handling that difficulty is to seek
inspiration from the art world, an alternative Olsson has frequently used.

In the theories of art paintings are often conceived of as lines and points put on a
canvas, the canvas being the material plane that catches and sustains what is being
depicted. The crux in the invisible domain of human thought-and-action is that no
such material plane can be found. So what is it in that realm that corresponds to

the canvas, that onto which thought is being projected? For Olsson it is the "taken-for-granted", sometimes referred to as "the sixth sense of culture", that functions as the plane or screen (the cartographer's *mappa*) that holds together and sustains the dematerialized points and lines into which thought can be distilled:

> the self-evident truth is that without the resistance of the projection screen there would be no shadows to catch, hence nothing to see, hence nothing to share. Understanding the structure of this social surface is a challenge of the utmost order, for it is the screen, not the screened, that constitutes the taken-for-granted. (Olsson 2007: 197.)

So it is that in Olsson's cartographic imagination the taken-for-granted is a transparent canvas which "serves the same function as the philosopher's Cave Wall – the culturally prepared projection screen onto which invisible relations are casting their visible marks" (Olsson 2007: 126). And finally, after a long journey through the "humus of western culture" (Olsson 2007: xii) it is the taken-for-granted of cartographic(al) reason *itself* that arrives at the destination which was set by its own destiny:

> For even though the base map of the Human Territory poised between the non-crossable limits of the spiritual Mindscape and the material Rockscape is an extraordinary rhetorical success, it comes with a real problem. *It is not true!* To be precise, the mapmaker's major mistake is that (s)he is treating the processes of deification and reification as if they were separable, while in fact they are two sides of the same coin of socialization. What on the base map is shown as two boundaries – two forms of silence – is consequently not two but one. Likewise with the mappa, the projection screen whose structure is not a reflection of the world as it is but a consequence of how the canvas has been gessoed. Surely the world is not a flat thing, but something far more abstract. And therein lies the real challenge, for it is in the nature of the semiotic animal to put more trust in the Q.E.F. than in the Q.E.I. A difference which makes a difference. (Olsson 2007: 361.)

So it is that cartographic(al) reason is also a reason on the verge of turning itself into a stiff corpse. In theory and practice it has the tendency of "thingification" deeply entrenched, not least through spatial concepts and theories of space, but the attempts to fit the world "into the geometrical mode has reached a critical point" (Olsson 1998: 149). Hereinafter, also after Olsson's *Abysmal*, it should be:

> evident that from the beginning to end cartographical reason serves as the handmaid of power, sometimes concealed with its face covered, sometimes naked with its genitals bared. (Olsson 2007: 54.)

Yet, although designed for other times and for other places cartographic(al) reason is nevertheless likely to continue to haunt them humans every now and then, here and there. And for Olsson the trustworthy truth is still that the map remains "the most powerful of all power-filled tools" (Olsson 2007: 364).

Postscript: prolegomena to human geo-graphiology

If I now, finally, were to project Olsson's key question – what it means to be human – onto the reference plane of the discipline of geography, it would bounce back slightly transformed as: *What does it mean to understand and explain humans as geo-graphical beings?* This is a question that actualizes a strange and paradoxical blind spot in geography, and even more so if attention is turned to the specifics of *human* geography. What I have in mind is the lack of a systematic discipline theory in geography of humans as geo-graphical beings.

In the discipline of geography the question about what it means to be human cannot be approached as if it pertained to humans in general, nor can it be premised on a particular social science conceptualization of humans as social, cultural, political, economic, or as any other such kind of beings. When housed in geography humans must instead be theorized as geo-graphical beings who live their lives on the Earth. It follows that their ontology is one of earthlings, which suggests a reclaiming of the etymological meaning of *homo* ("from humus, by analogy [hu]man from earth", Olsson 2007: 58). A distinguishing feature of these earthlings, as shown by Olsson in numerous ways, is that they are able to translate and transform their earthly existence by the means provided by various planes of reference that their "imaginationings" are able to bring forth. "Nature" and "culture" (or "society") are foremost among them but many more subsist the Territory of Humans, including those most dear to geographers (place, space, landscape, even geography itself) and those of religion, science and the arts which appear in *Mappa Mundi Universalis* (Olsson 2007).

Boiled down to its essentials, the Earth is what it is, and "if there were no imaginations, then there would be no humans" (Olsson 2007: 388). It follows that humans as geo-graphical beings can be conceived as ontological transformators in-between *geo* and *graphein*, the latter term etymologically tied to carving, drawing, scratching, writing, mapping, language, and the like. The future discipline that would focus on the study of humans as such beings, and that would be devoted to the understanding and explanation of their whereabouts in-between geo and graphien, is however yet to be born. Perhaps its birth has been delayed because human geography has too long been aligned with geography and its study object, the Earth itself. In accordance with the etymology, the rationale of geography has been to describe, to represent, the Earth, and foremost through the geometric modality of the map. In other words, geography is too remote from human imaginationings in-between geo and graphein *per se*. What is needed to fill that hiatus is therefore a newborn discipline, an enterprise for which I offer a proper name in advance: *human geo-graphiology*. Such is my bold thoroughfare for prospective human geo-graphiologists, ripe and ready for further (in)definite descriptions.

In the meantime, and in my judgment, Olsson's path breaking geographies – of what it means to understand what it means to be human – will be absolutely invaluable for anyone who cannot resist the temptation to wander further in/on/to/ at the ultra-thin human abysmal abode between the graphein and the geo. Perhaps, in some unknown future, someone will even write a fifth critique that lays bare the taken-for-granted plane onto which Olsson's geographies and cartographic(al) reason have been projected. Its provisional title: *Ode to Earthlings*.

Acknowledgements

Thank you so much for help, support and inspiration: *Gabriel Bladh, David Crouch, Edward H. Huijbens, Kristina Tryselius, Marcus Doel, Chris Philo*. And last (but absolutely not the least): *Gunnar Olsson*.

Bibliography

Derrida, J. 2002. *Writing and difference*. London: Routledge.

Farinelli, F. 1998. Did Anaximander ever Say (or Write) any Words? The Nature of Cartographical Reason. *Ethics, Place and Environment*, 1(2), 135–144.

Farinelli, F. 2001. Mapping the Global, or the Metaquantum Economics of the Myth, in *Postmodern Geographies: Theories and Praxis*, edited by C. Minca. Oxford: Blackwell, 135–149.

Farinelli, F. 2003. *Geografia. Un'introduzione ai modelli del mondo*. Torino: Einaudi.

Gren, M. 1994. *Earth writing: exploring representation and social geography in-between meaning/matter*. Göteborg: Publications edited by the Departments of Geography, University of Gothenburg, Series B, no. 85.

Latour, B. 1988. *The Pasteurization of France*. Cambridge: Harvard University Press.

Olsson, G. 1968. *Distance, human interaction, and stochastic processes: essays on geographic model building*. Ann Arbor, Michigan: The Department of Geography, University of Michigan.

Olsson, G. 1975. *Birds in Egg*. Ann Arbor, Michigan: Michigan Geographical Publications, no. 15.

Olsson, G. 1977. Servitude and inequality in spatial planning: ideology and methodology in conflict, in *Radical Geography: alternative viewpoints on contemporary social issues*, edited by R. Peet, London: Methuen & Co. Ltd, 353–361.

Olsson, G. 1980. *Birds in Egg/Eggs in Bird*. London: Pion Limited.

Olsson, G. 1990a. *Antipasti*. Göteborg: Bokförlaget Korpen.

Olsson, G. 1990b. Lines of Power. *Nordisk Samhällsgeografisk Tidskrift*, 11, 106–116.

Olsson, G. 1991a. *Lines of Power / Limits of Language*. Minneapolis: University of Minnesota Press.

Olsson, G. 1991b. Invisible maps – a prospectus. *Geografiska Annaler*, 73B(1), 85–92.

Olsson, G. 1993. Chiasm of thought-and-action. *Environment and Planning D: Society and Space*, 11, 279–294.

Olsson, G. 1994. Heretic Cartography. *Cultural Geographies*, 1(3), 215–234.

Olsson, G. 1998. Towards a Critique of Cartographical Reason. *Ethics, Place and Environment*, 1(2), 145–155.

Olsson, G. 2002. Glimpses, in *Geographical Voices: Fourteen Autobiographical Essays*, edited by P. Gould and F. Pitts, New York: Syracuse University Press, 237–268.

Olsson, G. 2007. *Abysmal: A Critique of Cartographical Reason*. Chicago: University of Chicago Press.

Part B
EQUAL SIGNS

Chapter 3

GEOGRAPHY 1984

Gunnar Olsson

Department of Geography, University of Michigan[1]

For the individual researcher struggling with a
detailed and often technical problem, it may sometimes
be difficult to place his own piece of work into a
broader and more meaningful context. It may therefore
be beneficial sometimes to pause for a moment to isolate
some of the major trends in today's thinking. Having done
this, it becomes exceedingly tempting to extrapolate
those trends into the future. Although such attempts
always have a dangerous flare of science fiction that of
course should be resisted, Scandinavian mid-summer nights
have a strange way of making people let their conventions
go and rather seek momentary joys and satisfactions than
the fulfilment of dull and commonday responsibilities.
Being in that mood, it is never difficult to find luring
tempters, which in this particular case assumed the
form of Peter Haggett and David Harvey, who suggested
that this paper be included in the Discussion Paper
Series from the geography department at Bristol. Asking
the reader to keep these reservations in mind, we will
proceed to a brief and admittedly biased exposé of theory
and model building in modern human geography.

Long-range perspective

As discussed in more detail elsewhere,[2] considerable
parts of the traditional work in human geography and

regional science have at least implicitly been based
on a general ideal of minimization of effort. Nowhere
has this come out more clearly than in the works in
social physics by Zipf, Stewart and others.[3] Their ideas
about human behavior and least effort can of course
be directly related to such fundamental optimality
principles in physics as Fermat's principle of least
time in geometrical optics, and Maupertii and Hamilton's
principles of least action and conservation of energy
in mechanical systems.[4] In the simplified two-dimensional
case that geographers usually have dealt with, it has
been natural to assume that this principle of least
effort exerts itself as a minimization of physical
distance.[5] Restating the spatial optimality principle
in a somewhat different terminology, Bunge has talked
about a nearness effect or a nearness principle, by
which he means for instance that countries place areas
as near themselves as possible, or, more generally, that
interacting objects tend to place themselves as closely
together as possible.[6]

But an overriding principle of this very general
kind is not worth much, until it has been simplified via
special theories into operational and testable models.[7]
The parameters of these models can then be interpreted
and predictions may be based on observed systematic
changes in parameter values. To illustrate the occasional
pregnancy of this approach, one can point to the social
gravity model. Thus we have been able to arrive at
fairly reliable estimates of the relationships between
distance and the intensity of human interaction simply
by analyzing systematic variations in the b-values of
the gravity regression model or, put another way, by

3.

accumulating data on the slopes of different regression
lines.[8] More specifically, the social gravity model
has made it possible to prove in an exact manner that
people's distance sensitivity has become smaller and
smaller over time while it is also a function both of
the interactors' personal characteristics and of their
relative position in the central place hierarchy.

The comments above offer a rough caricature of
theoretical geography and its achievements up till only
a few years ago. Very often the overriding principle
of the discipline was taken to be that of minimization
of distance as measured in the traditional sense.
At the same time, it had been clear right from the
beginning that the principle of minimum distance can be
carried over into a number of secondary concepts such
as specialization, agglomeration, and scale economies.
These concepts have played a major role in all classical
location theories and they explain how Christaller
and Lösch could derive their regular and periodic
settlement patterns. Essentially the same problem has
been discussed more recently by Curry, who has concluded
that the interplay between the forces of distance and
specialization results in a kind of undulating spatial
economic surface.[9] It is comforting that Curry's surface
resembles the type of central place lattice that also
Dacey has been experimenting with.[10]

The type of periodicities in the spatial economy that
Curry, Dacey and others have observed can be specified
mathematically by two-dimensional polynomials or Fourier
functions.[11] Essentially the same results can be obtained
with a recently developed optical device, by which laser
beams are sent through digitized maps or pictures.

4.

Since they are highly relevant for pattern recognition in remote sensing, these new modes of analysis have attracted the military's attention and considerable research funds seem to be available.[12] But the increasing popularity of harmonic and spectral analysis in geographic research does not only reflect a submission to filthy lucre. Because of their close relationships to Curry's and Dacey's ideas it is in fact likely that the accumulation of data on wave-lengths and frequencies in spatial series ultimately will lead to theoretical implications which we may not realize today.[13]

In the writer's opinion, Curry's explicit suggestions that the spatial economy can be analyzed in terms of periodic functions represents a new approach to traditional geographic theory and a sophisticated extension of the principle of least effort. But the classical models are not only based on the principle of least effort or of minimzing distance. Typically, they have also involved the simplifying assumption of a homogeneous plain over which the decision makers in the model can move themselves and their economic activities. Reality, however, is only remotely similar to such a theoretical plain and it seems that the progress in theoretical geography has been seriously hampered by the long discussions of this obvious fact. One of the few constructive suggestions in this polemic between "theoreticians" and "empiricists" has been made by Tobler, who has worked explicitly on the design of map transformations by which complex realities can be turned into homogeneous plains.[14] Conceptually, this approach reminds us of von Thünen, who deformed his theoretical pattern of concentric land-use zones by introducing a

river on which goods could be transported more cheaply
than over land.

But there are also other ways in which reality may be
transformed to fit into the assumptions of the classical
models. Although these techniques are not nearly as
sophisticated and valuable in the long run as those
suggested by Tobler, they may have been more helpful
in the short run. Political geographers, for instance,
often hypothesize that boundaries create irregularities
in theoretically symmetrical interaction patterns. These
hypotheses have then been tested by various distance
models and observed jumps or discontinuities in otherwise
smooth distance functions have been attributed to the
effect that political, topographic, economic or social
boundaries have on spatial human interaction.[15] It has
been most common to quantify these effects in terms of
a- or b-values in the gravity regression model. Even this
very simple and well-known formulation may therefore be
a valuable tool in Tobler's search for the "appropriate
coordinate system" or in his attempts to draw maps based
on functional instead of physical distance.

It indicates the generality of Tobler's idea that
his search for the appropriate coordinate system can be
related not only to studies of spatial interaction, but
also to the current works in spatial perception. Although
they usually deal with another set of variables, the
perception group is thus interested in the relationships
between objective and functional space. There is little
doubt, therefore, that the perception approach offers
a way in which existing location theories can be made
more realistic.[16] But the vigorous group engaged in
this type of research must now prove the practical

usefulness of their seemingly good ideas by translating
them from often cumbersome English into simplified but
operational and testable models. What seems most urgent
is to formulate an extremely simple but reliable model
of perception – a model of the psychology of space with
the potential for contributing to spatial theory as much
as the gravity regression model already has done. There
is a clear need for such a model containing only ore or
two parameters, whose systematic variations are easy
both to compute and to interpret. Instead of developing
this type of model, however, the students of spatial
perception and behavior have gone to two extremes. The
group centered around White, Burton and Kates has thus
provided a number of general ideas about basic variables
relating to man's perception of his environment but even
though these ideas have already constituted the basis
for important political and planning decisions, they
have yet to be translated into exact models which can
be manipulated and tested on empirical data.[17] The works
belonging to the other extreme are best brought out in
Wolpert's and Gould's studies of individual decision
making in agriculture.[18] Although this group definitely
works with theoretical and symbolic models, the suggested
formulations are often of a type that cannot be solved,
either because the computers are still too small to
handle them or because we lack appropriate input data.
This does not deny, however, that the use of symbolic
models in perception seems promising indeed and one
can only regret that those practicing verbal studies so
definitively outnumber those devoted to the mathematical
approach. What actually can be achieved with fairly

simple but exact techniques was recently illustrated by
Gould in his exciting study of mental maps.[19]

The studies discussed thus far are conceptually
closely related to each other and they represent a
trend in human geography that probably will remain
important for many years to come. More specifically, it
has been suggested that the works by Tobler, Curry, and
the perceptionists actually fall into the same general
class of transformation studies. These writers have all
suggested new and promising approaches to geography,
while they still have focused mainly on traditional two-
dimensional space and on two-dimensional transformations.
This has perhaps been unfortunate as it may be more
fruitful to deal with a complex multi-dimensional
behavioristic space, whose exact location on earth may
be of minor importance. The limitations of the Euclidean
two-dimensional approach have consequently become more
and more recognized, particularly by the people already
mentioned. Judging from recent publications, it is
thus clear that topics such as satisficing behavior,
information, perception, and decision making are
being given increased attention, while discussions of
geography in terms of two-dimensional Euclidean geometry
have become almost non-existent. At the same time,
the advocates of this new wave in geography do not
say that distance is unimportant per se, only that it
must be measured in terms that are meaningful from the
functional point of view. Instead of placing things as
closely together as possible in two-dimensional space,
this implies that it may be relevant to minimize only in
a more complex multi-dimensional non-Euclidean space.
Specification of such multi-dimensional transformations

appear to lie at the heart of our discipline and they seem to offer a framework within which concepts like perception and decision making can be simplified and tested in operational models.

To avoid misinterpretations, it should be pointed out that these distances and transformations are not the same as those employed in traditional multivariate techniques like factor analysis.[20] In those techniques one is still operating within the linear Euclidean space, i.e. a space that Tobler's experiments indicate not to be overly realistic. Everything suggests, therefore, that geographers have a need for a geometry which topologists have suggested but not yet seem to have invented – a geometry with holes in. Travelling from Detroit to London, for instance, most people probably conceive of the airports in these two cities as points with no or very little space between them. In summary then, it appears as if the geographer of tomorrow will have to concern himself with the analysis of something that may be conceived as one of Curry's periodic surfaces carried over into the nth dimension of such a strange and unknown geometry. At present, it has not been possible to specify the properties of this Henry Moore-like structure. This is a great challenge and it can be predicted that parts of the future work in theoretical geography will be devoted to the search for mathematical transformation functions that can be used to described complex realities in such terms. There is little doubt that this would bring us beyond both Curry's and Tobler's current work and it is likely that its ultimate success would depend on mathematical perception studies.

9.

Finally, while analogies constitute a classical way
of drawing faulty conclusions, it will not be suggested
that this proposed long-range development of geography
has any similarities with the history of physics. Despite
this, it is interesting to recall that the formal law
of transformation from one space-time system to another
was know already by Lorentz and that it in fact was an
intrinsic property of Maxwell's equations. The usefulness
of the Lorentz transformation is limited, however, since
it is linear and expresses the physical equivalence only
of systems which are in relative motion with constant
velocity. Einstein's theory of gravitation, on the
other hand, is formally based on a generalization of
these transformations into non-linear ones; Einstein's
transformations consequently express the transition
from one system of reference to another accelerated and
deformed system. To prove this, we must only assume that
time can be defined in relative terms, which are valid for
a given coordinate system, but not for another system in
relative motion.[21]

Short-range perspective

The first part of this paper has furnished a broad and
very general framework into which some new approaches
to theoretical geography can be fitted. By reviewing a
biassed selection of existing studies, attempts were
finally made to predict some long-range developments
within the discipline. No matter how intriguing such
long-term predictions may be, however, they are still
extremely dangerous and they have a strong tendency
to become disproved by the course of actual events.

10.

It is appropriate, therefore to change the emphasis
slightly and turn to the future of geographic research
with a shorter time-perspective in mind. In doing
this, attention will be drawn to some existing models
which appear to have the potential for furthering the
understanding of the periodic multi-dimensional Henry
Moore-like economy that already has been discussed. More
specifically this section will be centred around two of
the more important dichotomies in modern model building;
that of the deterministic versus the probabilistic and
that of the static versus dynamic approach. Of these
dichotomies, the latter seems to be the critical one
and the model builder's choice between a deterministic
and a non-deterministic formulation is at least partly
contingent upon his choice between static, cross-
sectional, and truly dynamic models.[22]

Any discussion in such terms as static and dynamic
is, however, in _its_ turn dependent on our conception
of causality and the notion of time.[23] From the
history of physics, it should therefore be recalled
that Newton's definition of time was symmetrical and
reversible. Newton consequently conceived of time
simply as another independent variable, which appeared
in his equations of motion in such a way that if its
sign was changed from plus to minus, the equations
remained invariant. It follows from this that if all
velocities were reversed, then the system simply will
go backwards via the same paths as it earlier had moved
forwards. But Newton's conception of time was obviously
an abstract idealization, which conforms poorly with
the second law of thermodynamics or the principle of
increasing entropy.[24] As a consequence, it is now usually

11.

(although not always) accepted to perceive time as being irresistible and irreversible.[25]

If our general philosophy is stated in such terms, the impact that Hägerstrand has had on theoretical geography becomes more clear.[26] At least implicitly he seems to have conceived of time as having a definite direction which made his introduction of chance and probability into a set of deterministic laws meaningful. This is important because an unrestricted belief in chance appears highly unrealistic and it can in fact be argued that society is governed by a set of "regulated accidents", in which chance has been permitted to play its game with laws of nature and human behavior. Some philosophers and physicists have suggested that this constitutes a paradoxical situation in which observable events seem to obey the laws of chance, while the underlying probabilities in themselves obey some causal law.[27]

It is within this framework that most of Dacey's work should be judged. This holds true in particular for his formal studies of point patterns in which the points often have been interpreted as central places that topography or some institutional forces have dislocated from their theoretical positions on the hexagonal lattice.[28] The same framework can also be applied to Medvedkov's work, where he has employed the entropy concept from information theory as a technique for quantifying the agreement between empirical and theoretical settlement patterns; the hexagonal pattern was treated as the "signal", while the "noise" was attributed to disturbances in the isotropic conditions.[29] To make Medvedkov's calculations meaningful, however, one must assume that human settlement ultimately will reach

a state of spatial equilibrium. Given the continuous
change in external conditions, this may not be realistic,
and several alternative assumptions have therefore
been made about the generation of two-dimensional point
distributions. Several of these generating processes
have then been specified in stochastic models of the type
employed in quadrant counts or cell counting analysis.[30]
Dacey, for instance, has described and interpreted the
location of central places in homogeneous areas by means
of double Poisson distributions, while he has found
that data from inhomogeneous areas are better described
by the negative binomial.[31] Rogers has used essentially
the same technique in his studies of retail location in
cities,[32] while Hudson employed it in his analysis of
rural settlement.[33] Finally, largely drawing upon previous
works in plant and animal ecology, Harvey has furnished
an excellent discussion of the use of cell counts in
diffusion studies.[34] Provided the employed distribution
functions are related to their physical urn models, he
has noted how the cell counting analysis offers a way in
which time and space can be treated concomitantly in the
same model.

Exactly as in the widespread simulation technique,
the cell counting models have been built around a core
of deterministic laws which then have been mixed with or
given motive power by the use of probabilities.[35] Several
stochastic models are well suited to handle problems that
involve this type of regulated accidents and they have
been applied also outside the field of point analysis. One
reason for attempting this is the recognition of spatial
lag effects, by which we refer to the fact that once
something has been located in space, it tends to become

fixed and very difficult to move.[36] As a consequence, it is
virtually impossible to achieve a society in which all
economic activities are optimal from the spatial point
of view. Already Hotelling had commented on the same
question in his well-known discussion of the equilibrium
approach to location theory.[37] It is conceivable that
some of these problems could be solved by a reformulation
of the old theory in terms of differential equations
with built-in lag effects and specific assumptions
about depreciation. Such a reformulation would not be
basic, though, since it would still rely on the usual
assumption about economic rationality, perfect knowledge,
locational flexibility, etc. It has already been noted
how these assumptions have been severely criticized by a
number of geographers, who argue that location processes
cannot be properly understood unless classical economic
theory has been intertwined with pertinent parts of
behavioral theory. As mentioned, discussions of spatial
behavior in terms of Simon's concept of satisficing man
have become fairly common and increased attention has
been paid not only to the flow of information but also
to the decision makers' perception and subsequent use
of this information. Despite this increased awareness,
the exact knowledge about individual behavior in space
is still very restricted. It is evident, though, that
chance and uncertainty play important roles and it is
often suggested that the model builder should make more
use of probability formulations by which this degree of
uncertainty or ignorance can be specified.

When it comes to the spatial flow of information on
which most decisions are based, it is likely that the
development of such stochastic formulations will draw

upon already existing models of spatial interaction
in general, and of human migration in particular. In
most migration models, it has namely been assumed that
migration is the observable response to information
largely flowing in the opposite direction.[38] A number
of models have been suggested to describe this process
and although most of them are related to the basically
deterministic gravity regression model, some have
been extended into probability formulations. Such
extensions of simple models of human interaction were
recently discussed elsewhere,[39] and it was noted that
Porter's tentative migration law can be connected both
with Harris' derivation of the gravity model and with
Stouffer's model of intervening opportunities and
competing migrants.[40] More specifically, these three
models all consider two sets of probabilities, where one
specifies the number of opportunities already reached
(e.g. settled migrants or satisfied trip-makers), while
the second specifies the proportion of unsatisfied migrants
or trip-makers who have to go beyond a certain distance
to reach an acceptable opportunity.[41] Nevertheless, the
probabilistic modelling of information flows leaves a lot
to be desired and it is likely that the work within this
area will increase considerably.

This remark applies also to models of spatial perception
or of the subjective filtering that individuals undertake
of the information they receive. Thus far, most attention
has been given to Markovian models, which again have
been applied mainly in interaction and diffusion studies.
Brown, for instance, used a Markov chain model to describe
the spread of propane tanks, while Rogers employed the
same type of formulation in his analysis of interregional

migration flows.[42] Rogers' approach differs somewhat from that by Wolpert and McGinnis et al., who have suggested that migration should be viewed as a Markov process rather than as a Markov chain.[43] As recalled, the main difference between a Markov process and a Markov chain is that the transition probabilities in the former change over time, while they stay constant in the latter. Markov processes are therefore well suited to handle problems that involve search procedures and learning processes.[44] Particularly marketing behavior has been described in these terms, most recently by Golledge in his adaptation of the Bush-Mosteller learning model.[45] It deserves finally to be mentioned in this context that a decision maker's actions are conditioned not only by the experience and learning that he accumulates during this search procedure, but also by his anticipation of the future. This indicates of course that the transition probability matrix very well may change over time as in a Markov-learning process, but that the probabilities in themselves must be regarded as subjective. An alternative approach to the perception decision making problem may therefore go via Bayesian statistics. It certainly seems reasonable to use this approach, even though geographers have not yet explored its possibilities to any significant extent.[46]

This very brief review has demonstrated that geographers actually have been dealing with several probability models of space and human behavior.[47] Considering the possibilities that seem embodied in the stochastic approach, however, the number of people involved in this research is extremely small. At the same time, traditional geography teaching has rarely provided the type of skills that the new ideas require,

even if this deficiency is now being remedied in some
key departments. As the profession gradually becomes
better equipped, one may therefore expect the number
of theoretically relevant studies to increase.[48] In due
course, we can also hope to fill some of the theoretical
gaps that require extensive field work and systematic
studies of changing parameters. Once again to take
the example of the Markovian models, we are thus faced
with the problem of specifying transition probability
matrices that are meaningful both theoretically and
empirically. Viewed in isolation, this particular problem
may of course appear less important, but considered
within the framework that has been suggested in this
paper, it becomes one amongst many critical building
blocks. Since the matrices usually change over time and
space, there should be little doubt that historical and
regional geography may become important for the future
developments in theory and symbolic model building.
Although specific operational models are not to be
found in Pred's excellent study of the urbanization
and industrialization in the United States, his volume
represents a step ahead in this respect.[49]

Overview of the overview

This paper was written with the twofold purpose of
isolating promising trends in today's theoretical
geography and to extrapolate some of these into the
future. In doing this, it seemed appropriate to employ a
longer time perspective for the development of theoretical
concepts than for the specific model building which is more
dependent on unpredictable changes in research technology.

In both these cases, however, we have occasionally
touched upon the relationships between geography and the
philosophy of science. Since the philosophy of science in
its turn has been extremely dependent on developments in
physics, it is finally tempting to note some provocative
similarities between that discipline and geography.

Thus, as we have mentioned earlier, classical
theoretical geography may have shown some conceptual
similarities with the classical Newtonian approach
to physics. In the course of testing the classical
location models, it became evident that these basically
deterministic formulations were too simplified to offer
any meaningful insights into spatial processes. This
caused a shift in interest away from the pure locational
approach to an increased emphasis on the psychology and
spatial perception of decision makers. As a consequence,
geographers have become more concerned with multi-
dimensional space and with the search for very general
non-linear transformations. These transformations
are conceptually similar to Tobler's search for the
appropriate coordinate system, although they will probably
have to go beyond his current work into something that
vaguely may be envisioned as a recurring multi-dimensional
non-Euclidean probability surface.

But the similarities between current geography and
the history of physics can be carried beyond the purely
conceptual stage into some of the operational models
that geographers are now experimenting with. This is
particularly true about the probability models and it
is interesting to recall that the quantum theorists'
rejection of the deterministic Newtonian approach was
contingent upon their conception of time as having only

18.

one direction. If geographers conceive of time in a similar way, it is feasible to treat reality simply as one particular realisation of a process that may as well have produced another result. It is important, therefore, that we try to derive the probability matrices which form the bases for the game that nature or human beings are playing. Provided this is done, it becomes very appealing to approach not only economic geography, but also historical, regional, and even physical geography from a probabilistic point of view. In this way one can treat "unique" cases occurring in time and space as parts or samples from more general processes. This approach becomes all the more attractive, if we consider the presence of contingency, i.e. the fact that one particular outcome of a stochastic process influences the probabilities for other particular events occurring close in time and space.

Footnotes

1. The ideas present here are essentially those discussed in seminars at Northwestern University, May 5, 1967, and at Bristol University, May 18, 1967. The writer is grateful both to those who arranged these meetings and to the Swedish Council for Social Science Research, which has supported parts of his work. Finally, thanks are due to Stephen Gale, one of my graduate students, for providing the challenge that every teacher needs.
2. Olsson, Gunnar: <u>Distance and Human Interaction: A Review and Bibliography</u>. Philadelphia, Pa., Regional Science Research Institute, 1965.
3. Zipf, George K.: <u>Human Behavior and the Principle of Least Effort</u>. Cambridge, Mass.: Addison Wesley, 1949. Stewart, John Q.: "Empirical Mathematical Rules Concerning the Distribution and Equilibrium of Population", <u>Geographical Review</u>, Vol.37, 1947, pp. 461–485.

19.

Stewart, John Q. & William Warntz: "Physics of
Population Distribution", Journal of Regional
Science, Vol.1, 1948, pp. 93-123.
Ajo, Reino: "Contributions to 'Social Physics': A
Programme Sketch with Special Regard to National
Planning", Lund Studies in Geography Ser.B, Human
Geography, No.11, 1953.
4. Rosen, Robert: Optimality Principles in Biology.
London: Butterworths, 1967.
5. Cf. the classical
von Thünen, Johann Heinrich: von Thünen's Isolated
State (ed. by Peter Hall). Oxford: Pergamon Press, 1966.
Weber, Alfred: Theory of the Location of Industries
(Friedrich translation). Chicago: University of
Chicago Press, 1929.
Christaller, Walter: Central Places in Southern
Germany (Baskin translation). Englewood Cliffs,
N.J.: Prentice Hall, 1966.
Lösch, August: The Economics of Location (Woglom-Stolper
translation). New Haven, Conn.: Yale University Press,
1954.
Isard, Walter: Location and Space Economy. New York:
John Wiley, 1956.
6. Bunge, William: Theoretical Geography. Lund: Gleerups
Förlag, 1966.
7. For discussions of the relationships between general
theories and specific models, cf. e.g.
Forrester, Jay W.: Industrial Dynamics. Cambridge,
Mass.: MIT Press, 1961.
Chorley, Richard J.: "Geography and Analogue Theory",
Annals of the Association of American Geographers,
Vol.54, 1964, pp. 127-137.
Olsson, Gunnar: "Teori, modell och planering", Choros,
No.3, 1967.
8. Claeson, Claes-Fredrik: "En Korologisk publikanalys.
Framtällning av demografiska gravitationmodeller med
tillämpning vid omlandsbestämning på koordinatkarta",
Geografiska Annaler, Vol.46, 1964, pp. 1-130.
9. Curry, Leslie: "The Random Economy: An Exploration in
Settlement Theory", Annals of the Association of
American Geographers, Vol.54, 1964, pp. 138-146.
Curry, Leslie: "Central Places in the Random Spatial
Economy", Paper presented at a conference on new trends

20.

in Cultural Geography, Columbus, Ohio, November 1966.
Also in Journal of Regional Science, forthcoming.

10. Davey, Michael F.: "The Geometry of Central Place Theory",
     Geografiska Annaler, Vol.47, Ser.B, 1965, pp. 111-124.

11. James, William R.: "The Fourier Series Model in Map
     Analysis", Technical Report, No.1, ONR Task No. 388-078,
     Contract Nonr-1228 (36), Geography Branch, Office of
     Naval Research, April 1966.

     Krumbein, W.C.: "A Comparison of Polynomial and Fourier
     Models in Map Analysis", Technical Report, No.2, ONR
     Task No.388-078, Contract Nonr-1228 (36), Geography
     Branch, Office of Naval Research, June, 1966.

     It is interesting that these techniques have suggested
     new approaches to the analysis of scale factors in
     spatial series; cf. e.g.

     Casetti, Emilio: "Analysis of Spatial Association by
     Trigonometric Polynomials", Canadian Geographer,
     Vol.10, 1966, pp. 199-204.

     Harvey, David: "Processes, Patterns, Scale Problems in
     Geographic Research". Unpublished Discussion Paper,
     British Section of the Regional Science Association,
     March 1967.

12. This type of funding may be questionable since the
     ongoing works in pattern recognition are extremely
     relevant also for the analysis of information obtained
     from more conventional sources than high-altitude
     aerial photographs or orbital images. Cf. e.g.

     Simpson, Robert B.: "Radar Geographic Tool", Annals
     of the Association of American Geographers, Vol.56,
     1966, pp. 80-96.

     Risley, Edward M.: "Developments in the Application of
     Earth Observation Satellites to Geographic Problems",
     Professional Geographer, Vol.19, 1967, pp. 130-132.

     Tobler, Waldo R.: Problems and Prospects in
     Geographical Photo-interpretation. Paper presented
     at a Symposium on Automatic Photo-interpretation,
     Washington, D.C., June 1967.

13. But cf. the early suggestions along the same line in
     Ajo, Reino: "An Analysis of Automobile Frequencies
     in a Human Geographic Continuum", Lund Studies in
     Geography, Ser.B, Human Geography, No.15, 1955.

     Ajo, Reino: "An Approach to Demographical System Analysis",
     Economic Geography, Vol.38, 1962, pp. 359-371.

Ajo, Reino: London's Field Response, I and II", <u>Acta Geographica</u>, Vol.18, 1964.

14. Tobler, Waldo R.: <u>Map Transformations of Geographic Data</u>. Unpublished Ph.D. dissertation, Department of Geography, University of Washington, 1961.
    Tobler, Waldo R.: "Geographic Data and Map Projections", <u>Geographical Review</u>, Vol.53, 1963, pp. 59–78.

15. Mackay, J. Ross: "The Interactance Hypothesis and Boundaries in Canada", <u>Canadian Geographer</u>, Vol.11, 1958, pp. 1–8.

16. Pred, Allan R.: <u>Locational Behavior</u>. Lund: Gleerups Forläg, 1967.

17. White, Gilbert F.: <u>Choice of Adjustment to Floods</u>. University of Chicago, Department of Geography, Research Paper, No.93, 1964.
    Burton, Ian: <u>Types of Agricultural Occupance of Flood Plains in the United States</u>. University of Chicago, Department of Geography, Research Paper, No.75, 1962.
    Kates, Robert W.: <u>Hazard and Choice Perception in Flood Plain Management</u>. University of Chicago, Department of Geography, Research Paper, No.78, 1962.
    Burton, Ian & Robert W. Kates: "Perception of Natural Hazards in Resources Management", <u>Natural Resources Journal</u>, Vol.3, 1964, pp. 412–441.
    Lowenthal, David (Ed.): <u>Environmental Perception and Behavior</u>. University of Chicago, Department of Geography, Research Paper, No.109, 1967.
    Kasperson, Roger E.: <u>Political Behavior and the Management of Natural Resources: An Inquiry into Cognitive and Decision-Making Processes</u>. Paper presented at the Meeting of the Association of American Geographers, St. Louis, April 1967.

18. Wolpert, Julian: "The Decision Process in Spatial Context", <u>Annals of the Association of American Geographers</u>, Vol.54, 1964, pp. 537–558.
    Gould, Peter R.: "Man Against Environment: A Game Theoretic Framework", <u>Annals of the Association of American Geographers</u>, Vol.53, 1963, pp. 290–297.
    Particularly Wolpert's work draws heavily upon the notion of satisficing man as discussed in e.g.
    Simon, Herman A.: <u>Models of Man</u>. New York: John Wiley, 1957.

19. Gould, Peter R.: "On Mental Maps", <u>Michigan Inter-University Community of Mathematical Geographers</u>, Discussion Paper, No.9, 1966.

20. Harman, Harry H.: Modern Factor Analysis. Chicago:
    University of Chicago Press, 1960.
    Berry, Brian J.L.: "A Method for Deriving Multi-Factor
    Uniform Regions", Przeglad Geograficzny, Vol.33,
    1961, pp. 263-282.
    Berry, Brian J.L. & D. Michael Ray: "Multivariate
    Socio-Economic Regionalization: A Pilot Study in
    Central Canada", in T. Rymes & S. Ostry (eds.):
    Regional Statistical Studies. Toronto: University of
    Toronto Press, 1966.
    Berry, Brian J.L.: Essays on Commodity Flows and the
    Spatial Structure of the Indian Economy. University
    of Chicago, Department of Geography, Research Papers,
    No.111, 1966.
21. Reichenbach, Hans: The Philosophy of Space and Time.
    New York: Dover, 1958.
22. For a classification and discussion of "laws" or models
    in these terms, cf.
    Bergman, Gustav: Philosophy of Science, Madison, Wisc.:
    University of Wisconsin Press, 1957.
23. Plaut, H.C.: "Condition, Cause, Free Will, and
    the Direction of Time", British Journal for the
    Philosophy of Science, Vol.11, 1960, pp. 212-221.
24. Reichenbach, Hans: The Direction of Time. Berkeley,
    Calif.: University of California Press, 1956.
25. Narliker, J.V.: "The Direction of Time", British Journal
    for the Philosophy of Science, Vol.15, 1965, pp. 281-285.
26. Hägerstrand, Torsten: Innovationsförloppet ur
    korologisk synpunkt. Lund: Gleerups Förlag, 1953.
27. Born, Max: Natural Philosophy of Cause and Effect. New
    York: Dover, 1964.
28. Dacey, Michael F.: "Imperfections in the Uniform Plane",
    Michigan Inter-University Community of Mathematical
    Geographers, Discussion Paper, No.4, 1964.
    Dacey, Michael F.: "The Geometry of Central Plane Theory",
    Geografiska Annaler, Vol.47, Ser.B, 1965, pp. 111-124.
29. Medvedkov, Yuriy V.: "The Regular Component in Settlement
    Patterns Shown on a Map", Soviet Geography, Vol.8,
    1967, pp. 150-168.
30. Olsson, Gunnar: "Lokaliseringsteori och stokastiska
    processer", in Tor Fr. Rasmussen (ed.): Forelesninger
    i regionale analysemetoder. Oslo, 1967.

23.

McConnell, Harold: "Qadrat Methods in Map Analysis",
Department of Geography, University of Iowa,
Discussion Paper, No.3, 1966.
31. Dacey, Michael F.: "Modified Probability Law for Point
Pattern More Regular than Random", Annals of the
Association of American Geographers, Vol.54, 1964,
pp. 559-565.
Dacey, Michael F.: "Order Distance in an Inhomogeneous
Random Point Pattern", Canadian Geographer, Vol.9,
1965, pp. 144-153.
Dacey, Michael F.: "A Compound Probability Law for a
Pattern More Dispersed than Random and with Areal
Inhomogeneity", Economic Geography, Vol.42, 1966, pp.
172-179.
Dacey, Michael F.: "A County Seat Model for the Areal
Pattern of an Urban System", Geographical Review,
Vol.56, 1966, pp. 527-542.
Dacey, Michael F.: "A Probability Model for Central
Place Locations", Annals of the Association of
American Geographers, Vol.56, 1966, pp. 549-568.
32. Rogers, Andrei: "A Stochastic Analysis of the Spatial
Clustering of Retail Establishments", Journal of the
American Statistical Association, Vol.60, 1965, pp.
1094-1103.
33. Hudson, John C.: Theoretical Settlement Location.
Unpublished Ph.D. dissertation, Department of
Geography, University of Iowa, 1967.
34. Harvey, David: "Geographic Processes and the Analysis
of Point Patterns: Testing Models of Diffusion by
Quadrant Sampling", Transactions of the Institute of
British Geographers, Vol.40, 1966, pp. 81-95.
35. Pitts, Forrest R.: "Problems in Computer Simulation of
Diffusion", Papers and Proceedings of the Regional
Science Association. Vol.11, 1963, pp. 111-119.
Morrill, Richard L.: Migration and the Spread and Growth
of Urban Settlement. Lund: Gleerups Förlag, 1965.
Morrill, Richard L.: "The Negro Ghetto: Problems and
Alternatives". Geographical Review, Vol.55, 1965, pp.
339-361.
Malm, Roger; Gunnar Olsson & Olof Wärneryd: "Approaches
to Simulations or Urban Growth", Geografiska Annaler,
Vol.48, Ser.B, 1966, pp. 9-22.
36. Devletoglou, N.F.: "A Dissenting View of Duopoly and Spatial
Competition", Economica, Vol.22, 1965, pp. 140-160.

24.

37. Hotelling, Harold: "Stability in Competition", Economic Journal, Vol.39, 1929, pp. 41-57.
38. Hägerstrand, Torsten: "Migration and Area", in David Hannerberg et al. (eds.): Migration in Sweden. Lund: Gleerups Förlag, 1957.
39. Olsson, Gunnar: "Central Place Theory, Spatial Interaction, and Stochastic Processes", Papers and Proceedings of the Regional Science Association, Vol.18, 1967.
40. Porter, R.: "Approach to Migration Through its Mechanism", Geografiska Annaler, Vol.38, 1956, pp. 313-343.
    Harris, Britton: "A Note on the Probability of Interaction at a Distance", Journal of Regional Science, Vol.5, 1964, pp. 31-35.
    Stouffer, Samuel A.: "Intervening Opportunities and Competing Migrants", Journal of Regional Science, Vol.2, 1960, pp. 1-26.
41. Also cf. the interesting interpretation of these probabilities in
    Porter, Herman: Application of Intercity Intervening Opportunity Models to Telephone, Migration, and Highway Data. Unpublished Ph.D. dissertation, Department of Geography, Northwestern University, 1964.
42. Brown, Lawrence A.: "The Diffusion of Innovation: A Markov Chain-Type Approach", Department of Geography, Northwestern University, Discussion Paper, No.8, 1963.
    Rogers, Andrei: "A Markovian Policy Model of Interregional Migration", Papers and Proceedings of the Regional Science Association, Vol.17, 1966, pp. 205-224.
43. Wolpert, Julian: "Behavioral Aspects of the Decision to Migrate", Papers of the Regional Science Association, Vol.15, 1965, pp. 159-169.
    Wolpert, Julian: "Migration as a Adjustment to Environmental Stress", Journal of Social Issues, Vol.22, No.4, 1966, pp. 92-102.
    McGinnis, Robert; George C. Myers & John Pilger: Internal Migration as a Stochastic Process. Paper presented at the meeting of the International Statistical Institute, Ottawa, August 1963.
    McGinnis, Robert & John E. Pilger: On A Model for Temporal Analysis. Paper presented at the meeting of the American Sociological Association, Los Angeles, August 1963.

25.

44. Coleman, James J.: <u>Models of Change and Response Uncertainty</u>. Englewood Cliffs, N.J.: Prentice Hall, 1964.
Atkinson, Richard C.: Gordon H. Bower & Edward J. Crothers: <u>An Introduction to Mathematical Learning Theory</u>. New York: John Wiley, 1965.
45. Maffei, Richard B.: "Brand Preference and Simple Markov Processes", <u>Operations Research</u>, Vol.8, 1969, pp. 210-218.
Herniter, Jerome D. & John F. Magee: "Customer Behavior as a Markov Process", <u>Operations Research</u>, Vol.0, 1961, pp. 105-122.
Herniter, Jerome D. & Ronald A. Howard: <u>Progress in Operations Research, Vol.2</u>. New York: John Wiley, 1964.
Golledge, Reginald G.: "A Conceptual Framework of a Market Decision Process", <u>Department of Geography, University of Iowa</u>, Discussion Paper No.4, 1967.
46. Curry, Leslie: "Seasonal Programming and Bayesian Assessment of Atmospheric Resources", in W.R.D. Sewell (ed.): <u>Human Dimensions of Weather Modification</u>. University of Chicago, Department of Geography, Research Paper, No.105, 1966.
47. For a fuller discussion of probability concepts in geography, cf.
Curry, Leslie: "Chance and Landscape", in J.W. House (ed.) <u>Northern Geographical Essays in Honour of G.W.J. Daysh</u>. Newcastle-upon-Tyne University, Department of Geography Publications, 1967.
48. LaValle, Placido; Harold McConnell & Robert G. Brown: "Certain Aspects of Quantitative Methodology in American Geography", <u>Annals of the Association of American Geographers</u>, Vol.57, 1967, pp. 423-436.
49. Pred, Allan: <u>The Spatial Dynamics of Urban-Industrial Growth, 1800-1914: Interpretive and Theoretical Essays</u>. Cambridge, Mass.: MIT Press, 1966.

Chapter 4
Emerging from the "Egg"[1]

Reginald G. Golledge

Drinking "glug". That was my first experience with Gunnar Olsson. It was New Year's Eve at the home of Dr Leslie J. King in Columbus, Ohio. Gunnar had concocted that dreadful mixture and warned me not to imbibe too much. I didn't take his advice, and thereby learned a lesson. I don't remember much about the next few days, except that Gunnar had returned to Ann Arbor. I breathed a sigh of relief.

As soon as I could stand and walk again, I acquired a copy of "Distance and Human Interaction," Gunnar's published PhD dissertation. At the time of the Theoretical and Quantitative Revolution in Geography, this publication epitomized what was beginning in Geography – deeper thinking than just empirical generalization; objective evidence of spatial regularity in human behavior; "explanation" via mathematical models; discovering patterns in apparent chaos; and thinking in terms of process rather than form.

"Birds in Egg" and "Eggs in Bird" followed. Incidentally, Gunnar, I still have "the egg" that symbolized these productions! And, while these seminal works gave hints at to what was to come, they represented mere cracks in the eggshell of thinking and reasoning that encapsulated Geography at the time. While certainly other cracks in the shell could be traced to the emergence of Behavioral Geography, and to Social Theory and Political Economy, it was Olsson who finally broke the shell and emerged as philosopher–scholar and conscience of the discipline.

At a time when I was low in spirits and was banging my head on the ceilings of language and theory, Gunnar extended a lifeline that mixed personal friendship with academic curiosity. And this was done in the absence of "glug". We arranged commuting weekends between Ann Arbor (Michigan) and Columbus (Ohio) in which some faculty, graduate students, and staff personnel from each geography department alternated in traveling to each campus for weekends of thinking and talking. No agendas. We really wanted to know and understand what our peers were thinking and doing. In many ways, these exchanges and other informal visits kept my mind and body on track during a period of personal difficulties. And during these weekend visits, it was not unusual to find friends and associates

---

1   This chapter was originally written by Reginald Golledge in February 2008, in response to an invitation to participate in a *Festschrift* for Gunnar Olsson. Sadly, Golledge passed away in 2009. He was a very close friend of Gunnar Olsson, and we have decided to include his original contribution here (with only some minor editing).

like Peter Gould, Les King, David Harvey, John Hudson, and Michiganders such as Tobler and Nystuen, all sitting on Gunnar's floor and talking to each other. And what an amazing collection of thoughts poured out of these meetings. They produced academic "highs" with only a glass or two of wine as a tongue loosener.

Gunnar's extensions (not excursions, for those are but short lived) into philosophy, art, planning, design, poetry, and religion typified his almost complete emergence from his formative geographic eggshell. But not completely. Spatial thinking and reasoning still occupied a prominent place in Gunnar's academic repertoire. And, while sometimes relaxing in lawn chairs in his Uppsala backyard and savoring the specially preserved "last cherry" from a backyard tree, I felt I'd "lost" Gunnar. "Lost" in the sense of being unable to follow the deep and convoluted thinking he was professing at the time. For example, once Gunnar sent me a paper entitled "Very Preliminary Notes for Some Beginning Concepts on a Mandala of Internal Relations"; I'm still trying to figure out what he meant!

But it all seemed to come together somehow, and I always felt that I had gained some additional insightful powers when it was time to move inside for some of Birgitta's wonderful meals. But, I must confess to spending many restless nights trying to reconstruct what it was that we had talked about that made me feel so good. Maybe it was the sound of my eggshell cracking!

There are many experiences I could elaborate that have meaning mainly to Gunnar and myself (such as the hearing aid situation), but it's probably more profitable to tell a few anecdotes that involved others as well.

Anecdote #1: 1974 in New Zealand

After an IGU Regional Meeting in the city of Palmerston North, New Zealand, Gunnar, Les King, Yorgos Papageorgiou, and I undertook a trip around New Zealand's South Island – hosted by New Zealander Les King. After a fabulous journey through the Southern Alps with their crashing waterfalls, rock rapids, and beautiful beech and pine forests, we emerged on the West Coast on December 6th – my birthday. A picnic meal on a windswept and flotsam-ridden beach was crowned with deep fried oysters, whitebait fritters, and, to my surprise, bottles of carefully hoarded champagne, carried from overseas in anticipation of this surprise event. Surprised? I should say so!

Then, down the West Coast stopping at a beach consisting of perfectly ovaloid shaped pebbles that were either pure white or obsidian black. And the trip continued to get better, drinking million year old water from the Fox Glacier – to staring at huge trout at Lake Wakatipu. And seeing the look of horror on Les's parents' faces when Papageorgiou, after stuffing the leg of lamb destined for a Sunday lunch with lemon wedges and garlic cloves, cut it vertically (i.e., into steaks) rather than the traditional slices when serving it.

Anecdote #2: 1980s

On several occasions, Gunnar visited me after I moved to Santa Barbara. On every visit, some worthwhile anecdote could be told. I'll settle for the one when Gunnar and Peter Gould were both visiting. Gunnar had recently acquired a gold earring – much to the confusion of my 6 year old daughter Brittany. At dinner one night, Gunnar modeled a black cocktail dress that he had acquired for his daughter. This had such an impact on Brittany that in "Show and Tell" next day at elementary school, she painted a picture of a bearded Swede with a golden earring, who wore a black dress and slept in her brother's room! Word from the school got back to us *very* quickly! Perhaps Brittany understood Gunnar more so than any of us did!

Anecdote #3: Uppsala

The first time I visited Gunnar and Brigitta at Uppsala, I took crunchy peanut butter and English muffins. On following visits, I took only ideas. Not all my anecdotes involve humorous events. All who have visited Gunnar in Uppsala know of the peace, the serenity of his home environment, a state that lulls one into opening one's thoughts and thinking processes without seeming to do so.

I recall on several occasions, while walking through the beautiful sensory experience of the Uppsala forest with Gunnar, that I realized we had slipped into deep philosophical discussions while ostensibly looking for wild blueberries and strawberries. At home, in the forest, or in an office environment, Gunnar has always exuded an aura of peace and contentment – even in the case of the gastronomic hearing aid – accompanied by humor and good fellowship. Many times I've valued this presence more than anything else. Now I know this is produced by the satisfaction Gunnar was getting from his thinking and reasoning processes. He was doing exactly what he very much wanted to do. Whilst many of us have had an "Aha!" experience, Gunnar seems to be able to have these daily as things click into place in his ontology of thinking.

Rather than academic and professional, many of my experiences with Gunnar have been human and behavioral. Since this has been my own area of expertise, I have looked at memories of Gunnar and the role he played in encouraging me to emerge from a shell of disciplinary strictures to a freeform thought process. At times, it was difficult to see any of Gunnar's original matrix of geographic thought in his academic meanderings. But what could be more revealing of an umbilical connection to his parent discipline than his latest book?[2]

I find it impossible to measure or describe what Gunnar Olsson has meant to the discipline of Geography and, specifically, to my own intellectual development.

---

2   Referred to by Golledge as "Cartography and geographic reasoning", but eventually published as "Abysmal: A Critique of Cartographic Reason" (Olsson, 2007).

In both cases, it has been substantial, ever-present, and freely proffered. My life has been enriched in many different ways through my friendship with Gunnar, and I believe that I am but one of a multitude.

Chapter 5

INFERENCE PROBLEMS IN LOCATIONAL ANALYSIS

Gunnar Olsson

The University of Michigan

A. Introduction

Contributions to this and other recent volumes indicate that some quantitative geographers have shifted their attention from the modelling of large-scale aggregates to studies of group and individual behavior. As a result, the earlier stress on the geometric outcome of the spatial game has lessened in favor of analyses of the rules which govern the moves of the actors who populate the gaming table. Thus, the new studies aim at a better understanding of those cause and effect relationships which are relevant to the decision makers themselves, i.e. to those whose actions eventually will determine the success of various planning programs. With such pragmatic planning ideas in mind, the behaviorists wish to complement the traditional work in quantitative geography by establishing explicit linkages between individual behavior and spatial patterns. Restated and simplified, the behavioral approach suggests a different solution to the geographical inference problem of form and process; while the spatial analyst attempts to infer individual behavior from knowledge of a given spatial pattern, the behaviorist argues for reasoning the other way around.

15

To assess the epistemological merits of the suggested
solutions to the geographic inference problem, the initial
purpose of this paper is to discuss current location theories
from the viewpoint of the philosopher of science. The secondary purpose is to investigate how well the ideal approach of
the behaviorists outlined in the first half of the paper
actually compares with their operationalized models. The
latter investigation is prompted by the realization that it is
one thing to make attractive methodological statements but
quite another to translate such predilections into testable
formulations.

B. The Truth Status of Geographic Theory

A major proposition of the Vienna Circle is that scientific
statements become lawlike by being logically consistent and
empirically true, i.e. by being acceptable in terms of both
syntax and semantics. A theory may then be defined as a set
of deductively connected laws.[1] It follows that once the
syntax of a specific theory is accepted as true, its additional
value can be assessed in terms of semantics or by its ability
to predict empirical events. To test a statement derived
from a theory is therefore to seek the instantiation of what
is presently considered a law.[2] An important vehicle in this
verification procedure is the notion of correspondence rules
by which the calculus of the theory can be interpreted in
terms of real world observations, or, if an alternative view
is adopted,[3] the calculus of the real world can be interpreted
in terms of the theory. In case the correspondence rules
cannot be specified or if their application indicates differences between the theoretical and observational languages,
then the truth status of the theory is in question. Con -
versely, a theory is held to be true when all its extra-logical
terms have factual reference.[4]

Even though geography rarely has been discussed within the rather stringent framework of epistemology,[5] it may still be suggested that predictions based on geographic theory tend to be dubious. This is certainly the opinion of Pred, who argues that the breakdown of location theory is due to discrepancies between the motives which govern the decisions of the actors in the theory and the actors in reality.[6] Others have conveyed essentially the same message about the validity of existing theories but in a lower key and relating it to other causes. For instance, Curry has criticised Lösch's notion of the unbounded plain so severely that he advocates that "little remains of existing theory to allow its refashioning."[7] Dacey, finally, is less definitive, despite his claim that it is "inconceivable that any pattern of central places corresponds exactly to the specified geometry."[8]

There can be many reasons why theoretical predictions do not sufficiently agree with empirical observations. One reason relates to the aspiration level or the degree of generality at which a particular theory aims. This observation draws on the fact that theories derived within a hypothetico-deductive system by definition can be ordered into a hierarchical structure of statements.[9] An important characteristic of those structures is that axiomatic statements on one level may be testable theorems of a higher level theory. This suggests that a particular theory should not be rejected simply because it contains unrealistic axioms and therefore may provide unsatisfactory high level predictions. At the same time, it should be noted that the possibility of syntactical mistakes cannot be ruled out until the theory has been completely axiomatized. It reflects some of the biases of the work in theoretical geography that the only attempts at axiomatization relate to the geometric properties of Christaller's and Lösch's theories.[10]

The previous arguments suggest that the low predictive power of geographic theories perhaps may be due to weak linkages in the interlocking system of hierarchically ordered statements. If this is true, then it is difficult, perhaps impossible, to say anything conclusive about the validity of existing constructs. Since the theories have not yet been fully formalized, it is uncertain exactly which aspiration level is being sought. The root of the uncertainties is supposedly in the employed axioms or perhaps rather in ambiguities introduced when these were provided with semantical meaning and turned into assumptions. It is the purpose of the next section of the paper to discuss the validity of this supposition.

C. The Assumptions of Geographic Theory

It is sometimes helpful to make a superficial distinction between the spatial and the behavioral axioms of location theory. More exactly, the former postulates relate to the properties of the area over which the actions occur, while the latter concern the motives and behavior attributed to the actors themselves. The behavioral assumptions are usually the same as those of the theory of the firm, while the spatial assumptions commonly are those of the unbounded homogeneous surface. By combining spatial and behavioral postulates, theorems like the hexagonal arrangement of central places may be derived.

To point out that the assumptions of location theory are unrealistic is almost trite. It is far more important that the lack of realism becomes critical only above a certain aspiration level, i.e. when a higher level hypothesis is refuted by observations which would not refute a lower level hypothesis.[11] Taking the pragmatic view of many economists,[12] it is enough, therefore, to decide whether the assumptions

lead to sufficiently accurate predictions for the purpose at
hand -- it would be foolish to stop making everyday predictions
on the basis of Newton's laws only because of Einstein's subsequent work.   Likewise, provided the location analyst is content with devoting himself to pattern analysis and interpretive
descriptions of large-scale spatial regularities, he may possibly be satisfied with existing constructs.   If, on the other
hand, the aspiration level were changed to include analyses
also of those micro-units, groups or individuals whose actions
give rise to the large-scale regularities, then the situation
would be different.   The reason is, of course, that the
axioms in the first case have become testable theorems in the
second.   There is little doubt that on this new aspiration
level, the axioms of the traditional theory are unacceptable.[13]
Instead, the essential problem seems to become that of understanding the internal structure of goal conflicts and adaptive
systems.   It is for reductions to this level of explanation,
i.e. to the level at which inner and outer environments are
treated as adapting to one another,[14] that some behaviorally
oriented geographers are striving.

The question now arises as to whether the imagined new
breed of geographer really is new or whether it is only the
old one dressed up as Hans Christian Andersen's Emperor.   It
would probably seem so to geometrically inclined students like
Hudson, who recently claimed that "one central problem of geographic theory is that of relating individual behavior to that
of (spatial) distribution."[15]   Few would object.   The difference between the spatial and behavioral approaches to this
geographic inference problem becomes clear, however, when
Hudson later refers to a "given system of nodes (within which)
the individual must acquire a set of spatial relations so as
to navigate ... in an efficient manner."   Thus, the prime
concern is not to derive a spatial pattern from axiomatized

behavior but rather to make inferences about behavior from the
knowledge of spatial patterns.

Since there are no clearcut one directional cause and
effect relationships between geographic form and process, the
spatial analysis approach of Hudson and others will almost
certainly provide valuable insights.   Nevertheless, it is
significant that most geographic geometricians have found it
necessary to tamper with the classical spatial assumptions and
work in transformed non-Euclidean space.[16]   Most importantly,
the choice between different assumptions is conceived more as
a matter of expediency than as a problem involving explicit
statements of aspiration level or purpose.   This view has been
stated most succinctly by Tobler, who asserts that the theory
can be made "more realistic by relaxing the assumptions, but
(that) this generally entails an increase in complexity."
Difficulties arise, however, when this proposition later is
being executed through the removal of "the differences in geographic distribution by a modification of the geometry or of
the geographical background."[17]

The terms "geographic distribution" and "geographical
background" may convey the impression of areal variations in
physical landscape and population densities, i.e. of phenomena
which have nothing or very little to to with individual or
group behavior.   Judging from the rest of the paper, however,
Tobler must have something more far reaching in mind.   Thus,
the idea is illustrated by references both to the logarithmic
migration maps of Hägerstrand[18] and to cartograms like 'A New
Yorker's Idea of the United States of America', i.e. to maps
explicitly derived from the theory of cognitive behavior.
This means, of course, that the aspirations in fact have been
increased to a level where the spatial axioms take on the role
of testable theorems derivable from the theory of cognitive
behavior.

What is crucial in the comments on map transformations is not that people seem to do something they claim not to be interested in. Instead, it is the consequences that the approach has for the testing of classical location theory. More exactly, it remains to be seen how the data underlying the estimation of transformation functions can be separated from the empirical information with which the theoretical predictions are to be compared; clear specifications of the employed correspondence rules should help to clarify this issue. Even then, however, it may be questioned whether the approach actually is as expedient as sometimes suggested. Speaking exactly to this point, Curry has recently observed that "if we could gain the level of sophistication necessary to transform, we could probably write theory in terms which would not require it."[19]

In short, it appears unclear which exact linkages in the alleged chain of deductive reasoning suggested that map transformation would be a valid approach. In less clever accounts than the ones by Tobler, it even seems that large-scale data occasionally have been used as a basis for inferences about small-scale behavior. It is tempting to speculate that such logical peculiarities stem either from the traditional reliance on the map as a given, or from the almost metaphysical belief that the same model can be applied to both physical and human phenomena.[20] More importantly, though, the previous discussion suggests the spatial postulates of location theory to be special cases of behavioral theorems; the questions asked by the spatial and behavioral analysts therefore tend to belong to different scientific aspiration levels. This suggestion supports the common sense conclusion that the reliability, explanatory power, and the potential planning applications of any social science theory depend on its treatment of individual and group behavior.

21

Since the behavioral assumptions of location theory recently have been treated elsewhere, it now seems superfluous to
discuss those in the same detail as the spatial assumptions.[21]
Likewise, it is rather pointless to elaborate on the lack of
realism in terms of actual decision processes. Notwithstanding, it is possible that the postulates still will lead to
sufficiently good results, provided that only large-scale predictions are aspired. If, on the other hand, the goal is to
understand the finer workings behind large-scale regularities,
then it is doubtful whether the traditional approach with its
firm grounding in classical utility theory and normative economics will provide reasonable explanations. Examples of
other, supposedly more fruitful approaches include Hägerstrand's
ideas about information, diffusion, and migrations, Curry's
notions of shopping lists, inventories, queueing, and central
places, Wolpert's work on the spatial attributes of stress and
goal conflicts, and Pred's explorations into the behavioral
matrix.[22] The obvious conclusion is that large-scale patterns should be deduced from explicit statements about individual behavior rather than the other way around. More specifically, spatial patterns should be viewed as reflections of
habits and institutionalizations which in turn can be accounted
for by individual decisions governed by continuous learning
processes.

The attempts to derive spatial patterns from realistic
assumptions about individual or group behavior are appealing
because they aim at higher level explanations. Occasionally,
however, some issues relating to the axioms, theorems, and
aspiration levels still remain unclear. For instance, Curry
has recently suggested that the problem of writing theory is
to obtain postulates which are not so directly linked to the
final results that added insight is not gained.[23] Provided
this means that one should not use the same data sets for

estimating parameter values as for testing theoretical predictions, or that logical deductions may lead to the discovery of new types of scientific facts, then there can be no disagreement.   If, on the other hand, the term "gaining added insight" refers to something less tangible, then the statement becomes less clear.   This is particularly so in view of the tautological nature of the deductive method.[24]

D. Inference from Geographic Models

The discussion thus far has centered on some epistemological issues relevant to the geographic inference problem of how to connect spatial patterns and human behavior.   It has been proposed that this basic problem of form and process can be approached from two basically different directions.   Thus, one may arrive at conclusions about individual behavior through analyses of given spatial patterns, or one may draw conclusions about spatial patterns from detailed knowledge of individual behavior.   Both approaches involve difficult inference problems, some of which are related to the choice of axiomatic system or aspiration level, while others are embodied in the lack of one-directional cause and effect relationships. Given this situation, continued epistemological reevaluations of locational analyses seem mandatory.

Evidence that such reevaluations are important is provided by the fact that even the very superficial comments of this paper have helped to isolate some attractive features of the behavioral approach.   The question therefore arises as to whether the logically appealing syntax can be matched by meaningful semantics.   To illuminate this question, the discussion will now turn to a review of some models recently employed by students frequently associated with the behavioral school. It should be noted that only operationalized and tested formulations will be treated; to extend the comments into

23

suggested, hypothetical and nontested constructs would not add anything substantive to the comments already made.

As an introduction to the model review, it may be helpful to rephrase the geographic inference problem in terms of large-scale patterns with small variances and small-scale processes with large variances. It follows that large-scale systems, i.e. systems from which portions of the internal variation has been filtered out, tend to be more deterministic, while small-scale systems are more probabilistic. For this reason, the efforts of rewriting well tested deterministic models like the gravity, regression, and rank size formulations in probability languages becomes interesting from the inference point of view. These attempts to get around rather than solve the problem have been discussed in detail elsewhere,[25] and it is therefore now sufficient to add a reference to some subsequent works on the entropy concept by Wilson.[26]

Provided the same correspondence rules apply in both cases, the rewriting of classical models in probability terms will make large-scale regularities interpretable and derivable from explicit and quasi-realistic assumptions about individual or group behavior. However, the value of these reformulations should not be overestimated; the fact that the final results have been arrived at via another route has not appreciably changed the character of the models. As a consequence, the regulating forces remain deviation counteracting rather than deviation amplifying.[27] On the practical level, however, the translation of the same model to another language with important syntactical and semantical differences may have considerable utility. Thus, it is a common problem in planning situations to determine the scale below which probability techniques must be used; the current discussion of the efficacy of population planning as compared to family planning offers an excellent case in point.[28]

A related but methodologically very different answer to
the problem of connecting large-scale spatial regularities
with small-scale generating mechanisms is provided by the Monte
Carlo simulation technique. Thus, the main characteristic
of the simulation technique is that evolving spatial patterns
are viewed as resulting from an interplay between deterministic
and random factors. More specifically, the general development is determined by distance functions translated via the
relative frequency interpretation of probability into the
operational form of mean information fields, while the exact
development is influenced by a large number of chance factors,
operationally represented by the drawing of random numbers.
This means, of course, that the resulting patterns may be
viewed as a set of 'regulated accidents', in which chance and
contingency have played their game with laws of nature and
human behavior.

It has sometimes been suggested that conceiving reality
as the result of regulated accidents is paradoxical because
the observable events are said to obey the laws of chance,
while the underlying probabilities in themselves obey some
causal law.[29] On the other hand, the same observation has
been extended into the epistemological tenet of complementarity. The issue can be discussed either in terms of axioms and
theorems as in the first part of this paper, or in terms of
the idea that causal laws at one level of aggregation normally
result from averages of statistical behavior at a deeper level,
which in turn can be explained by deeper causal behavior, and
so on indefinitely.[30]

The diffusion work after Hägerstrand[31] has focused almost
entirely on refining details of the original model. With
few exceptions, the efforts have been devoted to the construction of specialized computer programs,[32] experiments
with different mathematical distance functions,[33] and the

25

derivation of biased or unbiased mean information fields,[34]
while little has been done with the more basic issues of
testing and interpretation of underlying theory and functional
relationships. This means that most diffusion students --
Hägerstrand himself not included -- in fact have neglected the
behavioral approach to the geographic inference problem.
Thus the employed procedures involve implicit reasoning from
the large-scale regularities of the mean information field to
the behavior of the individuals as this is governed by the
random number matrix.   It follows that more insight may be
gained through detailed experiments with different parameter
values based on observed systematic spatial and temporal variations in resistance, distance sensitivity, communication net-
works, etc.   Proceeding to the testing of subsequent model
generations, this suggests that more attention should be paid
to sensitivity analyses and less to evaluations of spatial
end products.

In practice, spatial applications of the Monte Carlo
technique rely almost exclusively on large-scale aggregate
data, which then are treated as the joint product of deterministic and random variables.   As a consequence, inferences
about individual behavior can be made only indirectly via a
reasoning from spatial patterns to generating mechanisms.
The same characterization generally applies to the cell counting technique,[35] even though Dacey in some of his county seat
models has attempted to deduce spatial distributions from
explicit assumptions about underlying processes.[36]

The inferential problem involved in the cell counting
technique is best illustrated in studies employing the negative binomial distribution.   It is well known that this distribution can be generated in at least six different ways,
some of which are complete opposites.   For instance, a spatial point pattern can be described by the negative binomial

if it consists of randomly located clusters generated through a two-stage diffusion process in which the parent points have been distributed randomly over the area, and the secondary points have been assigned among the initial nuclei independently of one another but in such a fashion that the growth over time is logarithmic. The opposite to this generating mechanism is the urn scheme for heterogeneous Poisson sampling, according to which a negative binomial may be obtained for the total area, provided the area can be divided into regions within which the points have been randomly distributed but in such a manner that the mean number of points per cell varies between the regions according to the gamma function. The obvious conclusion is that very little can be said about generating mechanisms solely on the basis that the morphology of a point pattern can be described by the negative binomial.[37]

It could perhaps be tempting to conclude from the discussion of the negative binomial that it is possible to reason from process laws to morphological laws but not in the other direction. However, not only are the causal linkages in geography too intricate to allow such a conclusion, but it may also be shown mathematically that at least one single mechanism -- that of space filling -- can give rise to either clustered, random or regular spatial patterns. In such situations of conflicting results, one solution is to fit the same data to another set of distributions connectable with only one of the previous interpretations.[38] Despite its value in the special case, this approach is clearly more expedient than elegant. It will hardly bring the solution of the geographic inference problem much closer.

To varying degrees, the models discussed have started from a given spatial pattern and then proceeded to indirect inferences about generating processes and individual behavior. This conclusion has been possible to substantiate only because

27

the cited models have been refined to the extent of operationalization and empirical application.   Unfortunately, the
rest of the model work in behavioral geography still awaits
rigorous testing and does therefore not permit the same degree
of conclusiveness.   On present evidence, however, it is not
unlikely that current theoretical explorations will end up
with testable formulations which are based on the traditional
approach to the geographic inference problem rather than on a
new logically more attractive one.

The suspicion that techniques for generating spatial
patterns from individual behavior still may be far away, relates closely to the attempted adaptations of classical learning theory to the geographers' need.[39]   Thus, the focus has
been more on how individuals learn to act efficiently in an
existing spatial system than on how their actions cause
existing spatial patterns to change.   Basically, the same
holds for the notion of subjective preference functions[40] and
for most studies of mental maps.[41]   Ignoring the extremely
thorny measurement problems as being beside the point in the
present context, the most interesting property of the latter
studies is their amenability to trend analysis.   Perhaps it
is on this level that the relationships between the allegedly
new approaches and the traditional work in spatial analysis
become most evident;   trend surface analyses and map transformations appear in fact to be the dual of one another.

E. Summary and Conclusions

This paper was based on the premise that the limited predictive power of geographic theories is due to a preoccupation
with spatial patterns and a neglect of small-scale generating
processes.   The paper has attempted to evaluate this premise
by comparing the spatial and the behavioral approaches to the
geographic inference problem of form and process.   To

establish some general guidelines, attention was first given
to the overlaps between geography and the philosophy of the
social sciences. The subsequent conclusion was that the
behavioral axioms of location theory belong to a higher level
of the hierarchical structure of the hypothetico-deductive
system than do the spatial axioms. As a consequence, the
behavioral approach can provide more detailed explanations and
is therefore preferable, particularly if the findings are to
be extended into planning applications.

It is one thing, however, to isolate attractive methodological approaches and quite another to translate these predilections into operational and testable models. In order to
assess what has actually been achieved rather than merely
talked about, the latter half of the paper reviewed a number
of models recently used by students more or less identified
with the behavioral school. More specifically, attention was
given to the rewriting of deterministic models in probability
terms, the use of Monte Carlo simulations, the cell counting
technique, and the geographical amendments to psychological
learning theory. It was found that practically all studies
had started from given spatial patterns and then proceeded to
indirect inferences about generating processes and underlying
human behavior. Although suggestions about alternative and
epistemologically more attractive approaches do exist, it has
been difficult to find cases where such models actually have
been applied to empirical data. Recalling the positivists'
quest for combinations of logical consistency and empirical
truth, this leaves the assessor bewildered. On the one
hand, it is possible to point to a number of low order spatial
formulations with considerable empirical reliability. On the
other hand, one may imagine some logically attractive behavioristic formulations which unfortunately still await empirical
evaluations.

29

The final conclusion must be that the behavioral approach
to quantitative geography may or may not alter the current
rather peculiar state of the art.   Speaking *for* improvement is
the growing recognition of studies from quantitative psychology
and non-normative economics as well as the epistemological
bases of most behavioral work.   Speaking *against* substantial
and quick change is the existence of multi-directional causal
relationships as well as the shortage of suitable highly
disaggregated data.

NOTES

1.   Merle B. Turner, *Philosophy and the Science of Behavior* (New York: Appleton-Century-Crofts, 1967), 226; May Brodbeck, *Readings in the Philosophy of the Social Sciences* (New York: Macmillan, 1968), 583.

2.   Thomas S. Kuhn, *The Structure of Scientific Revolutions* (Chicago: University of Chicago Press, 1962).

3.   Henry Margenau, *The Nature of Physical Reality* (New York: McGraw-Hill, 1950); Norwood Russell Hanson, *Patterns of Discovery* (Cambridge: Cambridge University Press, 1958).

4.   Carl G. Hempel, *Aspects of Scientific Explanation* (New York: Free Press, 1965), 217-222.

5.   For notable exceptions see: Fred Lukermann, "On Explanation, Model, and Description," *Professional Geographer,* Vol. 12 (1960), 1-2; Fred Lukermann, "The Role of Theory in Geographical Inquiry," *Professional Geographer,* Vol. 13 (1961), 1-5; Reginald Golledge and Douglas Amedeo, "On Laws in Geography," *Annals of the Association of American Geographers* (1968), 760-774; Gunnar Olsson, *Distance, Human Interaction, and Stochastic Processes: Essays on Geographic Model Building* (Ann Arbor: University of Michigan, 1968); David Harvey, *Explanation in Geography* (London: Edward Arnold, forthcoming).

6.   Allan Pred, *Behavior and Location. Part I* (Lund: Gleerup, 1967).

7.   Leslie Curry, "The Geography of Service Centers within Towns: The Elements of an Operational Approach," in Knut Norborg (ed.), *Proceedings of the I.G.U. Symposium in Urban Geography* (Lund: Gleerup, 1962), 33.

8.   Michael F. Dacey, "Imperfections in the Uniform Plane," *Discussion Papers of the Michigan Inter-University Community of Mathematical Geographers,* No. 4 (1964), 1.

31

9.   Richard Bevan Braithwaite, *Scientific Explanation* (New York: Harper, 1960).

10.   Michael F. Dacey, "The Geometry of Central Places," *Geografiska Annaler,* Vol. 47, Ser. B (1965), 111-124.

11.   Braithwaite, *op. cit.*

12.   Milton Friedman, *Essays in Positive Economics* (Chicago: University of Chicago Press, 1953).

13.   Herbert Simon, "Theories of Decision-Making in Economics and Behavioral Science," *American Economic Review,* Vol.49 (1959), 253-283; Richard M. Cyert and James G. March, *A Behavioral Theory of the Firm* (Englewood Cliffs, N.J.: Prentice-Hall, 1963).

14.   Herbert Simon, *The Sciences of the Artificial* (Cambridge, Mass.: M.I.T. Press, 1969).

15.   John Hudson, "A Model of Spatial Relations," *Geographical Analysis,* Vol. 1 (1969).

16.   William Bunge, *Theoretical Geography* (Lund: Gleerup, 1966).

17.   Waldo R. Tobler, "Geographical Area and Map Projections," *Geographical Review,* Vol. 53 (1963), 59-78.

18.   Torsten Hägerstrand, "Migration and Area," in David Hannerberg, Torsten Hägerstrand and Bruno Odeving (eds.), *Migration in Sweden* (Lund: Gleerup, 1957).

19.   Leslie Curry, "Quantitative Geography 1967," *Canadian Geographer,* Vol. 11 (1967), 265-279.

20.   Michael J. Woldenberg, "Spatial Order in Fluvial Systems: Horton's Laws Derived from Mixed Hexagonal Hierarchies of Drainage Basin Areas," *Bulletin of the Geological Society of America,* Vol. 80 (1968), 97-112.

21.   Gunnar Olsson and Stephen Gale, "Spatial Theory and Human Behavior," *Papers and Proceedings of the Regional Science Association,* Vol. 21 (1968); David Harvey, "Behavioral Postulates and the Construction of Theory in Human Geography," *Seminar Papers of the Department of Geography,* University of Bristol, Ser. A, No. 6 (1967).

22. Torsten Hägerstrand, *Innovations förloppet ur korologisk synpunkt* (Lund: Gleerup, 1953); Leslie Curry, "Central Places in the Random Economy," *Journal of Regional Science,* Vol. 7 (1967), 217-238; Leslie Curry, "A 'Classical' Approach to Central Place Dynamics," *Geographical Analysis,* Vol. 1 (1969); Julian Wolpert, "Migration as an Adjustment to Environmental Stress," *Journal of Social Issues,* Vol. 22 (1966), 92-102; Pred, *op. cit.*

23. Curry (1967), *op. cit.*

24. Hempel, *op. cit.*

25 Gunnar Olsson, "Central Place Systems, Spatial Interaction, and Stochastic Processes," *Papers and Proceedings of the Regional Science Association,* Vol. 18 (1967), 13-45.

26. Allan G. Wilson, "Notes on Some Concepts in Social Physics," *Working Paper from the Centre for Environmental Studies, London,* No. 4 (1968).

27. Walter Buckley, *Sociology and Modern Systems Theory* (Englewood Cliffs, N.J.: Prentice-Hall, 1967).

28. Kingsley Davis, "Population Policy: Will Current Programs Succeed?" *Science,* Vol. 158 (1967), 730-739.

29. Max Born, *Natural Philosophy of Cause and Chance* (New York: Dover, 1964).

30. William Kneale, "Scientific Revolution for Ever?" *British Journal for the Philosophy of Science,* Vol. 19 (1968), 27-42.

31. Hägerstrand (1953), *op. cit.*

32. Forrest R. Pitts, "MIFCAL and NONCEL: Two Computer Programs for the Generalization of the Hägerstrand Models to an Irregular Lattice," *Working Papers from the Social Science Research Institute, University of Hawaii,* No. 4 (1967).

33. Richard L. Morrill, "The Distribution of Migration Distances," *Papers and Proceedings of the Regional Science Association,* Vol. 11 (1963), 75-84.

33

34.   Duane F. Marble and John D. Nystuen, "An Approach to
the Direct Measurement of Community Mean Information Fields,"
Papers and Proceedings of the Regional Science Association,
Vol. 11 (1963), 99-109;   Richard L. Morrill and Forrest R.
Pitts, "Marriage, Migration, and the Mean Information Field:
A Study in Uniqueness and Generality," *Annals of the Association
of American Geographers*, Vol. 57 (1967), 401-422.

35.   David Harvey, "Geographic Processes and the Analysis
of Point Patterns: Testing Models of Diffusion by Quadrat
Sampling," *Transactions of the Institute of British Geographers*,
Vol. 40 (1966), 81-95;   Gunnar Olsson, "Lokaliseringsteori och
stokastiska processer," in Tor Fr. Rassmusen (ed.), *Regionale
Analysemetoder* (Oslo: Norsk Institutt for By og Regionforsk-
ning, 1967);   John Hudson, *Theoretical Settlement Geography*
(unpublished Ph.D dissertation, University of Iowa, 1967).

36.   Michael F. Dacey, "A County Seat Model for the Areal
Pattern of an Urban System," *Geographical Review*, Vol. 56
(1966), 527-542.

37.   William Feller, "On a General Class of Contagious
Distributions," *Annals of Mathematical Statistics*, Vol. 14
(1943), 389-400;   David Harvey, "Some Methodological Problems
in the Use of the Neyman Type A and the Negative Binomial Probability Distributions for the Analysis of Spatial Point Patterns," *Transactions of the Institute of British Geographers*,
Vol. 44 (1968), 85-95.

38.   Gunnar Olsson, "Complementary Models: A Study of
Colonization Maps," *Geografiska Annaler*, Vol. 50, Ser. B (1968),
1-18;   R.G. Swinburne, "Vagueness, Inexactness, and Impression,"
British Journal for the Philosophy of Science, Vol. 19 (1969),
281-299.

39.   Reginald G. Golledge, "Conceptualizing the Market
Decision Process," *Journal of Regional Science*, Vol. 7 (1967),
239-358;   Reginald G. Golledge, "The Geographical Relevance of
Some Learning Theories," *this volume* (1969);   Reginald G.
Golledge and Lawrence A. Brown, "Search, Learning, and the
Market Decision Process," *Geografiska Annaler*, Vol. 49, Ser. B
(1967), 116-124.

40.   Gerard Rushton, "The Scaling of Locational Preferences,"
this volume (1969).

41.   Peter R. Gould, "On Mental Maps," *Discussion Paper of the Michigan Inter-University Community of Mathematical Geographers*, No. 9 (1966); Peter R. Gould, "Problems of Space Preference Measures and Relationships," *Geografiska Annaler*, Vol. 50, Ser. B (1968); Peter R. Gould and R.R. White, "The Mental Maps of British School Leavers," *Regional Studies*, Vol. 2 (1968), 161-182; Roger M. Downs, "Approaches to and Problems in the Measurement of Geographic Space Perception," *Seminar Papers of the Department of Geography*, University of Bristol, Ser. A No. 9 (1967).

Chapter 6

Gunnar Olsson: A Very Short Introduction

Michael Dear

> You cannot write anything about yourself that is more truthful than you yourself
> are. That is the difference between writing about yourself and writing about external
> objects. You write about yourself from your own height. You don't stand on stilts or
> on a ladder but on your bare feet.
>
> Ludwig Wittgenstein (1980 [1938]: 33e.)

For Gunnar Olsson, *homo sapiens* as writer (a) is naked; (b) possesses eyes in the back and front of the head; (c) elevates uncouthness to an art; (d) speaks in tongues; and (e) is strange, frightening, blasphemous, wrong, stupid, and incomprehensible. The rationale for this characterization unfolds in this essay, through Gunnar's own words and my recollections of our encounters in person, in print, and imagined.

The project

> [My] study objects ... remained the same: the power-filled relation between the
> individual and society, the crises of representation, the concepts of intentionality
> and purposeful political action. The specific questions also remained: How do
> I know the difference between you and me, and how do we share our beliefs
> in the same? (Olsson 2002: 260.) ... How do we become so obedient and so
> predictable? (Olsson 2002: 265.)

> It is exactly this issue of logic and rhetoric that has been the core of my entire
> life. (Olsson 2007: x.)

> Surely the most common purpose of human action is to topple truth, not to
> preserve it, to falsify rather than preserve what is now the case. (Olsson 2007: x.)

I first met Gunnar Olsson when I was a graduate student at the University of Pennsylvania in the early 1970s; he was visiting Philadelphia for a Regional Science conference. I already knew Gunnar by reputation; he was a general in the Regiment of Quantitative Geography though manifestly chafing at disciplinary confines (my favorite among his early mutant essays is Olsson, 1974). He was instantly approachable, throwing a carefree arm over my shoulder, guffawing heartily, and readily recruiting me as co-conspirator in his transgressions. In the looping cadences of his precise English, he made me feel that what I said was of value and that I had something to contribute. He treated me as an equal, though I knew better.

At later conferences, I observed people gravitate to Gunnar in spontaneous knots of intellectual combustion. So I joined the Gordian groupings. Sometimes, there would be informal salons in a hotel room, in a bar or a friend's apartment. At first, approaching him felt a bit like seeking an audience with some visiting potentate, except you could come and go as you pleased, imbibing heady drafts of intoxicating ideas and then stepping outside for some air. Truth be told, in those gatherings I teetered between a black hole of incomprehension and the exploding galaxies that Gunnar unraveled by going boldly where no geographer had gone before. Rarely had I met anyone so intellectually voracious, so enraptured by a life of the mind, so delighted by gate-crashing any and all boundaries. He could take the narrowest sliver of an idea and enlarge it with his warm, brilliant light.

Language

> [W]hen we encounter a new reality, our first reaction is to baptize it. We give it a name. This practice is necessary. Yet it is extremely inhibiting, for if we subsequently proceed as if word and object were the same, then we are liable to confuse use and mention and to commit ourselves to the preservation of things denoted. (Olsson 1979: 297.)

> [B]y committing ourselves to the scientific language, we effectively imprison meaning. (Olsson 1979: 302.)

> [H]ow can I think what I have not already thought and how can I say what is not already in my language. The best I can do is to move freely among schools of thought and modes of expression. (1980: preface, unpaginated, 1.)

Birds in Egg appeared in 1975 out of the University of Michigan as a fat duplicated typescript bound in soft covers that were decorated by daughter Ulrika's drawings, which were destined to become as famous as her father's words. This epic gained a rapid notoriety, not least because it represented such a loud departure from his previous research. My copy soon began to disintegrate after repeated handlings. The volume gained a new lease on life when it was re-published by Pion Ltd. in their Research in Planning and Design series, edited by Allen J. Scott. (Pion was already well-known as publisher of the *Environment and Planning* series.) The original text was supplemented by an extended series of commentaries and essays that formed a second book within the volume, entitled *Eggs in Bird*.

The legendary John Ashby, one of Pion's two directors, was renowned for his love of books, typefaces, paper, everything that related to the physical production of books. If you can, take a closer look at the book. It's possible to start reading the book at either end: *Birds in Egg* is one front cover; but flip the book and you begin again with a second front cover entitled *Eggs in Bird*. The two parts of the volume meet in the middle in a "Circular Embrace," or "Paz-way to desire," according to which direction you approach the book. The front cover of *Birds* has a drawing in red by Ulrika; the cover of *Eggs* is the same drawing in blue. Now look at the binding of the book: the stitching atop the spine of *Birds* is red and white, to match red Ulrika; but flip the book and you'll notice that the stitching of *Eggs* is blue and white, to match blue Ulrika. A beautiful filigree touch by Ashby, the master book-maker. You may find these details a little too esoteric, but Gunnar loved Ashby's attention to detail. I did too, when later I became editor of *Society and Space* in the *Environment and Planning* series.

Birds in Egg/Eggs in Bird established Gunnar as a major post-quantitative revolution thinker. He often refers to it as the work of his that has most influenced him. You came away from the book with Wittgenstein's phrases ringing in your head like an unshakable tinnitus: the limits of my language mean the limits of my world. It's also the book in which Gunnar's beloved Marcel Duchamp makes a sustained cameo appearance, and exposes his obsession with Duchamp's *Bride stripped bare by her bachelors, even*. Twenty-seven years later, in *Abysmal*, Duchamp and his bride still crowd Gunnar's consciousness; he refers to the work as "an outstanding map of what it means to be human," (Olsson 2007: 374)." Duchamp, he confesses "is the only person apart from my wife with whom I have had daily discourse for almost four decades" (Olsson 2002: 259).

Social science, Geography

> [Examine] that ethical glue that simultaneously keeps individual and society together and apart. It is the decomposition of this invisible glue that I see as the major task of critical social science. (Olsson 1991: 54.)

> Nietzsche's insight was that the identity of an individual is intimately tied to the fact that he simultaneously *has* a body and *is* a body. ... To experiment with one's own body is therefore a sophisticated technique for investigating who you are. Your skin is escape-proof. (Olson 1991: 146) ... [H]e who questions his body, thereby questions his culture. (Olsson 1991: 150.)

> Is it at all possible to draw valid inferences about human behavior from a map of spatial distributions, to reason from form to process? The short answer is that such inferences are not valid, a conclusion which in 1968 hit the disciplinary ship of Geography below the waterline. (Olsson 2007: x.)

Not long after I became editor of *Society and Space*, I received for consideration a paper from Gunnar entitled 'The Social Space of Silence.' Still a relative neophyte, I approached the paper with some trepidation. It was a set of fractured, seemingly tangential paragraphs, sometimes in poetry other times in prose, adopting variable typefaces, filled with puns and dense allusions. Gunnar knew what he was doing; this was his way of fighting the totalitarianism of codification, by writing in fragments and creating collage. Unable to find my way into or out of the argument, I made what I now understand was the worst possible response: I suggested Gunnar clarify his argument by contextualizing it a bit more. Amicably, we parted company and the paper was not published. But this was not the end of the story.

A year or so passed, and the journal was about to publish an issue including several papers with a philosophical bent. Truly I wasn't feeling guilty about refusing Gunnar; such an emotion would paralyze a journal editor. But his paper had lodged in my mind, and it occurred to me that the forthcoming essays were a perfect complement to Gunnar's earlier musings. Juxtaposed, they could set the scene for each other, and maybe even enter into dialogue. So I wrote to Gunnar suggesting we include 'Silence' and he agreed. The papers were published together in *Society and Space* (Olsson 1987). But this was not the end of my story.

A shortened version of 'Silence' was subsequently published in *Lines of Power/Limits of Language* (Olsson 1991). In dutifully acknowledging its original publication source, Gunnar gently mentioned that the essay had been submitted for publication to *Society and Space*, but then "promptly turned down." Yet three years later, he continued, "the rejecting journal came back and asked for it." (Olsson 1991: 216). I was grateful that he did not identify me as the active agent in the essay's temporary banishment, because yes, I had been acting as a "repres(s)entative editor" auditing the audience (Olsson 1991: 110). I consoled myself that at least I had corrected my earlier blunder, even if somewhat belatedly. But this was not the end of our story.

My desired rehabilitation was completed by the publication of *Abysmal* in 2007. In it, Gunnar describes a period and a project (the completion of his *Mappa Mundi Universalis*, wrought with Ole Michael Jensen) that brought him "some of the best moments of my life" (Olsson 2007: 411). The obsession with the *Mappa* went deep with both its creators, but Gunnar traces the project's most immediate origins to the two papers, his and Jensen's, that appeared together in a 1993 issue of *Society and Space* (Olsson 1993 and Jensen 1993). This pairing was one of my last acts as journal editor; and knowing that it became an inspiration to further collaboration of such personal significance is the kind of intangible that gives both pleasure to editors and meaning to editorial labor.

Postmodernism, education

> The braiding of epistemology and ontology is inevitable ...; whereas modernism deals primarily with epistemological questions of a relational nature, postmodernism experiments with ontological issues of fundamental importance. As modernism searches for purposeful answers to well-defined

riddles, postmodernism explodes in cascades of paradoxical situations. (Olsson 1991: 205.)

As a speaking subject, I have no choice but to express myself in a language that is common and social. And thus it is that as modernism becomes postmodernism, the epistemological search for knowledge transcends itself into ontological issues of Being. (Olsson 1991: 206.)

[W]hereas research sometimes can be creative, research training is always conservative. … There is a contradiction between creativity and socialization. Just as the overriding aim of the former is the creation of the new, so the overriding aim of the latter is the preservation of the old. (Olsson 1991: 28.)

The point is not to produce pre-ordered copies but to allow some genuine originals the freedom they need to be what they are. (Olsson 2002: 264.)

I am personally convinced that the young geographer interested in the power of the gaze and the politics of the Other has more to learn from recent art history than from the disciplining masters… (Olsson 2001: 264.)

In Oslo a few years ago, I scrambled for a ticket to the opening night of Henrik Ibsen's *Gengångare* (*Ghosts*) performed by the Royal Swedish Theater under the direction of the great Ingmar Bergman. I had discovered Ibsen independently during my grammar school days in Wales; needless to say, he was not in the official school curriculum (neither was Tennessee Williams, August Strindberg and so many of my other favorite writers). So, I knew Ibsen's play pretty well and such was the power of the performance it mattered little that I did not understand the language. The production faithfully moved along its familiar groove of doom – although the sexual interplay between son Osvald and Regina the maid was much more demonstrative than is customary (well, it was Bergman after all). At the end of the play, Osvald's impending insanity (a legacy of his father's dissipation) renders his mother speechless as she disengages from her son's embrace and watches in horror as he slumps vacantly on a couch. The stage directions in my English-language version of the play (Ibsen 1964: 102) have Osvald, motionless, affectless, begging his mother to give him the sun. Curtain.

Except that he didn't slump across the couch. Instead, under Bergman's direction, the Oslo Osvald stripped off all his clothes and lay naked, center-stage in a foetal curl, illuminated by the morning sun's first rays. The play's climax was transformed into a tableau of birth and death. Moreover, this stunning dénouement shifted the play's crescendo from a mother's future of suffering without end, to the object of her despair: Osvald, with nothing to hide, no more ghosts to endure, in a profound and truthful expression of self within his skin. Gunnar would, I think, applaud Bergman's audacious transformation. Bergman had altered our way of seeing Ibsen's play literally by stripping open its true heart.

Gunnar has especially valued artists because they break the rules more readily and frequently than social scientists. Duchamp above all, of course, and most certainly Magritte and Lewis Carroll. Since *Birds*, engagement with art (especially painting and sculpture) has been a prominent feature of his work. His choice of illustration for his book covers are alone cause for speculation: for instance, the child's drawings in his earliest dash for intellectual freedom (Olsson 1980); his choice of Wassily Kandinsky's modernist geometries for his first sustained exploration of the cartographies of reason (Olsson 1991); and the delightfully corrupted Vermeer in collage with Kent Karlsson and Gunnael Jensson for his magnum opus (Olsson 2007).

Now, I would not dream of putting words into Gunnar's mouth. That would be the most unforgivable of transgressions! But he understands more than most the enduring, deeply radical challenge posed by postmodern thought, which I will telegraph simply as the unavoidable uncertainty overshadowing our truth claims. Perhaps Gunnar and I would shake our heads in unison as young and old dismiss postmodernism as an out-of-date fashion (while insisting, naturally enough, on the enduring significance of their own intellectual persuasions). The ontological challenge posed by postmodernism is inescapable, as much as our own skins. We can ignore it, assume it away, or bludgeon it with politics; but it remains.

Politics

> I gradually came to see Swedish politics not as the profitable farce it may well be but as an updated version of classical tragedy – everything beautifully right at the beginning; everything horribly wrong at the end; no one to blame in between. (Olsson 2007: x.)

> [As] a devout nonpolitician, with a conservative bent, I am spending a great deal of joyful time with leftists, albeit coming from a variety of congregations. (Olsson 2002: 267.)

> But if instead [these discourses] were to lure just one single student into the forbidden grounds, then this romantic comedy will proceed. (Olsson 1982: 264.)

Olsson's transgressive conservatism – if that is what it is – would easily appeal to intellectuals of many persuasions, including those pitching their tents on the multi-hued left. In the early days, I was never aware of his politics; I simply made assumptions given the company he was keeping and his country of origin (I know, I know…). But Gunnar was not a soap-box orator despite the clear outrage he felt and expressed about diverse human injustices. Now, as he ages imprecisely (like us all), Gunnar is perhaps more willing to let his veil trail. People tend to get more conservative as they grow older, or so it is said. Perhaps he has become more willing to articulate his view of politics, goaded by the obscene perversities of the present. Life's lessons are often hard, unless one is not paying attention.

As expressed in *Abysmal*, Gunnar's lifeworld – that of a self-described critical social scientist – contains a particularly jaundiced view of formal politics. Yet his work is an intricate tapestry of the micropolitics of everyday life and the potentials of democracy through human action and agency. He is drawn back again and again to one of our deepest terrors: why and how do we become so docile? For all that, *Abysmal* has few references to Michel Foucault and none to Henri Lefebvre. You can discover other unexplained lacunae by browsing the indexes of his books. Olsson's political project remains an unfinished symphony, but then, so is mine.

Encore!

> We live forward and understand backward. This explains why in retrospect I can detect exciting parallels between the development of my own work and that of Wassily Kandinsky. (Olsson 1991: 5.)

> ... Janus, my own favorite among gods. (Olsson 2007: 5.)

> The time has come for dancing as naked as we are. (Olsson 1979: 303.)

> Every reader – especially the specialized expert – is therefore bound to run into passages [in *Abysmal*] (s)he will find strange, frightening, blasphemous, wrong, stupid, and incomprehensible. (Olsson 2007: xi.)

The reader should not be surprised if my representation of the Olsson project differs from everyone else's in this volume. Mine is, after all, a personal reflection on the enormous canvas of his work. It tells you more about how I perceive the exciting parallels connecting my work to his (though not necessarily vice-versa).

Gunnar Olsson was/is inevitably a child of his times, even as those times continue to unfold. In the broadest terms, his continuing search over four decades for what it means to be human has intersected with a set of powerful intellectual shifts that propelled us from the strident certainties of Quantitative Geography through to the radical uncertainties After Postmodernism. The key to his search is language, but his method is madness, involving an embrace of the ill-mannered and the uncouth; these are the deliberate tactics of someone dedicated to casting off the conventions of discipline. Gunnar would prefer to see us naked, too.

Gunnar Olsson is a critical social scientist focused on the invisible glue that keeps individual and society together and apart. All knowledge begins from what we observe (hence the power of the visual), but Geographer Olsson rejects a simplistic reasoning from form to process, even though he approvingly quotes Edward Casey: "There is no creation without place" (Olsson 2007: 4). I do not think that Gunnar intends to deny the project of Geography, but merely to insist that our earlier conflations of space and society were naïve. Contemporaneity in time and space is not necessarily an indication of connectedness or causality. But I have this ineffable sense that he has lost some of his bearings. In an overtly autobiographical essay, 'Glimpses,' Gunnar identifies Franco Farinelli as "the only geographer with a significant impact on my own thinking" (Olsson 2002: 265). Leaving aside Farinelli's undisputed wisdom, this is a curiously depressing admission, especially given the long lists of indebtedness to geographical friends and colleagues that invariably decorate Olsson's books. However, just as I earlier refrained from psychoanalyzing his choice of artists, I will leave it to him to elaborate on this self-willed exile.

Gunnar Olsson has learned many languages, spoken in many tongues, and moved freely among myriad schools of thought and modes of expression. His polyvocality has taken in the visual and the mathematical, the spiritual and scientific, and the fragmented and formalistic. The key to his method is triangulation, a "modus vivendi" of mapping and cartographical reason (Olsson 2007: 184). Triangulation is also widely understood as a general analytical imperative favoring multiple perspectives. This is why, Janus-like, we need eyes in the back of our heads whenever we elect to travel with him, and his four constant companions: Duchamp, hard-wired into the Olsson cortex, cannot be left behind under any circumstances; Wittgenstein comes along for the conversation; Bergman casts a piercing Scandinavian eye over everything; plus Derrida, who simply enjoys skewering everyone's cherished intellectual pretensions. In conversation, this happy five – Olsson and his Four Musketeers – sets our universe spinning. It reminds me of Barbara Stafford (2007) whose visual bricolage of proliferating images and colliding perspectives are a method of undermining existing categories and logic, resisting easy reconstruction because the sheer number of images is just too great to order.

One last thing. I cannot tell you how many times during the writing of this essay that I inadvertently wrote *Asylum* when I meant *Abysmal*. Whether I was thinking of a place of refuge or place of confinement, I cannot say; maybe both. Nor am I sure how or why I made this slip, but it is the sort of puzzle that amuses and engages Gunnar. Perhaps one day he will explain it to me.

Acknowledgements

Thanks to Gordon Clark who in 1981 introduced me to the Wittgenstein quote that prefaces this essay; and to Christian Abrahamsson and Martin Gren for the invitation to participate in this project.

Bibliography

Ibsen, H. 1964. *Ghosts and Other Plays*. (Translated by Peter Watts) London: Penguin Classics.

Jensen, O.M. 1993. Red River Valley: geo-graphical studies in the landscape of language. *Society and Space: Environment and Planning D*, 5, 295–301.

Olsson, G. 1974. The Dialectics of Spatial Analysis. *Antipode* 6(3), 50–62.

Olsson, G. 1979. Social Science and Human Action or Hitting your Head against the Ceiling of Language, in *Philosophy in Geography* edited by S. Gale and G. Olsson, Dordrecht: Reidel Publishing, 287–307.

Olsson, G. 1980. *Birds in Egg/Eggs in Bird*. London: Pion.

Olsson, G. 1982. Epilogue: A ground for common search, in *A Search for Common Ground*, edited by P. Gould and G. Olsson, London: Pion, 261–264.

Olsson, G. 1987. The Social Space of Silence. *Society and Space: Environment and Planning D*, 5, 249–262.

Olsson, G. 1991. *Lines of Power/Limits of Language*. Minneapolis: University of Minnesota Press.

Olsson, G. 1993. Chiasm of Thought-and-Action. *Society and Space: Environment and Planning D*, 5, 279–294.

Olsson, G. 2001. Washed in a Washing Machine™, in *Postmodern Geography: Theory and Praxis*, edited by C Minca, Oxford: Blackwell Publishers, 255–281.

Olsson, G. 2002. Glimpses, in *Geographical Voices: Fourteen Autobiographical Essays*, edited by P. Gould and F. Pitts, New York: Syracuse University Press, 237–268.

Olsson, G. 2007. *Abysmal: A Critique of Cartographic Reason*. Chicago: University of Chicago Press.

Stafford, B. 2007. *Echo Objects: The cognitive work of images*. Chicago: University of Chicago Press.

Wittgenstein, L. 1980. *Culture and Value*. Von Wright, G.H. (ed.), translated by Peter Winch. Chicago: University of Chicago Press.

Chapter 7

SERVITUDE AND INEQUALITY IN SPATIAL PLANNING: IDEOLOGY AND METHODOLOGY IN CONFLICT

GUNNAR OLSSON

I

In these notes I will draw attention to some curious problems we encounter when we use a family of well-known social science models as a basis for social, economic and regional planning. In this sense, I will deal with the problem of extending descriptive social science into prescriptive social engineering. In the process I will provide yet another illustration of the fascinating usefulness of the Hegelian concepts of self-consciousness, lordship and bondage.[1]

My main point will be that the mathematical and ideological foundations of the Pareto model—which includes such formulations as the gravity, rank-size and Clark models as special cases—make us suspect that whatever planning we base on it will be counter-productive. In fact, I have gradually and rather painfully come to the conclusion that if we continue along the methodological and manipulative path we have been following thus far, then we run the risk of *in*creasing those social, economic and regional inequalities, which the planning initially was designed to *de*crease; our good-natured attempts to rectify current injustices will be self-defeating, not because some vicious bureaucrat designed them that way, but because we have failed to understand the deep structure of social research and action.

Most of my specific references will be to regional planning in Sweden, but my main point applies more generally. My emphasis of the Swedish experience is more due to my own background and to my self-conscious attempts at understanding myself and my own society than to the uniqueness of my examples. And yet I know of few countries where descriptive social science has been translated more literally and with less modifications into prescriptive social engineering.[2]

Reprinted from *Antipode* 6, no. 1 (1974): 16–21, by permission of the publisher and the author.

It is from the slavishness of this translation that my main argument stems. To appreciate this remark, it should be recalled that the stated purpose of regional planning in Sweden has been to achieve equality in the ideal sense that somebody who lives in the valleys of the socio-economic undulating surface should have the same opportunities as somebody who lives at its peaks. I am not, of course, arguing against this ideal, for how could one say that someone who happens to be born in a far away village in northern Sweden should not have the same rights as someone who was born in Stockholm? My only quarrel is that this laudable piece of welfare ideology has been put into practice by means of a scientific methodology which reflects just the opposite thinking. To be more specific, the planning has been based on a variant of the social gravity model, which has the same mathematical form as the Pareto function. In this sense, the descriptive social gravity model encapsulates exactly those relations of *in*equality that characterize both Pareto's optimality principle and his Machiavellian theory of the elites.[3] I feel rather strongly that this mismatch of ideology and methodology has contributed to discontent and alienation which is becoming more and more visible. The proof of the pudding is in the election results; under the heat of methodology, ideology lost its flavor.

To substantiate this argument, I will limit myself to only one of several examples. More specifically, I will draw attention to the far reaching revamping of administrative and political districts carried out during the last decade.

The *reason* why this reform was deemed necessary was that the old division reflected a society in which the horse was the main mode of transportation and the local church was the prime purveyor of public service. With the advent of rapid technological change and with the widespread acceptance of welfare ideology, these administrative units became too small to afford the high level of education, health and social services which the population rightly was demanding. The *purpose* of the reform, then, was to create spatial units large enough to sustain the considerable burdens of the welfare state.

The *ideology* underlying the reform was a refined version of the classical ideal of human equality. The refinement lay in the argument that exactly as society guarantees that all people have the same rights regardless of whether they are rich or poor, so it must guarantee that everybody has the same rights regardless of where he or she happens to live. The goal was to abolish the spatial element of social and economic inequality.

This all sounds very good, until it is realized that the *methodology* through which the new degree of equality was to be implemented is of a rather different kind. Most importantly, the new and larger areas were designed to cause as little disturbance to the people as possible. As a consequence, the delimitation was made to reflect observed spatial interaction patterns as these revealed themselves through a variant of the gravity model. It was in fact data fitted to this particular version of the Pareto function that played a major role in the determination of the new boundaries. Essentially the same procedures were involved in the creation of the new educational system and in the reorganization of health care.

In summary, all these reforms were necessitated by technological and ideological changes, which in turn required adjustments of spatial boundaries. The purpose was to adjust society to change, but the analytical techniques by which the solutions were determined were more geared towards the preservation of status quo.

To be more specific, the spatial structure of the present Swedish welfare state has deliberately been built to reflect the structure of existing interactions patterns as these exerted themselves in a Pareto-type model. Thus, what happened was that a group of academicians—mainly geographers—went into the field of census taking, observed how people interacted over space, translated these observations into the positivistic language of a variant of the gravity model, determined from these fits where the boundaries between service areas actually fell, and then, finally, convinced the political decision makers that these boundaries were efficient boundaries to which the administrative and political areas ought to be adjusted. The approach clearly involves a reasoning from *is* to *ought*. In the process, the initial purpose of creating a just society became altered to that of finding a set of efficient solutions to a problem of geometric partitioning.

Unnoticed to spectators and performers, the play was changed in the middle of the act. The *ought* of justice disappears in the wings, invisibly stabbed by the *is* of methodology. Exit man with his precious visions, hopes and fears. Enter the Thiessen polygons with their crude distance minimizations and cost-benefit ratios.

II

I have just described how a profound misunderstanding of scientific methodology contributed to the institutionalization of a spatial organization which is most likely *not* to increase the level of equality. Instead it will reinforce that particular conception of man which lies at the heart of virtually all aspects of Pareto's voluminous, ambitious and influential work. This is truly ironic, for the conception of man embedded in these writings is completely opposite to the conception of equal and dealienated man which the Swedish politicians, planners and social scientists presumably were trying to institutionalize. To illustrate the deep structure of this miscarriage I must now provide a critique of the method which forms the bedrock of Pareto's models.

The starting point of the critique is that both Pareto and those who performed the analytic studies in Sweden allegedly wrote on what people actually do and not on what they ought to do. Pareto himself argued that observed behavior of this kind falls into a class of activities which he named "logical action." Into this same kind of action, he classified the activities of economic profit maximization, Machiavellian politics and scientific work.

But to call observed behavior "logical action" and to write about profit maximization and manipulative politics under the label of rational behavior is nothinq less than a sophisticated technique for rationalizing status quo. The reason is of course that in the minds of modern man the concept of logic has assumed the position which once was occupied by God himself.

But this deified view of logic is as mistaken as any absolutist religion; like all other dogmas, it is designed for escapists and serfs and it serves the interests of manipulators and masters. Instead of perpetuating the belief that the reasoning rules by which we structure our thought and action represent objective and unassailable a priori principles, we must learn that they are neither ethically nor aesthetically neutral. If we perform this intellectual striptease, then the first secrets we will uncover is that grammar tells us what kind of object anything is. As a consequence, we will understand that our analytical languages do not only provide labels which we attach to the objects we are talking about, but that they serve also as instruments, through which we determine and influence the relations among the phenomena we are dealing with. Given this conception of reasoning, the role of any language is not to describe reality so much as to shape it. And that holds for all languages no matter whether they are verbal or mathematical, natural or artificial, poetry or sculpture, touch or glance.

It was these crucial aspects of the internal relations between thought, language and action that Pareto and his many followers overlooked. To call some empirically observed actions logical and others non-logical is therefore highly misleading, for it deifies the existing by giving the impression that the particular categorial frameworks of Aristotelian logic are eternal and God-given. What is lacking in this view is an acknowledgment that all concepts are man-made tools, which become intelligible only in the context of our own personal and social lives.

It seems that if we dare not admit that our analytical languages have these characteristics, then we run the risk of imposing on reality a strictness which it neither has nor ought to have. If we in our roles as social scientists, citizens, and social engineers do not recognize this hallmark of positivistic methodology for what it is, then we may well be left with a society which mirrors the techniques by which we measure it and echoes the language in which we talk about it. At the end lies a society of human puppets with no dreams to dream and nothing to be sorry for. Instead of implementing plans which will aid man in his striving for becoming, we are entangled in so-called descriptive, objective and analytic techniques, which will produce just the opposite. Instead of building a world for the constant groping of autonomous man, we are on the verge of confining ourselves within spatial and social prisons which will serve only to increase our sense of loss and futility.

Even though spatial planning of the type I have discussed in these notes contributes only a small share to the formation of our future, it nevertheless contributes. My reason for this assessment is that the descriptive gravity model reduces to the mathematical Pareto function and thereby to the notion of logical action. But at the bottom of the latter is nothing less than Pareto's famous optimality principle, which defines an efficient equilibrium solution as a solution in which no person can be made better off unless someone else at the same time is made worse off. For instance, a distribution in which one person receives all benefits is Pareto efficient, because there is no rearrangement that will increase the utility of other individuals without at the same time worsening the situation for the one who initially owned everything. This is a problem of distributive justice which will become even more pressing once we push ourselves deeper into a no-growth society in which only small amounts of surplus will be available for redistribution.

By suggesting that there is a close methodological and mathematical connection between the Pareto function and the social gravity model on the one hand and Pareto's optimality principle and his theory of the elites on the other, I may be pointing at something very important. This is that Pareto never spoke about the ends which the members of society distributively *should* pursue. Instead, he always equated society's end with the ends which the Machiavellian elite actually *is* pursuing, through its manipulation of things, money and people.

Pareto's *ought* is therefore the *is* of the elite. This is crucial, because it means that the establishment of so-called empirical and logical relations is not limited to the manipulation of the symbols in a mathematical equation. Instead it represents relations between groups and classes of people. If we are not self-conscious about this deep structure of methodology, then man will end up as being doubly thingified, first by the producing-consuming forces of industrial technology, and then by the iron claws of scientific practice.

And so it is that Pareto and his influential followers came to equate the pursuance of logical reasoning and empirical truth with the advocacy of an elitist theory of society. To illustrate, the spatial and organizational set-up of the Swedish health, education and welfare system seems far more geared towards the growth and maintenance of its own bureaucracy than towards the interests of those sick and disadvantaged which it is supposed to serve. In this light, it is not surprising that one critic has labeled Pareto "the Karl Marx of the Bourgeoisie," while another has called him "the greatest rationalizer of authoritarian conservatism in our time."[4] How one should interpret the fact that Mussolini made him a senator is less clear, for at that time he might have been old enough not to see the difference between being honored and being used.

The irony is of course that it is a portion of Pareto's work—disguised from most people's recognition by its neat mathematical form—that lies at the heart of most spatial planning, no matter whether it is in Sweden or elsewhere. What is most remarkable of all is that the analysts have been so entrenched in their own social and professional relations that they never discovered that in the Pareto function, in the optimality principle, and in the theory of the elites is embedded exactly the system of inequality and servitude, which the plans were meant to alleviate.

III

In retrospect, it appears that the majority of spatial analysts—among whom I certainly include myself—have confined ourselves so thoroughly within our inherited concepts, within our categorial frameworks, within our particular mathematical language, and within our artifacts that we thereby have helped to perpetuate the functional inequalities of the past. In fact what we seem not to have realized is that in order to acquire a new world, we must at the same time acquire a new analytical language, less dogmatic than the old, but no less abstract and no less difficult.

Perhaps we should be blamed for having spent so much energy on memorizing the Isards and the Garrisons and so little on trying to understand Hegel, Marx and Wittgenstein. But such a criticism would surely be misdirected, for if we had never learned from the former, we would never have understood the latter. It is in this context that I admire so much the profundity of that Hegelian argument which says that any development comes through the subordination of one self to another; "for in shaping the thing [the consciousness] only becomes aware of its own proper negativity, its existence on its own account, as an object, through the fact that it cancels the actual form confronting it. . . . Thus precisely in labour where there seemed to be merely some outsider's mind and ideas involved, the bondsman becomes aware, through this rediscovery of himself by himself, of having and being a 'mind of his own.' "[5]

What most of us failed to pursue in our previous studies was exactly that element of self-consciousness which leads to the recognition that to engage in any form of language—of which formal model building is only a special kind—is to engage in an activity that goes far beyond the mere labeling and categorizing of phenomena; to speak, to estimate statistical functions, and to build scientific models, is nothing less than to act. The essence is in action, for even the attempt to escape it is itself action.

Put somewhat differently, it now appears far from clear whether what the analysts and planners are telling us conveys more information about the world they are talking *about* or about the language and the value system they are talking *in*. The root of this problem lies in the inability of our standardized model languages to capture the dialectical concepts of indeterminacy and qualitative change. I would even suggest that if we can not find ways to internalize these fleeting concepts into our thought and action, then we can neither hope to understand the deep structure of the past nor to form a freer, more dignified and less alienated future.[6]

IV

There is of course much to suggest that the vision of human freedom I have been painting in these notes is nothing but a romantic dream. Perhaps the most compelling of these pessimistic warnings is that in the coming years of increased stress and material shortages, anything resembling the concepts of freedom of thought and action will be luxuries limited to the class of charismatic leaders. Accompanying the recurring crises will almost certainly be demands for strong leadership, which can sustain itself only by resort to the higher principles of "rigorous" thinking and absolute values, no matter how simplistic and dehumanizing they may be.

If this is the human prospect—and there are many who say it is—then it is with considerable anguish that I publish these notes. The reason for my agony is that what I have written here illustrates so well how scientific methodology can be made the handmaiden of authoritarian ideology. Thus, I have noted that what we initially may take as empirical generalizations and theoretical deductions are as much reflections of those notions of equality, which we may hope to implement through our planning. As I demonstrated, methodology overtook ideology and in the process it became ideology itself. And yet it is through the labor for his master that the slave comes to self-consciousness and thereby to the realization of his own existence and freedom.

It is in this manner, by employing analytical techniques and social engineering devices which are founded on rather peculiar assumptions about categorization and linearization, that we in effect have come close to creating a society for human beings who themselves are peculiarly categorial and linear. If regional planning in Sweden, the Soviet Union and the United States have nothing else in common, it is exactly this simplified and dehumanizing conception of man. Instead of creating a world for becoming, we are creating thingified man; by treating the relations between people as if they obediently followed the multiplication table, we are ridding ourselves of that challenging ambiguity, which alone makes life worthwhile.

And that is the kind of world I see evolving, not in a distant future, but already in the present. For as a consequence not only of material shortages, but also of our well meaning attempts at planning, increasing numbers of people now find it harder to live. As an example, I call it nothing less than a fearful tragedy that the values and organization of health care have made it a rare privilege to die with one's eyes open and in the company of the few whom one loves. What I am conjecturing from my own limited experiences is that one of man's most fundamental rights is under serious attack, not because anyone wanted it this way, but because our methodological blinkers have kept us from seeing the deep structure of human and social relations. This right is not the right to live as one chooses, but to die in peace and dignity. What technology once bestowed upon us by letting our children live, it is now on the verge of reclaiming by not letting our tired die.

V

Since I see traces of all these prospects even in the best land I know, it is difficult to think of the rest, where freedom is just another word and where the issues of life and death, justice and decency, are of another magnitude. The best I can do may merely be to quote from Samuel Beckett's *Krapp's Last Tape*:

> Here I end this reel. Box 3. Spool 5. Perhaps my best years
> are gone. When there was a chance of happiness. But I
> wouldn't want them back. Not with the fire in me now. No,
> I wouldn't want them back.

how i wish it applied only to the holder of the secret tape

Notes

1. For other aspects of the descriptive-social-science-prescriptive-social-engineering problem see my paper "Some Notes on Geography and Social Engineering," *Antipode* 4, no. 1 (1972): 1–22. The best account of self-consciousness, lordship and bondage is in Hegel's *The Phenomenology of Mind*, pp. 228–40 of the Harper Torchbook edition, 1967. For an excellent treatment of the relations between Hegel's *Geist* and Marx's interpretation of it, see Richard J. Bernstein, *Praxis and Action* (Philadelphia: University of Pennsylvania Press, 1971), esp. pp. 14–28 and pp. 34–43.

2. The ideology and methodology of the Swedish reforms have been described in various publications such as Statens Offentliga Utredningar 1961:9, *Principer för en ny kommunindelning*, Stockholm, 1961; Arne Jakobsson, "Revision der Gemeindeeinteilung in Schweden," *Raumforschung und Raumordnung*, Jahrgang 22, 1964, pp. 177–92; Sven Godlund, "Population, Regional Hospitals, Transport Facilities, and Regions. Planning the Location of Regional Hospitals in Sweden," *Lund Studies in Geography*, Ser. B, no. 21 (1961); and Statens Offentliga Utredningar 1972:23, *Högre utbildning—regional rekrytering och samhallsekonomiska kalkyler*, Stockholm, 1972.

3. Pareto's original ideas are translated and most readily available in Vilfredo Pareto, *The Mind and Society*, 4 vols. (New York: Harcourt, Brace and Co., 1935). The most important commentaries and positive extensions of Pareto's ideas were first presented by Talcott Parsons in *The Structure of Social Action* (New York: McGraw-Hill, 1937), esp. Part I. For criticisms of both Pareto and Parsons see e.g., Alvin W. Gouldner, *The Coming Crisis of Western Sociology* (New York: Avon, 1971), esp. pp. 138–57. The broader setting within which the various brands of scientific naturalism developed is well presented in Edward A. Purcell, Jr., *The Crisis of Democratic Theory* (Lexington: University of Kentucky Press, 1973).

   The mathematical similarities between the gravity model and the Pareto function have been noted often, but usually not explored in any detail. For the best attempts see Jean-Michel Goux, "Structure de l'espace et migration," in J. Sutter, ed., *Les déplacements humaines* (Monaco: Entretiens de Monaco en Science Humaines, 1962); Richard L. Morrill, "The Distribution of Migration Distances," *Papers and Proceedings of the Regional Science Association* 2 (1963): 75–84; and Peter J. Taylor, "Distance Transformation and Distance Decay Functions," *Geographical Analysis* 3 (1971): 221–38.

4. I have borrowed the catchy characterizations of Pareto from Barbara S. Heyl, "The Harvard 'Pareto Circle'," *Journal of the History of Behavioral Sciences* 4 (1968): 317; and from H. Stuart Hughes, *Consciousness and Society* (New York: Vintage, 1961), p. 82.

5. Apart from reading Hegel's, Marx's and Wittgenstein's original works, I have benefitted greatly from the comments in Walter Kaufmann, *Hegel: A Reinterpretation* (Garden City: Doubleday Anchor, 1966); Bertell Ollman, *Alienation: Marx's Conception of Man in Capitalist Society* (Cambridge: Cambridge University Press, 1971); Hanna Fenichel Pitkin, *Wittgenstein and Justice* (Berkeley: University of California Press, 1972); as well as from the book by Richard Bernstein listed above under note 1. The quotes from Hegel are from the mentioned edition of the *Phenomenology*, p. 239.

6. There has been of course a continuing reevaluation and extension of the Swedish experiments, not the least by those geographers who themselves have played a decisive role in the shaping of the new society. However, none of them seem to have raised any of the fundamental issues I have dealt with in these notes. For some of the recent commentaries see e.g., articles by Bylund, Godlund, Hägerstrand, and Tornqvist in Statens Offentliga Utredningar 1970:14, *Urbaniseringen i Sverige*, Stockholm 1970; and ERU, *Regioner att leva i* (Stockholm: Allmanna Forlaget, 1972). Also see the detailed account provided by Allan R. Pred, in his "Urbanization, Domestic Planning Problems, and Swedish Geographic Research," C. Board et al., eds., *Progress in Geography*, vol. 5 (London: Edward Arnold, 1973).

Chapter 8

Between the Castle and the Trial:
The Spaceless Spaces of Planning

Tom Mels

CHOIR My definitions of ideology and planning are closely related. Both activities are nourished by the legitimating interplay of mystification and domination. It follows that planning is an ingredient of that ethical glue whereby the is of the past and the ought of the future are bound together

– Gunnar Olsson.

Das Gericht will nichts von Dir. Es nimmt Dich auf wenn Du kommst und es entläßt Dich wenn du gehst.

– Franz Kafka.

Spaceless spaces are the geographical expressions of a denial in which emptiness figures as a substitute for living meaning and matter. Planners should know all about them. Spaceless spaces: an outrageous contradiction in terms, a paradoxical oxymoron, but also the exact expression of an ideology in which denial of denial and other defense mechanisms are rife, and words seldom mean what they mean.

On an allegorical path to a critique of modern power, through the spaceless spaces of planning, there is a Castle and a Trial, a Lock and a Process. Along comes the haunting specter of K., a living dead looking for direction in the disorienting ciphers on Gunnar Olsson's maps.

Searching for keys

Although Gunnar Olsson's writings on power, language and representation have been widely discussed within geography, his critique of planning has gained far less attention. Planning is interesting not only because it remains a central theme of Swedish geography and part of the institutional and everyday backdrop of Olsson's oeuvre. It is also of interest because the fundamental problems of planning are very much at the heart of some of the core terms that make up Olsson's theoretical vocabulary: Ontological transformations (the shifts between mindscape and stonescape); the is-ought problematic (description becomes prescription); the equal sign (thrusting an = between unequal entities); cartographical reason; these and several other themes come together in the empirical case of planning.

Indeed, planning is one of the rare, more obviously empirical (rather than generic) examples from everyday life that recur in Olsson's work.

In this chapter, I will draw out a few lines of thought from Franz Kafka as a way of accessing and positioning Gunnar Olsson's critique of planning. First, my juxtaposition may help to understand and accentuate the *distance* between Olsson and the tradition of thinking about planning in Swedish geography. Swedish geography has a long record of involvement in applied research and planning. Much of this involvement could be described as mainstream in the sense that it usually proffered an active cooperation with official administrations rather than a critical detachment from state power and taken-for-granted practices. In many of the written histories of Swedish geography, this involvement is valued as a part of the discipline's success story. Renowned figures such as Torsten Hägerstrand and William William-Olsson come to mind, along with a heavy twentieth-century tradition of regional planning. Seen from this angle, it is not difficult to understand why Gunnar Olsson's writings on planning – a term which in his work appears largely in a pejorative sense as synonymous with a baleful social engineering – have provoked a certain degree of resentment and indeed distancing. A writer who associates himself with the critical humanism of Kafka cannot at the same time socialize comfortably with planning-geography.

Second, and perhaps less evidently, the juxtaposition may also help to bring out Olsson's engagement *within* this tradition. Unlike what he would see as the silent and grey masses of academic bureaucrats, the outsider-hero is often as easily discernible as a colorful bird, at least if outsideness is tactically positioned as an occasionally earsplitting Professor of economic geography and planning. Part of this is a question of seeing how the defying exception not only defines the rule, but also sometimes facilitates efforts to alter it. While at times, Olsson has described his own position as a K. in the trials and castles of geography (for example Olsson 1994) it is also evident that decades of service within the discipline have not passed unnoticed. While I will not have space to expand on this point, Olsson's influence on his own students' appraisal of planning is evident (I think of Jette Hansen-Møller, José Ramírez, Ole Michael Jensen, and other former doctoral students, but also of Olsson's more extensive efforts as a lecturer). Hence, it would be silly to regard his work as some transitory and silenced aberration on the outer edges of the field. It would be more reasonable to see it as a significant part of planning-geography in Sweden.

These two approximations – experiments in an interlocutory and partial reading of Olsson – emphasize the complex relationships of Olsson's work to various socio-historical and intellectual surroundings. These two loops through his work will then conjoin in a focused (and again politicized) attention to the import accomplished by virtue of this work's own, internal content-form dialectic. The reason for this is not primarily because this is such an obvious quality of Olsson's style, but because I believe that it allows for a better grasp of its external dynamics.

The chapter opens with a very brief description of the traditional links between planning and academic geography in Sweden. With a little help from Kafka it then moves on to (dis)locate Gunnar Olsson's work within and beyond this context: in Kafka's and Olsson's worlds alike, we are constantly both inside and outside the spaceless spaces encountered. The essay ends by asking questions about what kind of politics this position involves in postmodern or late modern times, and to what extent it overestimates the internal homogeneity of planning.

Norms: lineages of planning-geography

To understand human action is never to blame. It is instead to recognize that every actor is so entrenched in his role that he takes the shadow play to be reality and reality to be the play. It is indeed an integral part of all internal relations (and thereby of all ideologies and all mythologies) that we obey their commands without hearing them, and without knowing where they come from (Olsson 1991: 20).

It is not far-fetched to claim that these lines were inspired by personal experience of planning-geography in Sweden. The extensive alliance between geography and planning has been described in detail elsewhere (Buttimer and Mels 2006). Here, I will only be concerned with drawing out some important features of the long history of this bond because it may help to contextualize Olsson's work.

One defining moment in the early twentieth century was the modern welfare state's need for resource inventory and national surveys. This Linnaean project was well in line with ongoing engagements by geographers with for example national planning policies and the rationalization of agriculture and industry. Cartographic techniques were of central concern. For example, Sten De Geer's famous population mappings were of immediate practical use for the issue of regionalization, then one of the major tasks of national planning. De Geer recognized that such maps were invaluable to the planning and organization of space since they could be used for adjusting all sorts of institutional boundaries. His immediate practical engagements concerned the planning of counties, postal services, railways, commerce, roads, settlements, and the patterns of industrial location – an occupation which resonated with an understanding of geography as a science of spatial distribution. This approach highlighted some of the key challenges for many decades to come, including not just cartographic representation but also the analysis of mechanisms and processes underlying spatial form.

De Geer's student, William William-Olsson continued proceeded the combination of mapping and a deep involvement of research in current planning problems. William-Olsson's 1937 study of Stockholm revisited the form-process problem by inductive reasoning from urban patterns to expectations about general developments. He also constructed a theory of differentiation governing cities, and later efforts made him one of the early promoters of modern location theory.

At Uppsala, Gerd Enequist (Gunnar Olsson's mentor-to-be) used methods of cartographic representation of the existing structure of the Swedish economy, similar to William-Olsson's. Her work was of some importance to the Census and for comprehensive road planning. In 1950, Enequist was engaged in the *Towns and Hinterlands* conference, which would reinforce the position of planning-geography. The report from this conference featured contributions by Karl-Erik Bergsten, Sven Dahl, Sven Godlund, Torsten Hägerstrand, and Edgar Kant. New, more rational forms of planning were expected to be within reach through the application of positivist principles: Newton's law of gravity, Pareto's model on distribution, and Christaller's central place theory. In the decades to come, a host of empirically and statistically geared studies focusing on spatial differentiation would appear. Many of these were of immediately practical utility and most would aim at empirically determining the spatial differentiation of phenomena. Geostatistical material was converted into maps of existing patterns, which in turn would be the basis for policy proposals (Öhman 1994: 89).

Looking back upon his applied work of the 1960s and 1970s, Torsten Hägerstrand claimed that it was in the negative shock of experiencing the environmental, social, and political conditions in American cities that he found a justification for his engagement in Swedish planning: 'If we let industry and technology go as they please, then we quickly get that kind of society' (Mels and Buttimer 2006: 116. The quotation is from a 1978 interview). In a later recollection, he diminished the importance of his 'social and environmental idealism', and instead presented it as a more utilitarian question of opening up 'a new labour market for our advanced students' at Lund University (Hägerstrand 1983: 252). The success of this move 'eventually became a loss for the department, because a whole generation of graduate students disappeared into the world of planning. But on the other hand, these persons were able to demonstrate to the administration how useful geographers could be in practical matters. Within a decade they had all reached central positions in the new planning hierarchy' (Hägerstrand 1983: 252-253). Hägerstrand's own engagements featured massive projects such as the reorganization of municipal boundaries and the provincial government, the development of a national resource plan concerning land- and water-use, and the design of a national settlement strategy.

Much of this is well-known and roughly summarizes some features of and motivations behind the development of planning-geography (and much of this tradition is preserved today, albeit with different technological means of production and a broader array of methods). Yet it also fails to recognize the degree to which independent thought survived within academia. For one thing, there has long been a kind of self-reflexive undercurrent of scepticism accompanying geography's attitude to planning. William-Olsson would raise a critique of social engineering from the late 1940s onwards, declaring that he disliked the kind of central planning that was becoming mainstream in Sweden. From the results of his various mapping and forecasting efforts he claimed that 'we cannot accept places and regions delimited on political and physical grounds' (William-Olsson 1983: 159). According to William-Olsson's recollection, this challenge met with political resistance to the point where 'results were ignored because they pointed to a reality which was in potential conflict with politically stated goals'. His international experiences during the 1950s were not altogether different: 'Just as at home I had learned that unbiased social research was restrained if its results did not agree with predominant political interests, I now met criminal methods to suppress unwelcome knowledge' (William-Olsson 1983: 160).

Leading geographers such as Hägerstrand also became critics of planning and dehumanized quantification. His Swedish-language pieces from the 1970s and 1980s often made a particularly strong case. It is telling that Gunnar Olsson, despite his caustic critique of Swedish planning-geography, acknowledges Hägerstrand's classical essay *What about people in Regional science?* as a key text in 'a continuing re-evaluation of the Swedish experience, not the least by those who themselves played decisive roles in the shaping of the new society' (Olsson 1980: 207b). Far from being self-congratulatory, Hägerstrand admitted that: 'In many ways present-time critics are right when they say that we tried to sweep up after the moves of a capitalistic industry, involved in international competition. At the time, however, this seemed to be the sensible thing to do' (Hägerstrand 1983: 253).

One not very original conclusion from all of these episodes is that planning and geography in Sweden have extended and intimate connections. What is less often noted, but no less significant for my purposes, is that many of the geographers involved reflected on their role in official policy-making, and were at times overtly critical of where planning in Sweden was heading. Yet, at the same time, few if any opted out or forged a trenchant Olssonian or other critique. It seemed that most geographers involved in planning never lost hope of the possibility to make a difference from within admittedly unbending official machineries. Or perhaps, since opting out demands some kind of combination involving privilege, courage, insight, and despair, it may be less a question of faith, and more, to return to Hägerstrand, a sense of responsibility and utilitarian thinking. Opting out is neither the easiest way to satisfy the necessities of making a living, nor a safe method of securing a market for your students.

Within Swedish geography it is at this point between hope, faith or moral responsibility on the one hand, and no less moral objection to planning on the other hand, that the space of norms transforms into a space of exceptions.

Exceptions: locks and processes

Where can Gunnar Olsson's work be located within this context of Swedish planning and geography? I say *within* because it emphasizes Olsson's contribution to that tradition, locating his work inside the dynamics of the socio-historical totality to which it belongs. Yet this alone would not do enough justice to the distance of his work from that tradition and the inspiration it drew from *different* modes of thought. For Olsson the question was not that of the pragmatic mind looking for employment, adaptation to the given, strategic alterations from within, or a critique of planning in practice, but a far more categorical questioning of planning as a social phenomenon. As a modern-life K., Olsson cannot let go of the acute existential nightmare stemming from the awareness that "we obey their commands without hearing them, and without knowing where they come from". Olsson's critique was non-propositional, and possibly even dystopian, by displaying the tragedy of planning without offering solutions, while *at the same time,* like a piece of art, gesturing towards radical solutions and change. Therefore, perhaps a more satisfying way of approaching Olsson's otherness in relation to the geography-planning tradition could be reached from a second detour, this time to Kafka.

As my opening quotation indicates, Olsson renders planning as a particular form of power relations, "nourished by the legitimating interplay of mystification and domination". Throughout his work, this interplay combines with an on-going tension between individual and society. While this anti-dogmatic manoeuvre releases a host of associations to a broad tradition of critical academic thought, the dialectic of mystification-domination and society-individual is also ubiquitous in the literary worlds of *The Castle* and *The Trial*. Moving between these stories leaves one with the feeling that, regardless of the individual's victimhood (*The Trial*) or active search for knowledge (*The Castle*), the main thrust remains the same: the preservation of the system-society through a mystification-domination dialectic.

In *The Trial*, we are introduced to Joseph K. as someone whose seemingly monotonous and taken-for-granted life is interrupted by an opaque accusation torpedoed from the anonymous recesses of an intractable bureaucratic structure (Kafka 1946). The more K. exhausts himself to clarify the nature of his case, the more entangled he becomes in a web of bureaucratic power. His consecutive defensive probes are all in vain in a situation where law is not simply reduced to trial and judgment without meaning, but also, by turning exception into rule, becomes indistinguishable from the life which it is supposed to regulate (Agamben 2005). As the story progresses the situation becomes increasingly claustrophobic and threatening, ending with the infamous and seemingly unavoidable execution of the protagonist. Death ultimately seals the pact between domination and mystification.

In *The Castle* our hero explores a different behavioral strategy in the faceless face of power (Kafka 1935). Here, K. is not simply the innocent victim of a dehumanized power-system beyond his control, but a land surveyor guided by a strong intentionality. The main issue is a quest, in which K. moves around as an active cartographer of space, who uses his personal freedom to obtain knowledge of the meaning of things. Abandoning his hearth and family, and equipped with some sort of cartographic reason, he decides to struggle his way deeper into a structure that has little interest in his hopes and services. The castle, veiled in mist and darkness, soon unfolds as an impenetrable space of authority – the domicile of higher powers and a stronghold apparently promising disclosure of ultimate meaning. The more desperately K. tries to gain access to and acceptance from the grey bureaucrats of the castle, the more the muteness of existence confronts him. The castle is a proverbial and exact expression of a lockless lock (*Schloss* in German is a castle but also a lock), a space that by its very nature will not yield to lucidity, nor offer any solace. If its mist-enveloped façade tells us anything, it may well be that we have to learn to live with a curtailed reality, and should not put too much hope in the liberating sun and wind of ideology critique.

In Kafka's hands, the surveyor and dutiful bank clerk alike embody the existential horror of being caught in a rejection, codified and circulated through an ever-same flow of domination and mystification, and ultimately ending in private catastrophe and the preservation of the system (now becoming synonymous with society, as in any totalitarian condition). The system-society itself certainly has no interest in accommodating the dreams and needs of the human subject and thwarts anything that does not resemble it.

From a superficial reading, Kafka's world seems to be exceptional: a world where abnormality defines the norm. But a sharper image is given by Adorno, who notes that this world effectuates a permanent and collective *déjà vu*:

> Kafka sins against an ancient rule of the game by constructing art out of nothing but the refuse of reality. He does not directly outline the image of the society to come – for in his as in all great art, asceticism towards the future prevails – but rather depicts it as a montage composed of waste-products which the new order, in the process of forming itself, extracts from the perishing present. Instead of curing neurosis, he seeks in it itself the healing force, that of knowledge: the wound with which society brands the individual are seen by the latter as ciphers of the social untruth, as the negative of truth. His power is one of demolition. He tears down the soothing façade to which a repressive reason increasingly conforms (Adorno, 1981 [1967]: 251–252).[1]

Critical theory appears here as a negative dialectic, a deconstructive montage, composed of the refuse, the waste-products of the familiar and the unknown. In Kafka's world, the system-society wants us to believe that the norm is natural and hence defines abnormality and defiance (as Foucault would have it), but there is simultaneously an awareness that abnormality defines the norm. Such a reversal brings us back to Olsson. For Olsson too, I would argue, forces his readers constantly to consider the untruth, unreality, and abnormality of what carelessly and conventionally passes for truth, reality and normality. This is what happens when the taken-for-granted of planning is taken-apart, turning *jamais vu* into *déjà vu*.

---

1   Adorno's exquisite essay on Kafka is one of those prismatic texts which not only offers exegetic vistas on a particular oeuvre, but also works as a presentation of how critical theory can be practiced. Read in that way, it at once brings out glimpses of Adorno's negative dialectics and gives an introduction to Kafka's genius and literary tactics of demolition.

A search for keys

My reading of the Prague writer and Adorno's negative dialectic coincided with an engagement in a research project on planning in Sweden. It was at that particular point that Gunnar Olsson's publications on the subject attained some urgency for me – not so much because they had anything to say about the superficial specifics of official procedures, but more because they offered critical insights into the modern and postmodern consciousness. When we came to speak about the subject some years ago, he gently denied that Adorno's Marxism was of much significance to his own work. Yet, despite important differences, I could not help but notice commonalities between Adorno's Kafkaesque attitude to modern life, his avant-garde fight against any relaxed acceptance of the system, and Olsson's rendering of planning. Considering this a little longer, I provisionally concluded that it wasn't necessarily Adorno, but the Kafkaesque *déjà vu* effect he described, that is carried along in Olsson's work: the constant dialectic of norm and abnormality, in which the individual is overpowered by the system-society.

In Olsson's work, several of Kafka's quandaries achieve an unusual force of expression: the non-propositional way of calling for solutions; the persuasion that human failure and intentionality are coupled; the sentiment that estrangement confronts existence everywhere; the subjection to bureaucracies lacking ordinary human feelings; the anxiety to reach fulfillment and impose order upon a fragmented world; the acute material and mental crisis of disorientation; the awkward dialectic of individual and society; the theme of a search for categories without negotiating any progress – all of which are vital elements in modern and postmodern critiques of planning and power. Writers such as Kafka, Olsson once claimed, 'come closest to the evasive spirit of the present', transmitting a sense of ambiguity and fragmentation and managing (in a phrase which could have been Adorno's) 'to write both in and of power' (Olsson 1991: 113).

As in Kafka's image of bureaucracy, Olsson makes a strong connection between space, power and planning. Surely such connections are general enough to fit a wide variety of scholars, including major chunks of French poststructuralism. Indeed, Kafka's work remains only one in a myriad of existential, artistic, literary and philosophical references in Olsson's work. At the very least, this suggests some common conjunctions in the history of critical thinking about power and planning, one in which Olsson has a special place. And it is that special place –an important justification of Olsson's denial – which the remainder of this chapter seeks to explore in some more detail.

Of power-knowledges and planning-geographies

These two interlocutory readings of Olsson have primarily worked on the side of the interlocutors, whereas less attention has been paid to the subject of interlocution itself. The latter will be the topic of the remainder of this chapter.

Olsson's vista on the role of academic geography in Swedish planning can be extracted from a few pages in *Lines of Power/Limits of Language*:

> laws of social gravity should enable us to construct a human world of justice and efficiency. (Olsson 1991a: 186.) The message was that In the disciplinary environment of post-war Sweden, there were tight ties between geography and planning. If not then, then at least now, I believe that this coalition between science and politics is a modern strategy for wielding power, not only over space but also over time, not only over oneself but also over others. I believe planning to be a political and bureaucratic phallus symbol, whereby the present penetrates the future. I believe that to plan is to preserve what now is by transforming fleeting intentions into unyielding stones of physical and institutional structures. Values of the strong today are ontologically metamorphosed into the facts for the weak of tomorrow. The result is a modern version of the castration complex in which some fear the loss of something they once had, while others experience the lack of something they never possessed. Thus, from the suppressor's elevated perspective, planning appears as a thoughtful way of extending a heritage of hopes and fears from one generation to another. From a different perspective, it shows itself as an efficient technique for raping the future. From either perspective, it is like other cases of penis envy: inseparable from authority. The authoritarian elements of planning play over hierarchical organization levels. The main issue concerns the welfare functions of different decision units and how these are traded off against each other. As a consequence, planning interferes deeply in the dialectic between society and individual, a fact that posits planning in the holey cross between the social sciences and politics; none of these activities poses a more central question than that concerning the relation between individual and collective, one and many, subject and object, I and you, us and them. Myth shares the same concern (Olsson 1991: 17–18).

In isolation, this dense passage can easily be categorized as one more highly abstract version of planning critique, taking place in a spaceless space notwithstanding its reference to post-war Sweden. Much of its meta-political references to power and knowledge (science), abused and obedient bodies, phallocentrism, fears and hopes, authority, hierarchy and oppression, individual and society, myth and ideology critique, are not very distant from, for instance Foucault's image of docile bodies in disciplinary society, or Adorno's reified and mist-enveloped late capitalism.

But its meaning reaches further depth when read from within its specific context. Although it runs counter to intuition, I would argue that the passage cannot be fully understood if one disregards its embeddedness in a long tradition of Swedish planning writing, not least within geography. As can be inferred from my argument above, such a reading is counterintuitive only if Olsson's critical stance is situated outside this tradition. Yet, given the author's position as a professor of geography, it is impossible to avoid the conclusion that he was assaulting sections of his own discipline from within. Perhaps somewhat ironically, at the time when *Lines of Power/Limits of Language* appeared, he was working at *Nordplan* – an institute that would not have seen the light of day without the planning tradition Olsson never failed to deter.

And there is more to the critique than geography's disciplinary space. It is significant, too, that some of his most elaborate pieces on planning were published in the relatively easily accessible form of *Antipasti* and not only in heavy academic publications for a bookish international audience (Olsson 1990). In context, the passage stands as a cryptogram which excludes the customary success-story of Swedish planning with the intention of bidding its negative to display. In a more deconstructive way, it embodies a blasphemous moment made in Sweden and *therefore* proves the limits of the image of totally unyielding domination it intends to describe.

Punch-lines for starters

> We must cease once and for all to describe the effects of power in negative terms:
> it 'excludes, it 'represses', it censors', it 'abstracts', it 'masks', it 'conceals'. In
> fact, power produces; it produces reality; it produces domains of objects and
> rituals of truth. The individual and the knowledge that may be gained from him
> belong to this production (Foucault 1991: 194).

Because it was written in Swedish, *Antipasti* is hardly known outside the Nordic world, but may for exactly the same reason be regarded as one of Olsson's most important publications on planning (Buttimer and Mels 2006: 93–96). As I suggested, the reason for this lies in a pedestrian politics of reception rather than in language or differences in textual sophistication. Indeed, much of what *Antipasti* had to offer was already presented here and there in English. However, through publication in Swedish the message aimed its sharp points at a readership which might otherwise fail to notice foreign prose (some of the essays were originally published in a sensationalist newspaper!). By those who are/ought to be offended it could for various reasons be rejected, exiled from the discipline, or displaced as an Other genre of writing – but not simply ignored.

Three essays gathered under the apparently ironic title 'Reason' in part three in *Antipasti* summarize quite well Olsson's view of planning in Sweden. The first essay describes Sweden as a society in which the power of welfare capitalism rests heavily on a rather diffuse sense of anxiety among the populace. The theme of a 'disturbed relationship between individual and society' is of central importance to the diagnosis (Olsson 1990: 69). Olsson, following Gregory Bateson, describes this disturbed relationship as a double bind situation in which contradictory messages or demands are imposed by a respected person or authority (for example the state-parent) upon a victim (the citizen-child). This, in turn, has given rise to a general postmodern state of narcissism: 'The disturbed narcissistic person has preserved the child's primitive phantasies concerning the omnipotent parent and the grandiose self' (Olsson 1990: 68).

In the second essay, Olsson focuses on planning as an art of forcing together divergent levels of decision making into one mould. It is structured by an atomistic and rigid practice of categorization, 'freezing the dialectic of change', in which the future is no longer undecidable and no longer offers plural possibilities (Olsson 1990: 81). This, of course, recalls the lengthy quotation above, and also evokes the geographer-planner's, legal expert's and architect's manipulation of spatial form in order to alter societal functions and symbolic relationships. For Olsson, 'the point is that we cannot say anything definite about the outcome and the long-term consequences of this manipulation. The result is that we have built an unambiguous prison in which it has become increasingly difficult to live in plural and therefore creative ways' (Olsson 1990: 85). Conformity and subordination are the result.

The themes of a disturbed individual-society relationship – not least through repetitive ontological transformations of mindscapes into stonescapes, or to be more precise, 'the values of the strong today' into 'the facts for the weak of tomorrow' – encountered in the first and second essay, are rephrased in the third and most elaborate essay, entitled 'the paradoxes of planning'. Here, Olsson commences with an allusion to Kafka (in its Swedish translation *The Trial* is entitled *Processen*, that is 'the process', which preserves much of the meaning of the original German title):

The fundamental paradox of planning is that purposeful management never leads to final solutions, only to more management. But even if it is not the declared goal of planning, it may be the cynical function of planning: to arrange employment for a growing body of bureaucrats. The product is the process. But processes remind of Kafka, whose point was precisely that consistently implemented programmes consistently fail (Olsson 1990: 90).

Again, the disturbed relationships between individual and society are elementary, accompanied by a self-perpetuating bureaucracy (above associated with a castration complex) with ever-same routines. In a more historically inclined analysis, Olsson recounts the legacy of what he describes as some of the chief intellectual architects behind Swedish social engineering: the philosopher Axel Hägerström and his acolyte professors of law, Vilhelm Lundstedt and Karl Olivecrona.

Hägerström promoted a kind of ethical relativism which annihilated the question of the good society altogether. Instead of pausing at questions of justice or solidarity, pragmatic problem-solving on the basis of an instrumental rationality should prevail. Lundstedt (who also pursued a political career) and Olivecrona delivered further spadework for the development of the powerful Swedish welfare state. Their shared view was that in real life, there is no real ground to argue for the right of the individual against society. For them, the way Swedish law worked in practice was exactly the way it should work: 'the task of the courts of law is not to regard the law in individual cases, but to always maintain the public interest. In this manner, the descriptive *is* and the prescriptive *ought* melt together in the state ideology which would come to dominate Swedish politics, legislation and administrative practice' (Olsson 1990: 95).

This mode of critique, describing how the hermetic logic of spatial structures (the dominating trials and mystifying castles of the Swedish welfare state and its planning) are ontologically transformed into social relations and vice-versa, gives further empirical detail to what was argued in *Birds in Egg*. Here, Olsson claims that social engineering in general, and Swedish regional planning and the spatial structure of the Swedish welfare state in particular, give rise to a practice in which observed behaviour is accepted and rationalized as the desired state of affairs. In doing so, planning becomes a counterproductive practice which reproduces the very social and geographical inequalities it seeks to destroy (Olsson 1980: Ch. 14). *Birds in Egg* traced the paradoxes of planning back to a general but fatal misunderstanding of scientific methodology (notably social scientists' application of the Pareto principle) in post-war geography. *Antipasti* makes the same point with reference to a lineage of philosophical and political tactics emanating from the time of the nascent Swedish welfare state.

Tracing these ideas a bit further, it may be added that Olsson's presentation owes much to earlier critiques of positivism. For instance, a similar is-ought argument was presented by Adorno's Critical Theory colleague Marcuse, whose 1941 *Reason and Revolution* took Comte's positivism to task as a preacher of resignation and consolidation of the public order (and, although this might be a coincidence, remember that the section on planning in *Antipasti* was entitled 'Reason'). The similarities with Olsson's and indeed several other Frankfurt School theorists (and hence also Husserl) seem too obvious to be discarded. What is clear though is that Olsson finds the roots for his critique in a scientific misunderstanding leading from optimistic ideals to actual tragedy, rather than the calculated evil of, say, late capitalism. Science features here as a special case of what Olsson, following Kierkegaard, sees as the human condition in general.

To their feeble defence, Olsson writes, those who practise this principle turn to utilitarian ethics and the keywords of 'welfare' and 'consequences'. The *welfare* principle posits that our actions should promote the highest possible happiness for as many people as possible. Olsson claims that tragically, this has resulted in a sharp politics of normalization, in which a normalized society increasingly dominates minority groups, differences and otherness, and rigidly forces them to adaptation and assimilation. Again: 'Values of the strong today are ontologically metamorphosed into the facts for the weak of tomorrow'. The principle of *consequences* submits that our actions shall be appraised by their outcomes, not by their intentions. This presupposes a full command of the direct and indirect consequences of particular actions: a self-deceiving belief in our ability to predict the future. 'The product is the process' thus reads 'The product is The Trial'. At this point in *Antipasti*, Olsson introduces the gendered themes of, among other things, castration complex, phallocentrism, and masculine policy-making, which would return in *Lines of Power/Limits of Language*.

Messages from a shipwreck

What can we make of all these diagnostic accounts on planning? What politics is involved? Olsson's anti-modernism seems evident and can be encountered in nearly all his arguments on planning. On the level of cultural politics, I would suggest that this is not the same as an idealization of the past. To my best knowledge, Olsson has never suggested that earlier, pre-welfare state periods offered better solutions to social inequalities. There is no choosing between *Gemeinschaft* and *Gesellschaft*. Neither has he offered any detailed future-oriented pointer to how the is-ought principle could be reversed. Closest are the imprecise and surprisingly gendered gestures of 'implementing plans which will aid man in his striving for becoming', 'building a world for the constant groping of autonomous man', and the post-modern relativism of 'preserving ambiguity' instead of imposing certainty (1980: 204b).

If modernity is about systems, categorizations, planning and order, in Olsson's politics there is an unmistakable agitation on behalf of the individual, the other, difference, the exiles and outcasts, and against the system and its extreme intentionality. To some degree it is an anti-modernism pointing inward, towards the internal world of living individual human beings rather than the outer structures and stonescapes of society (Olsson 1980: Ch. 16). Yet, these gestures also take account of the inescapable presence of structures. Not surprisingly, several readings of Olsson find a seemingly paradoxical mixture of *individualism* and *structuralism* in his work, roped together by philosophically and politically conservative *idealism*.

An oft-noted observation is that Olsson's dialectics primarily engages 'with the "inner" world of individuals as manifest through the language used to ascribe meaning to the external world. This version was to be contrasted with another in which dialectics is used to explain sociospatial inequalities and guide systems of redistributive justice' (Dixon et al. 2008, 2553). From his Marxian viewpoint, Noel Castree laments that Olsson's dialectics 'has risked lapsing into abstruse, idealist language games far removed from the "mundane" imperatives that structure peoples' everyday lifeworlds' (Castree 1996: 342).

Such readings call for a further political and societal contextualization of Olsson's work, referring to its appearance not just within the confines of modern planning-geography and Kafkaesque philosophy, but within a postmodern and increasingly dominant, neoliberal politics. Olsson's above-quoted, imprecise propositional gestures – with their emphasis on apparently free individuals – may be instructive, and not only because they reinforce arguments found in humanistic geography of the 1970s and the post-modernism of the 1980s and 1990s. More specifically, it may be argued that the position taken by Olsson is liable to aggrandize issues of freedom, particularities, and individual liberties while neglecting other kinds of social and environmental justice. Such a post-modernism could easily become part of what Harvey calls a 'construction of consent' because it is liable to incorporation into the logic of the now globally dominant neoliberal mode of politico-economic thinking (Harvey 2005). Indeed, *Antipasti's* critique of the Social Democratic planning ideology and the Swedish welfare state coincided with the privatization of that very welfare state. This neoliberalization was initiated by the Swedish conservatives and then adopted by subsequent governments (cf. Harvey's exploration of the relationships between conservatism and neoliberalism). It also coincided and partly tuned into a wave of highly mediatized campaigning against what was seen as the excessive presence of a universalistic welfare state. In Olsson's work one looks in vain for critical comments on these currents or a deconstruction of neoliberal conservatism, although this might well have been part of a late-nineties sequel to *Antipasti*.

A consideration of the process of neoliberalization in Sweden and elsewhere would certainly add important aspects to a critique of planning-as-domination. Yet, while much can be said in favour of such a reading of post-modernity and the rise of a Swedish neoliberalism, it is also misleading to think that Olsson tuned into this chorus. After all, the neoliberal doctrine apparently dreams away planning and state-intervention while at the same time conserving those institutions and structures designed to secure the power and freedom of economic elites. This prospect of 'universal freedom' has been unmasked as the authoritarian architect behind a new world (dis)order, with its own Trials and Castles. Such authoritarianism sits uneasily with Olsson's critique of mystification and domination. If anything, neoliberalism knows exactly what it takes to metamorphose the 'values of the strong' into 'facts for the weak'.

Instead, and following my earlier allusion to the Kafkaesque, I would consider the possibility of valuing his linguistic tactics differently, precisely because it embodies in content and form a revolt against the 'mundane' and taken-for-granted. It is one possible avenue for anyone seeking to understand how particular renderings of the world become accepted as true. For this particular form of ideology critique, the task is to find a way through mist-enveloped spaces in which representations attain a life of their own. It differs from, but does nothing to deny the importance of the dialectic between outer and inner worlds, or what Olsson calls ontological transformations. This is not necessarily an anti-social commitment to the inner world of individuals, but rather an engagement with the dialectic of the senses and the sixth sense culture.

So if individualism and conservative dreams, kept together through philosophical idealism, offer only fragments of possible answers to my question about Olsson's politics, perhaps a *structuralist*, even dystopian possibility offers additional insights. Olsson criticizes planning as a power-system without offering any (class-biased or other) solutions. Instead of solutions, I would suggest that he presents us with a picture of collective action in which not even the complete dissolution of official planning could put the on-going dialectic of individual and society to a harmonious end. This approach forecloses any genuinely humane reconciliation between planner and planned, subject and object, since under Olsson's guidance these pairs recurrently translate into structural relations of reification, deification, and domination. If this is correct, the paradoxes of planning are likely to hang about, rather than to wither away in some utopian future. And that would refute any talk of feasible alternatives derived from critical analysis of the present world. Paraphrasing Adorno, Olsson's critique would remain oppositional rather than affirmative, his low voice would reverberate like 'the surviving message of despair from the shipwrecked' (Adorno 2003: 42).

Funambulism

> I keep a close watch on this heart of mine I keep my eyes wide open all the time
>
> I keep the ends out for the tie that binds Because you're mine, I walk the line – Johnny Cash

The ties that bind bring me to what I would call the structuralist tendency of Olsson's account. It is of interest that some commentators have argued that Olsson has a weakly developed notion of structure. Much can be said in favour of such an understanding, especially since Olsson's work has little to offer when it comes to exposing the worldly structures of, for instance the state, capitalism or the city. One motivation for this may be that 'geography has committed itself to the surface features of the external', which has 'produced studies of reifications in which man, woman, and child inevitably are treated as things and not as the sensitive, constantly evolving human beings we are' (Olsson 1980: 47e). Olsson, the humanist with his quest for understanding what it means to be human, finds a preliminary answer neither in the reifications of traditional science, nor among the sanctified beings of religious faith, but in the 'semiotic animal, a species whose individuals are kept together and apart by the use of their signs' (Olsson 1993: 280). This move from visible stonescapes and power-knowledge to a more humanistic prospect of self-exploration does not rule out a powerful presence of structure. On the contrary, Olsson's positioning is one in which structural reifications and deifications are eternally present and constantly challenged. However, it does to a substantial degree shift attention from external pressures to the internal or formalist level of language, thought, the unconscious, and being.

As Chris Philo has noted, such a shift is not without problems. He describes Olsson's manoeuvre as one in which a strong emphasis on the internal worlds of human agents is *compromised* by an equally internal view of structure as the subliminal social imperatives of a 'collective unconscious' (Philo 1984: 228). Indeed, accepting the collective unconscious means accepting that there are serious restrictions to the autonomy of individual human beings. If this reading is correct, it can be extended into Olsson's more thematic ruminations on planning and social engineering. In the passages devoted to this latter issue, 'structure' or society is precisely a collective, internalized structure. It reproduces an ontological transformation of the is-ought kind invoked above: the translation of a thingified descriptive language of spatial science into a prescriptive social engineering, imposed upon the social order.

In the case of planning, I think that it is in fact possible to identify an ultra-structuralist storyline of bureaucracy and planning-geography in which there is no place for, even hatred of, being otherwise. Everybody within the planning-geography system, perhaps even including academics without any practical involvement in official planning, participates knowingly or unwittingly to conserve its total control. Olsson's readers are constantly reminded of how good messages and intentions turn into shipwrecks and tragedy without anyone being to blame. This seems to essentiallize the grip of the collective unconsciousness: 'Our well-meaning attempts to rectify current injustices would consequently be self-defeating, not because some vicious bureaucrat designed them that way, but because we have failed to understand the deep structure of social research and action. Rather than dissolving the institutional facts which imprison us, we sink deeper into their masochistic embrace' (1980: 201b). In passages such as this, we are back in Kafka's world: a structural storyline that leaves the keyless individual with a bloody forehead at the armoured gates of the Castle or as a detainee in a predestined Trial.

But how precisely could we dissolve 'the institutional facts which imprison us'? This becomes particularly problematic and seemingly unanswerable, not only because Olsson's work remains fairly non-propositional, but also given the shrewd way it nearly forces its readers to obey and question the authority of the author himself. Put differently, Olsson is consistently ruthless in his attack on precisely the kind of rhetoric and cartographic reason which he deliberately avails himself of in his own work. Olsson's accounts tend to leave the reader with a self-imploding recommendation to trust and believe a writer who personally deconstructs any such urge as a core tactics of power. Such a credibility crisis illustrates the difficulty of practising an ideology critique critical of its own ideology, of writing in and of structure. This is no real surprise in a geographical imagination where structure, power and the problem of self-reference are everywhere, of course, but neither does it offer any breaking away from the system (cf. Gren 1993: 184–185, 190). Switching from thinking to knowing, an abyss forgotten in defence of Reason, a mindscape turned into stonescape: the line of Olsson's funambulism is also a tie that binds. No escape, balancing a rope of power held rigidly taut in the Bar de Saussure, bound to the ties of possession and love. Trust me; because you're mine, I bind the line.

Spacing out

> The message was that The work that shatters individuation will at no price want to be imitated: for this reason, surely, Kafka gave orders for it to be destroyed (Adorno, 1981 [1967]: 254).

Any imitation of Gunnar Olsson's work partakes in the same mindset, the same system, the same ceremonies of discipline, and the same machinery that the work seeks to deconstruct. But if this pushes critical moments in geographies of planning into the private orb of a one-man show, it would make the whole effort seem depressingly pointless. Thankfully, planning and geography in Sweden are no longer completely connected in the faultless power-knowledge assemblage that Olsson has repeatedly conjured up. I would certainly concede that, in its close alliance with planning authorities, geography has learnt how to operate a wide variety of techniques of discipline. At times, planning-geography may even seem to be arranged as what Foucault called 'tactics' or 'the art of constructing, with located bodies, coded activities and trained aptitudes, mechanisms in which the product of the various forces is increased by their calculated combination' (Foucault 1991: 167). In Olsson's preferred simile, any engagement in planning and the search for a dialogue with those who hold the scepter typically attains the structure of a tragedy. It leaves us with the contra-finality of good intentions gone terribly wrong, but without anyone to blame.

Even so, would not that be to deny the agency of more than one knowledgeable, capable and indeed critical human geographer within the broad hierarchy of the Swedish planning system? Olsson may be said to overestimate the internal homogeneity of planning and lose sight of the ways in which planning results from a complex, on-going process of struggle over places, resources and futures – a struggle in which geographers do play a role.[2]

This may support the conviction that Olsson's diagnostic statements on planning-geography can only be read as an over-pessimistic structuralism. Like Kafka's stories, there seem to be no exits, only blind walls. The dutiful J.K. and the assertive K. only show the futility of individual variations in the ever-same catastrophic dialectic between society and individual. In Kafka, humour in one passage only opens up for an abysmal dejection in the next. We look in vain for any Marxian politics of breaking away from the hegemonic order through an ideology critique mapping the real. No utopian withering away of the state. No obvious mode of thinking and acting outside the symbolic order of the forever absenting Phallus. No escape from the castration complex before the system of social engineering.

Yet, such a conclusion would ignore the volte-face movements of the avant-garde that animate Olsson's spatialities – not as ostentatiously academic chimera, but as real interlocutors in the drama of human thought-and-action. Olsson's authorship also encloses an unmistakable portion of Kafkaesque humor, but seems to lend a more optimistic ear to the power of laughter. Sometimes, I like to think that in these arms, as well as in the consciousness of the unconscious summoned in various places, lies some possibility of solace. It clears the anarcho-humanist path for a spatiality of thinking, writing, speaking, and living in your own way.

---

2 For elaboration on recent developments in Swedish geography and the significance of critical thought herein, see chapters 5, 7, 8 and 9 in Buttimer and Mels 2006.

It is a poststructural political geography with a cause, but without any ultimate conclusion. It is a liberating embracement that dissolves the ultra-structuralism that opposes the possibility of authentic critical voices – including Gunnar Olsson himself – and the academy as a meaningful source of emancipatory thinking. And fortunately, those who are unconvinced by such politics need not necessarily be on the side of convention and control.

Abandonments

Now that I finally abandoned the plan of not writing these notes, I recognize it as distant echoes of some of the laughs, tears and obscurities of a graduate course I attended more than a decade ago (Olsson 1998). I would like to thank the editors for patience and for encouraging me to return and remark my marks.

Bibliography

Adorno, T.W. 1981. [1967] Notes on Kafka, in *Prisms*, T.W. Adorno. Cambridge: MIT Press, 243–272.

Adorno, T.W. 2003. *Philosophy of Modern Music*. New York: Continuum.

Agamben, G. 2005. *State of Exception*. Chicago: The University of Chicago Press.

Buttimer, A. and Mels, T. 2006. *By Northern Lights: On the Making of geography in Sweden*. Aldershot: Ashgate.

Castree, N. 1996. Birds, mice and geography: Marxisms and dialectics. *Transactions of the Institute of British Geographers* NS 21, 342–362.

Dixon, D.P., Woodward, K. and Jones III, J.P. 2008. Editorial: On the other hand… dialectics. *Environment and Planning A* 40, 2549–2561.

Foucault, M. 1991. [1975] *Discipline and Punish: The Birth of the Prison*. London: Penguin.

Gren, M. 1994. *Earth writing: exploring representation and social geography in-between meaning/matter*. Göteborg: Publications edited by the Departments of Geography, University of Gothenburg, Series B, no. 85.

Hägerstrand, T. 1970. What about people in regional science? *Papers and Proceedings of the Regional Science Association*, 24, 7–21.

Hägerstrand, T. 1983. In search for the sources of concepts, in *The Practice of Geography* edited by A. Buttimer. New York: Longman, 238–256.

Harvey, D. 2005. *A Brief History of Neoliberalism*. Oxford: Oxford University Press.

Kafka, F. 1935. *Das Schloss*: New York: Schocken.

Kafka, F. 1946. *Der Prozess*: New York: Schocken.

Kafka, F. 2007. *Der Proceß*. Fischer, Frankfurt.

Öhman, J. 1994. Den planeringsinriktade kulturgeografin. *Nordisk Samhällsgeografisk Tidskrift*, Uppsala, 83–96.

Olsson, G. 1980. *Birds in Egg / Eggs in Bird*. London: Pion Limited.

Olsson, G. 1990a. *Antipasti*. Göteborg: Bokförlaget Korpen.

Olsson, G. 1990b. Lines of Power. *Nordisk Samhällsgeografisk Tidskrift*, 11, 106–116.

Olsson, G. 1991a. *Lines of Power / Limits of Language*. Minneapolis: University of Minnesota Press.

Olsson, G. 1991b. Invisible maps – a prospectus. *Geografiska Annaler*, 73B(1), 85–92.

Olsson, G. 1993. Chiasm of thought-and-action. *Environment and Planning D: Society and Space*, 11, 279–294.

Olsson, G. 1994. Job and the case of the herbarium. *Environment and Planning D: Society and Space* 12, 221–225.

Olsson, G. 1998. *Det kartografiska förnuftet*. Graduate course, Uppsala University.

Philo, C. 1984. Reflections on Gunnar Olsson's contribution to the discourse of human geography. *Environment and Planning D: Society and Space* 2, 217–240.

William-Olsson, W. 1983. My responsibility and my joy, in *The Practice of Geography* edited by A. Buttimer. New York: Longman, 153–166.

Part C
CHIASMS

Chapter 9

–/–

Gunnar Olsson¶

To translate is to express a sense in another language-parole. It is to carry to heaven without death and to remove the dead body or remains of a saint. It is to convey an idea from one art form to another. It is to make new boots from the remains of old ones. It is to interpret sings.

The definitions come from the OED. The conclusions are my own: Much is aVOIDable, translation is not. Anywhere/anytime, aeneymy/anyf(r)iend.

Anyhow:

My · is made, now I must erase it. Thus, to translate is to doubly lie, to beget by not getting at the truth. Little wonder then that all social laws are laws of the double. More wonder that most social scientists are one-eyed cyclops unable to imagine perspective; misled by our singular vision we tend to confuse use and mention, word and object.

Where is he, that wondrous wandering wonderer capable of releasing Ulysses from his wake?

* * *

In the convent, the point is more conventional:

All understanding involves crucial elements of translation, of movement from one conceptual world to another. Thus, all understanding is by necessity metaphoric, for I must always grasp what I wonder about as something different; the I becomes an Other, the Other becomes a Me.

It cannot be said more clearly:

To wonder about understanding is to be involved in language. But, to be involved in language is to exploit the distinctive connection between name and object, thing and relation, appearance and essence.

Already Odysseus caught in the Cyclop's cave knew that there is a distinction between what I think-and-say and what I think-and-say about. Wondering about understanding must consequently not be limited to the semantics of the signs which signify what I am talking *about*. It must also involve the pragmatics and syntax of the categorizations and relations I am talking *in*. The challenge is not to be tuned to the vibrations of my vocal chords or to detect the spots I leave on the white sheet. It is rather to be aware of the silent forces of that subtext which sneaks away into the emptiness that ties the marks together. But the overwhelming practice is to concentrate solely on what appears on the lines. The blank spaces which separate and unite them go unnoticed, precisely because they separate and unite.

¶ The author is grateful to Allan Pred for conversations, Bourdeaux, pheasant paté and Brie.

So:
See not only what is on the lines, but
  also what is between them.
Read less of what I am sufficiently ignorant to write and more of what
you know so well that it must be passed over in silence! Deafen yourself
to the noise of the expressible! Listen instead for the whisper of the
taken-for-granted! But be most curious about the limits between
categories, for it is only in the act of crossing a boundary that you
mistranslate and consequently learn! Everything else is obedient
reproduction.

* * *

Virtually everything is reproduction. But virtue is in the constraining
constraints of the mimicking social scienses, vice in the liberating
possibilities of the creative arts. This makes it possible to predict the
future (as in Foucault's scientia sexualis) and impossible to presense it (as
in his ars erotica). And yet, it is part of logic itself that 1984 is in the
midst of the 1980s. I therefore write now as a way of anticipating this
new world at the decade's end.

The outlines of this new world are already present, for how could it
otherwise be recognized. It is a world in which the familiar industrial
mode of production is overtaken by a hitherto unknown state mode of
production. Although the transformation can be read off the patience
cards that already lie on the table, this is not to say that the present
determines the future. Rather it is often the reverse, because our hopes
and fears for the future let us see only certain aspects of the present.
But what is shown to the observer are merely masks. The future itself
remains as invisible as the ontological stuff it is made of.

The following are nevertheless 1990-oriented questions to the 1980s:

Which aspects of the present are taboo because they are too important
to reveal? Which types of legitimating unknowing is society's science in
the midst of creating? How is today's state-capitalism disguising its
fundamental contradictions? How do I notice and then interpret the
signs of the rainbow in the sky? Who understands the silent language of
the taken-for-granted well enough to translate it? How can I re-member
what others have forgotten and how can I forget what others de-member?
Is the trancelation of relations the deconstructivist's version of
Wittgenstein's throwaway ladder?

And then:

Is it too early to translate the language of state-capitalism or too late
to trance-end its imprisonment?

* * *

Anyone who understands me eventually recognizes these stripteasing
questions as nonsensical, when he has used them—as steps—to climb up
beyond them.

He must transcend these rhetorical propositions. But he will not then,
as Wittgenstein suggested in *Tractatus* 6.54, see the world aright. He will

instead have created another world surrounded by other limits, guarded by other silences, ruled by other emperors.

How will the rulers be dressed tomorrow?

* * *

Just as noone can fully grasp Marx's *Capital* without first having studied through and understood the whole of Hegel's *Logic*, so noone can fully understand state-capitalism without first having internalized the meaning of the sign /. This sign is a symbol of relations, of the unity between identity and difference.

In conventional reasoning, relations are not denoted by a slanted line but rather by the parallel lines of the equality sign. This sing is then interpreted in the Leibnizian spirit of salva veritatae and the Russellian matter of logical atomism; a proposition is held to be both true and informative only if the equality sign is flanked by a proper name on the one side and a definite description on the other or, alternatively, by two different definite descriptions of the same object. But even though the whole point of

$$E!(\iota x)(Qx)=(\exists b)(x)[(Qx)\equiv(x=b)]$$

and

$$U(\iota x)(Qx)=(\exists b)\{(x)[(Qx)\equiv(x=b)]\&(Ub)\}$$

is to define away the definite description $(\iota x)(Qx)$, this *Principia* trick (*14.02 and *14.01) of abolishing definite descriptions by not mentioning them does not alter the fact that understanding requires their *use*. In use, however, definite descriptions reveal themselves as what they really are: contextual, metaphoric and self-referential. This is indeed why Russell wanted to rid them from all analysis; in his own words, "every proposition and every belief must have an object other than itself".

Materialists must now ask the idealist question:

Which is the object of state-capitalism and its precursing postmodernism, if it has to be an object other than itself?

The emerging answer:

Perhaps there are no such objects, for the major characteristic of state-capitalism is that it is paradoxically locked into itself.

Next question:

How is it constructed, the self-referential reasoning net capable of catching the emerging world of masturbation?

* * *

By writing / instead of =, I signal my interest in dialectical, internal and self-referential relations, names that most analysts have learned neither to use nor to mention. The words themselves are banned by the church of Fundamenalists.

The ban was issued because all relations (including equalities) are of an ontological kind alien to the ruling ideology of presence. Thus, relations are by necessity invisible as the emptiness between the lines of my text, inaudible as the silences that turn meaningless noise into meaningful words; the untouchable is pariah, the pariah untouchable. In society's interest of communication, there is consequently a strong tendency to do away with relations by thingifying them. In this tragedy of the common, however, the interest is turned from the concept of the relation itself to the phenomena or things related. Rather than questioning the relations we are talking *in*, we stare ourselves blind on what we are talking *about*. Instead of wondering about = or /, we get caught in the sign – –. This sign is a symbol of things related.

Direct contact with / is culturally forbidden. This is why it intrigues me, for whatever is dangerous enough to be taboo is important enough to understand. The constructive is not to question the static means of representative samples but to unravel the dynamic variances of distributional tails. And yet, culture is founded on its limits, civilization on its madness.
Thus:

Relations are not only relations between measurable things but also between cultural words, not only words but concepts, not only concepts but meanings, not only meanings but other relations. It follows that relations are always related to other relations all connected into strangely looping spirals. Self-reference is the word for this peculiar concept coiled at the center of current thought and extending beyond its frontiers.

Self-reference is the key to the coming revolution of the social sciences. Where are the lock-smiths who know the code?

* * *

Relations like beauty, sincerity, trust, malice, disgust, and nausea are not in the things themselves but in culturally determined conceptions and behaviors; isolated things are as meaningless as connected relations are meaningful. This is why comparative studies are both so promising and so dangerous; promising because they lead to understanding of the I through the Other, dangerous because they are potentially emancipatory. Benjamin, Horkheimer, Adorno, and Marcuse all set examples. But so does everyone else who in exile experiences how one never learns home until one goes away. But each exit is an exit with no return as Homeland becomes the safe symbol of escape.

Illustrative examples are in Geertz's analyses of the Balinese cock fights. For a western anthropologist it is easy to see those ritual dances as being performed not by roused birds but by people who via their cocks tie and untie knots of family relations. It is more difficult to interpret your reading of this writing as an integral part of geography's death and initiation rites.

And yet, a rite of author(ity) is exactly what this dual relation is: Preparations for the cooking of the raw. But who knows the recipe?

Who are the cooks and who are the cooked? And who furnish the pot
and the tempting spices?
* * *

The relations which tie the one to the many and the many to the one
are at the same time determining culture and determined by culture.

Writing those words is easy. Reading them is incredibly difficult.
The reason is that whenever I talk *about* culture I must talk *in* culture;
culture is like its own language in that it is bound to use itself to
understand itself. And so it is that any social scientist is handicapped by
the methodological praxis which requires him to be more stupid than he
actually is. Thus, in the interests of discipline, verification, and
communication he relies mainly on the two senses of sight and hearing:
What counts is what can be counted; what can be counted is what can
be pointed to; what can be pointed to is what can be unequivocally
named. Accumulation of knowledge about the nameable is consequently
the point of the scientist's game. Power, though, is not in uttering the
nameable things of commodity fetishism and penis envy but in innering
their symbolic condensation of relations: Un Coup de Dés played with
loaded dice.

Imagine here a Foucaultian study of filth and human excrement. To
see and hear the shit is barely passable. Uh! To touch it is nauseating.
Woh! But smelling and tasting it penetrates so deeply into the ego itself
that it is almost unthinkable. And so it is that killing a thousand people
by target-seeking robots is acceptable. But killing one person with a
bloody throat-bite is so brutish that the thought itself takes its holder to
the asylum.
* * *

The interesting is not to note obvious facts of empirical behavior. It is
rather to wonder about the particular socialization processes whereby
individual and society are brought together. But this is to wonder about
the taboos associated with the limit between the Ego and the Other. Put
differently, the issue is how you and I distinguish ourselves from each
other by establishing impregnable boundaries between us.

Perhaps the question is:

If definitions require distinctions, do relations affirm them by
transcending them?

Or, more operationally:

How do I teach my children to tell the truth and yet realize that truth
telling can sometimes be evil and therefore forbidden? How does the
Family Circ(l)e turn human beings into swine?
* * *

Relations are often called mystical and thereby silenced. Perhaps this
practice reflects the fact that proper understanding of social relations is a
prerequisite for the understanding of power. But such an understanding

is too fundamental for society to afford. Adam's apple is a double
symbol of temptation-and-fall and of knowledge stuck in manly throats.
   Power is another word for the relation between the I and the non-I.
It follows that the process of liberation can never end, for its driving force
is in the emancipation and creation of the self. Once this is understood,
it is easier to see not only that all power involves issues of translation but
also that every power struggle is a struggle of independence; power
would not be power if it were not a relation, that is, if it were not of an
ontological kind different from the things in which it momentarily seeks
to hide. He who has power knows how to mislead by mixing ontologies,
pretending to be concerned with things while in reality knowing that
things are meaningless until tied into meaningful relations. The essence
of power is thus in the slanted /, its appearance in the repetitive – –.
   And so it is an integral part of all relations that we tend not to notice
them until they begin to malfunction. Neither do I notice the air I
breathe or the blood in my veins until the relation between them is
disturbed. When it is, however, then every doctor is preconditioned not
to wonder about the relation *per se* but rather about the things of oxygen
and pumping muscles; we ask not about the relation /, but about the
categories – –.
   In this movement from / to – –, questions of epistemology turn to
questions of ontology:
   Is the practice of defining our problems into existence a technique for
getting at truth or for defining them away? Is it the practice of those in
power to thingify relations and thereby block the road to deeper insights?
Is it in society's collective interest to mislead its individuals into seeing
appearance rather than essence? Are we stuck in the serpent's truth that
"God knows that when you eat of the fruit of the tree in the midst of
the garden your eyes will be opened and you will be like God, knowing
good and evil"? Is the sign "God" nothing but the proper name of the
definite description "the collective unconscious"? Is the Barefoot-Father-
with-the-Beard a fetish of that social glue which is important enough to
be taboo? Is the crucified son of flesh and bone merely another stage-
stop on the ontological journey from subsisting relations to existing
things? Is reification deification, deification reification?
   Yes!
   The reason for the yes is that there is no objective reality to reflect
upon, for what appears is itself essentially a reflection of the reflector's
subjective self-awareness of that reality. As the dialectics of flexuose
flexion runs its course, questions of ontology therefore turn to self-
reflective questions of mythology:
   Why is the serpent the symbol of self-reference? Is it because it knows
the secret of the collective unconscious?

* * *

When a social scientist deifies by reifying, he christens the sign – – as
"society and individual". There is much to indicate that the dialectical
interplay between these two categories currently is under serious strain.
Perhaps the malfunctioning is most illustrative in welfare states like
Sweden, where the crisis is less a matter of resources and more one of
demoncracy. For what other is democrary than a powerful set of
principles whereby one-and-many, many-and-one are forged together into
what is presented as functional efficiency and moral justice? Put
differently, the principles of demoncrazy shape and reflect how the
psychological concept of the ego is translated into constitutional law;
the high court positioned as frontier guard in the wasteland between I and
the Other, therefrom ruling over what is equal to what, over good and
evil, life and death.

Some claim that the social collective now has penetrated deeply into
the realm of the individual. Others note the complementary trends
toward privatization. In my interpretation, these same tendencies are
further indications that capitalism is in the midst of a rapid, decisive, and
irreversible transformation from industrial capitalism into a form of
centralized state-capitalism. Both individual and society have yet to
adjust to this fundamental change from one dominant production mode
to another. For this to occur it is necessary to develop new decision
procedures, perhaps even new personality types. Whether we like it or
not, that is also the direction in which we are heading. As in the previous
shift from the feudal to the capitalist mode, devils and witches are
invented as scapegoats for commodity fetishists.

Who are the witches today? How is it determined whether they sink
or float?

* * *

Now it can be thought-and-said:

There is time for a new Marx. This is due both to the atrocities
committed in his name and to the disrepute brought by his Parties to
dialectics. But it is mainly because objective reality no longer is the
same; as Marx himself foresaw, quantitative changes in capital have led to
qualitative changes in Capital. It is obvious, for instance, that the
modern suburbs of Stockholm differ drastically from those of Manchester
a century ago, not only in their outer form but also in their functioning.
And yet, the sense of human deprivation inherent in the repressive
domination of man by man may be just as intense now as then. Thus,
the human sacrifices continue, for the gods of social cohesion and rational
exchange have simply changed from the suit of the old capitalist into the
open shirt of the new social bureaucrat; what is inside the velour pants
is nevertheless the same as was inside the strip(p)ed trousers. Perhaps
most analysts were too busy trying to understand the old world to notice
how the new was changing. Perhaps he who once was turned on his head
now is being turned back on his feet.

The emerging / in individual/society seems most evident in the grass-root movements currently spreading throughout the developed world. Here it is striking how the protests now focus less on the conditions of work and more on the holy family itself; it is in the micropowers of daily life at home, nursery, school, commuting, and hospital that we concretely experience the modern forms of social imprisonment. It is in those spheres of immediate existence that society reveals its fundamental contradictions of unfairness. It is through changes in familial relations that state-capitalism both reveals and hides itself.

Timely questions:

Which identity crises are in the commodities of the culture industry? Who would Oedipus have been without Laios? How do you rid yourself of the superego, if the superego is not a person but a faceless collectivity? Where does it reside, the fearful authoritarianism of state-capitalism?

* * *

It is easy to turn to Habermas on the ensuing crisis of legitimation. It is nevertheless more important to wonder about counterfinality, that is, about how we came to live in a world that is opposite to the good intentions it grew out of. But counterfinality flies in the face of traditional thought, for it is a situation in which the truths of the premises have not been preserved in the conclusions. It raises the issue of how I tell truths about a world whose very nature it is to be a lie:

What reasoning tools do I employ when I realize that the social world does not obey the rules of conventional truth-functional logic? What do I do when I notice that state-capitalism has many traits in common with such enemies of our culture as paradox and tragedy? Who is to blame when everything is perfectly right in the beginning and everything horribly wrong in the end?

Noone is to blame, for noone has broken those behavioral rules of reasoning into which he has been socialized. The tragic hero as expression of the Eros of the western ethos!

So:

How are we socialized into thinking-and-acting in ways which are at the same time individually praiseworthy and in the interest of society's state-capitalism? How do I move from categorizing crosses and directional arrows to self-referential loops? How do people learn to live in institutionalized double bind without going crazy?

* * *

The transformation of industrial capitalism into state-capitalism is already evident in the socialization and reproduction processes whereby society and individual are being adjusted to each other. By fulfilling moral codes, we experience how the temptation to dream and transform is overcome. But to experience is not to understand, even though it is a necessary step toward that boundary between the I and the Other where understanding resides. This boundary is taboo, now as much as in Paradise itself. In the process of trespassing, the anxiety of relations is turned into the fear of things, issues of power into fig leaves. And yet, castration is the metaphor that (fe)male power seems most eager to suppress.

So:

How can I simultaneously anchor – – in / and / in – –? If I ever did, how would I then translate my insights into communicable expressions without destroying them? How can you and I as individuals eavesdrop on society when it thinks-and-talks about itself in-through itself? How do I capture the dialectic of society and individual without falling into the – – trap of sociology and psychoanalysis?

* * *

Caught in culture, the only way to produce is to reproduce by putting words out of conventional contexts, by making new boots from the remains of old ones. Thus it is in self-reflection that reason sees its own interest. To ask again is consequently not to repeat but to translate anew.

Therefore:

Which forces of social cohesion are illustrated and further entrenched in the Odyssean act of ontological juggling? Is his appearance essential or is his function merely to divert attention from the pickpockets that raid the appalauding audience? How is this desiring piece of writing itself a legitimating instance of the socialization processes of state-capitalism? Is not striptease the appropriate metaphor of a society that talks about its own silences, reveals the powers it exerts, and promises to rid itself of the laws that protect it?

What does it mean to engage in dialectically mediating history-specific communication?

* * *

The challenge is enormous. Not for geography or any other well ordered discipline, but for its individual members exploring the limits of culture. Mallarmé was eons ahead of Christaller:

NOTHING WILL HAVE TAKEN PLACE EXCEPT THE PLACE EXEPTÉ

PEUT-ÊTRE

UNE CONSTELLATION

Toute Pensée émet un Coup de Dés
Hazerdous hazard.

Chapter 10
Of Bats, Birds and Mice[1]

Michael J. Watts

Vilfredo Pareto provides us with the classic statement of [Marx's peculiar use of words] when he asserts that Marx's words are like bats: one can see in them both birds and mice.

Bertell Ollman, *Alienation* (1971: 3)

Ann Arbor, Michigan is often seen as a sister city to Berkeley, California: the mid-West's sibling of that unruly bastion of free speech, anti-war politics and student radicalism on the Left Coast. There is some truth to this. Certainly by the time Gunnar Olsson arrived on the campus in 1966 – several years after the founding of Students for a Democratic Society (SDS) and the drafting of the Port Huron Statement and in the immediate wake of the Gulf of Tonkin resolution and President John's massive troop build-up in Vietnam – Ann Arbor had established its credentials as a centre of dissent: a forcing-house in a widening network of opposition to what was then called, as President Eisenhower put it in his 1961 speech, the 'American military-industrial-congressional complex' and its imperial ambitions. Olsson's decade-long sojourn in the US became the setting – and the Michigan campus the ether – for what one might call his 'epistemological break'. He came as a distinguished member of a foundational group who ushered in the so-called quantitative revolution, fresh from a year at the Regional Science program at the University of Pennsylvania.[2] A decade later, the person who scrambled aboard the *Queen Elizabeth II* in New York in 1977 – almost missing his First Class passage back to Sweden after a hastily convened good-bye party by some of his Ann Arbor students – was hardly the same geographer who arrived on the Michigan campus armed with the gravity model, and a slide-rule.

In the course of the short arc of his Ann Arbor years, Gunnar discovered that the good ship geography had run aground on the reef of the inference problem. He ran up against the limits of (ordinary) language and of translation, he came to see the tragedy of social engineering (and hence of the Swedish model of

---

1 This chapter is devoted to Gunnar Olsson's "Ann Arbor years" (1966–1977), though I can only speak with confidence about the period between 1973 and 1978 during which I worked very closely with him (Gunnar returned to Sweden in 1977). The Michigan Department of Geography was closed formally in 1982.

2 Trevor Barnes has conducted some important archaeological work in his social history of the quantitative revolution and of the so-called space cadets (see Barnes 2010).

social democracy), and he began the process of mining the deep veins of modern power, which led him to chart the abyss of cartographic reason. It is a remarkable journey that took him away from Sweden and ultimately back to it. There is a song by the great American troubadour Tom Waits – a strange and brilliant figure in contemporary popular music who keeps company with the likes of novelist and art critic John Berger and composer Robert Wilson – which comes to mind when I reflect upon Gunnar's return from what he called the possibility of 'death in exile':

> What made my dreams so hollow, was standin' at the depot
> With a steeple full of swallows, that could never ring a bell
> And I've come ten thousand miles away, with not one thing to show
> It was a train that took me away from here, but a train can't bring me home

It is not at all clear that Olsson's personal trajectory – or something like it –could ever have been accomplished, or even been thinkable, within the stifling confines of the Uppsala or Lund professoriat. What America, and the Ann Arbor campus offered was a ten-year postdoctoral fellowship in which he inhabited that peculiarly fecund yet strange and contradictory world – freedom, flexibility, superficiality, pettiness, fear, obedience and comfort – that American universities could, and to a certain extent still do, offer.[3] But ten years is a long time. And 'ten thousand miles away' the Swedish *heimat* sung its siren-song. The dialectics of exile and return were never far away.

23 August 1973. An outdoor coffee shop in Ann Arbor on the west side of campus (let me be more precise: it was a pizza parlor with an outdoor patio. Then it was a small family operation called Domino's: now it has 9000 outlets in sixty countries). It was here I met Gunnar Olsson for the first time amidst the sweltering humidity of a late Michigan summer. I had been prepped for the meeting. A classmate from University College, London – Adrian Pollock – had arrived the year before and quietly snowed me in a small avalanche of letters – I was working in West Africa at the time – extolling the virtues of not just the American academy in general but of Michigan Geography in particular. Wild and wonderful things could be found in the bland 'salmon loaf' building on State Street which contained the Geography Department. Not unnaturally Gunnar asked me what I was interested in pursuing at Michigan. I was twenty-three years old, undereducated, and not long out of University College, London. I only knew that I needed to make sense of what I had witnessed in Africa – poverty, famine, Islam, and state power. Gunnar

---

3   The University of Michigan was a wonderful place to be in the 1970s. Not so much for its campus and city liberalism but because of a sort of cultural-institutional openness – I suppose what we might now call inter-disciplinarity – that bequeathed a labile and ever changing landscape of intellectual vitality: impromptu seminars, reading groups and free-floating salons abounded. By comparison Berkeley, when I arrived in 1978, seemed staid, institutionally rigid and stifled by a sort of silo-mentality. The 1960s, I discovered, had left its mark.

seemed to delight in the fact he knew absolutely *nothing* about African peasantries or military government or Muslim clerics.[4] What mattered, Gunnar informed me, was how I *thought* about these issues: I must aspire, he said, to always be radical and abstract. With some solemnity, he told me, this did not mean Marxism: it was something *much* more difficult. It meant understanding how I knew the difference between him and me, and why we are so obedient and compliant. This was not the introduction to postgraduate work than I had anticipated.

Inevitably we drifted into a discussion of epistemology – a word I barely understood – and still further into what seemed to me like a glittering cosmos of intimidating names and referents – Hegel, Quine, Russell, Marx (ah, a flash of recognition), Wittgenstein, logical types, ontology and denotation – and finally an exhortation to understand the logic of science. All of this was delivered with enormous seriousness and without condescension (I knew nothing after all, and Gunnar was always the first to note that newly-minted PhDs knew a lot about not very much[5]). What I recall more than anything was a great warmth and humanity, a sense, characteristic of all of Gunnar's dealing with students who fell into his orbit, of caring deeply about what you thought, about your passions and the need to pursue unhindered one's intellectual passions but, at the same time, to recognize the seductions and the grave dangers of convention and obedience – of being capable of meeting the challenge of being "abstract enough". I felt as if I might be entering the clergy or perhaps therapy. It turned out to be neither. It was above all a friendship, and a life of questions: it required a sort of dedication that was insular, selfish and often monastic (little did I know how much Gunnar would turn to religious texts in his long trek toward *Abysmal*). Genius he often used to tell his posse of Michigan graduate students was nothing more than hard work; critique could only proceed on the basis of the deepest and most profound knowledge of the object of critique. Marx's deep grasp of – and indeed respect for – the very best of the Enlightenment political economists and philosophers provided a model of such critique. And all of this he told me – we were now into twilight and the dregs of a bottle of wine – required us (I think he meant me) to learn how to *read*. For someone with Saharan dust still on his shoes, this sounded both other-wordly, utterly confusing, and totally compelling. The wine was done. We departed – he with that characteristically long stride and slouch heading off toward Geddes Avenue. I fled to the library in search of Hegel.

In some respects the number of students Gunnar graduated in his eleven Ann Arbor years – he served as Chair to 16 students – is out of keeping with both his

---

4   A decade later I had written a book which I had dedicated to Gunnar: when we met in Uppsala shortly after I had mailed him a copy he told me that what moved him most was his appearance on a dedication page in a book the substance of which was utterly foreign to him.

5   This was the first, but certainly not the last occasion on which he expressed some skepticism of clever English boys who had read everything, synthesized everything and understood nothing.

charisma and the scope of his intellectual footprint in the discipline. These days, admittedly in a much more industrial and Taylorist system of doctoral training, the average Geography professor might sign-off on two or three dissertations each year. His first PhD student was Stephen Gale; I was the last, minted incidentally at the very moment that the Michigan department was, to the abiding shame of the University, terminated.[6] But it is not the quantity I wish to dwell upon as such – as if this were a failing of productivity. It is rather that a number of students began but some never finished: some took their PhDs and headed away from academia. None of these things are perhaps surprising. My own cohort for example contained an extraordinarily gifted group of individuals: some finished their doctoral training, many didn't. Those who did went on to become, interior designers, self-made software millionaires, business consultants. The most gifted became a high school teacher. Many drifted into what I cannot say. Perhaps this heterogeneity is not unusual but the reality is that the residuum of the 1960s still lingered, and sometimes it felt like a pall that hung over the academe.

When I touched down in the US, the Vietnam war had not drawn to a close (the drama of the fall of Saigon was still two years distant). The reverberations of the anti-war movement still saturated the atmosphere of the campus. It took until the mid-1970s for the 1960s to draw to a close. The nightmare of Vietnam cast a very long shadow. There were casualties along the way. Gunnar often talked of some of his students who simply went mad. He rarely talked of 1968[7] (the *annus mirabilis* as *Time* magazine dubbed it) or the sixties campus protests: the BAM strikes of 1968, the violence surrounding the attempt by students to "liberate" the south campus in 1969, the struggles over the ROTC on campus, the anti-draft

---

6   To this day I have never given a penny to the University of Michigan in spite of their insistent pestering through the alumni network, and for almost two decades would not consider speaking on campus.

7   The events of that year bear repeating: the Tet offensive and the massacre at Mai Lai, the assassinations of M.L. King and Robert Kennedy, the May *événements* in Paris (including the strike of 10 million French workers), 'Socialism with a Human Face' in Prague, Warsaw and Belgrade, the Soviet invasion of Czechoslovakia, the high-tide of the Chinese Cultural Revolution, the Medellin convention that launched Liberation Theology, students' rebellions in Tokyo, Delhi, Berkeley, Rio and Berlin (indeed just about *everywhere*), the massacre of Mexican students in Tlateloco Square, Regis Debray's imprisonment in Bolivia (and marriage in jail), growing turmoil and civil strife in Ireland and Palestine, the debacle surrounding the Democratic Convention in Chicago, the so-called hot winter in Pakistan (and later in Italy), the early stirrings of feminist protest surrounding the Miss America Contest in Atlantic City, the end of American growth liberalism, the collapse of the British pound, and the first rumblings of a major economic crisis, and, lest we forget, the election of Richard Nixon and the move across Washington, from the Pentagon to the World Bank, of a beleaguered and morally contorted Robert McNamara. Just to blink was to miss something, said Christopher Hitchens (1998: 101). *Fortune* magazine, surveying the events of spring 1968 judged, with good reason, that American society had been 'shaken to its roots' (quoted in Horowitz, 1970: 185).

movement, and the faculty mobilization led by Marshall Sahlins among others against the war. I do not recall Gunnar ever talking about these issues. And to the best of my knowledge he has never written a word on the subject of America. It was a period of darkness in which the personal costs for many – some of his closest students included – were devastating.[8] 'Twas in another lifetime', sang Bob Dylan in 1974 looking backwards, 'full of toil and blood, blackness was a virtue, and the road was full of mud' (*Shelter from the Storm*).

Curiously it was during the heat of the war that Gunnar was drawn to southeast Asian himself, to Thailand as part of a research project directed by his Geography colleague Peter Gosling. It was, I think, a profound experience. It soured him forever on much of what passes as development, to say nothing of planning. It proved to be an echo of what he was beginning to grasp through the ethics problems he attributed to the social engineering by geographers, Torsten Hägerstrand among them, on Sweden in the 1960s ("everything beautifully right at the beginning ... horribly wrong at the end ... no one to blame in between" (Olsson 2007: x). The inference problem had come home to roost.

1968 was, as turns out, a pivotal year for Gunnar, in fact revolutionary to a fault. But not due to the evenements on the streets of Paris, Mexico City, Berkeley or London. The real weight and significance of '68 resided elsewhere. As a student Gunnar had been influenced by the Swedish economist Herman Wold who delved deeply into the philosophic foundations of correlation and regression. At stake was what we might now call the conundrum of situated knowledge (the knowledge of the vantage point): in Wold's case the ties between causality, inference, reasoning rules and the deep structure of human thought and action. What concerned Gunnar was what this might mean for the gravity model – he once told me that by the mid-1960s he knew more about the gravity model than anyone in the world – and how spatial variation could be understood – as a cause or an effect of human interaction? How to grasp spatial form and social process? What linked the two was planning or what he called social engineering, a utopian project which as the sixties wore on he saw as tragedy in the guise of social democracy, as discipline, fear, and obedience dressed up as liberty, fraternity and equality. It was in 1968 that "it was proven that the traditional two stage approach to geographical description and interpretation is fallacious. The various stochastic models...showed that the same spatial form can be generated through radically different processes" (Olsson

---

8   There is another aspect to the attrition of students and to the very different life courses of Michigan students. It is a delicate matter. To open the door of dialectics and the prison-house of language to young doctoral students and to invite them to walk through, is a very risky business. Not only because they have to then navigate the choppy waters of Departmental politics and egos of faculty too insecure to admit what they do not know. But equally because it left many of us without foundations, unmoored – or at least it felt that way – and often immobilized by the sheer difficult and complexity of what we confronted, of what this meant for something called research. In Geography 810 – our own prison-house – this anxiety was a constant spectral presence.

2002: 252). The most perfect description of spatial form cannot be used to infer how and why it came about. The good ship Geography had run aground. The inference problem – born as it were in 1968 – was too radical, too unsettling to fully absorb. Many fled from it or continued as if nothing had happened. It cast a very different light on the challenges to authority witnessed on Hash-Bash day (a 1960's carnival to celebrate legalized pot on the university diag) or the political theatre of draft-card burnings on the steps of the Harlan library.

The turning point of 1968 represented the coming to fruition between 1968 and 1975 – between the shipwreck of the good ship geography and the birth of *Birds in Egg* – of seeds sown much earlier. It was a period described by Gunnar as "the best years of my life". Olsson's struggle with the reasoning contained within the gravity model was in evidence by the mid-1960s but it began in 1960 and was stimulated in 1961–62 by Julian Wolpert who visited the Uppsala department (both Wolpert and Olsson attended Wold's seminar and the same course in computer programming). Without a doubt the most formative period was during 1963–64 in which Olsson spent time on an ACLS Fellowship at the Regional Science Department in Philadelphia. Time once more with Wolpert but most especially with Walter Isard, and Michael Dacey and in two NSF summer institutes – the first in Regional Science at Berkeley, the other in spatial statistics at Northwestern – in the good company of his founding generation of quantitative geographers (David Harvey, Akin Mabogunje who almost succeeded in recruiting him to Ibadan, and Leslie King among them). All of these roads were to lead, ironically via the *pur et dur* of Michael Dacey's geographical science, to the inference problem and the 'crisis' of 1968.

Olsson's epistemological break fell into focus in the years between the late 1960s and the early 1970s. In 1967 he is publishing on interaction and stochastic process; he is writing about correspondence rules and differential logics with Stephen Gale in the early 1970s; the rupture was just about complete by the time I arrived in Ann Arbor in 1973 (about the time of the appearance of his *Antipode* piece on dialectics and planning). The struggle over form and process – or geography and history, identity and difference – proved to be the forcing house for both Olsson (*the* central intellectual problem of his life as he sees it) and his immediate cohort, most especially John Hudson, Leslie Curry, David Harvey, Reg Golledge and Leslie King. For Gunnar the inference problem was at base a *translation* problem: it led him into in the first case philosophy of science (how was consistency to be maintained in positivist reasoning, what were its reasoning rules?), into other forms of logic consistent with the tensions between theory and observation, between the phenomena themselves and the language in which we tried to pin them down (this led to a brief flirting in his case with many-valued logics and fizzy-set theory) and finally deep into the prison-house of language itself. The insurmountable translation problem – as he came to see it – was connected to the present and the past: to Swedish social engineering, and to Aristotle's "On Interpretation", the law of the excluded middle and the sea-battle tomorrow. As he put it in an interview with Trevor Barnes:

The outcome of those investigations was, of course, that even though we cannot do without the copula, there are several alternative ways of defining it, each mode expressible through drastically different languages; dialectics one of them, conventional logic another.

For others in his immediate *groupiscule*, the road from the inference crisis led elsewhere: for Harvey into the world of Marx, for Golledge into the world of seeing and believing, for Hudson into the museum of history and the making of landscapes, and for King out of the world of research altogether.

In my mind it is a large, unencumbered space, light carpets, a beautiful leather couch (my then-wife regularly fell asleep on it). The living room floor at 2128 Geddes Avenue. I spent a not inconsiderable amount of my graduate life there, along with Adrian Pollack, Jonathan Mayer, Peter Hoag, Cathy Baker, Tony Black – all card-carrying members of Geography 810. Others spent time within its circumference including a steady stream of professors who stood at an angle to much of what was going on in Michigan social science: anthropologists Aram Yengoyan and Mick Taussig, philosophers Frithjof Bergmann and Henryk Skolimowski, linguist Peter Becker It was on this floor too that Harvey, King, Golledge and Hudson fought the inference problem. And there was the wine, always very good wine.

It was during the first half of the 1970s that *Birds in Egg* was crafted, itself largely a product of the year-long introductory graduate class Gunnar regularly taught and a seven-year university wide course entitled 'Thought and Action'. It was the former that was my baptism into the Church of Olsson. To experience a lecture by Gunnar is quite indescribable. He paced rather like a caged panther in a zoo: he gestured and pondered. It often felt like being taken on a deep dive: oxygen was depleted, we became light-headed, suffocation seemed to approach. At a certain point there was desperate need to get back to the surface for air (I have often had this experience in listening to John Coltrane, another individual testing the limits of language). Each lecture laboriously written out on yellow note-pad paper in that distinctive labyrinthine script. Always a performance, but also a sermon. An Olsson lecture was, after all, as likely to be filled with references to Abraham as much as Althusser, the icons of the Orthodox church as much as Marx and Hegel. I recall vividly receiving a letter from Chris Smith in 1976, then teaching at the University of Oklahoma, describing a talk by Gunnar to the geographers: at the end of the lecture a man on the front row was sobbing; another stood up and left muttering: "back to the drawing board". As Trevor Barnes (in this volume) describes his first brush with Olsson in London in 1978, it was an incandescent and brilliant experience.

For me, and I suspect for others, there was a paradox at the heart of Olsson's teaching, and it is this: he is an utterly compelling and a commanding presence but I sometimes understood almost nothing. He often produced a sense of intellectual vertigo. Much I did not understand but I felt it's truth. Some simply seemed confused or quizzical. Nobody doubted the gravity of what was uttered. I came to realize much later than this paradox proved the point – his point, a

Hegelian point: "truth," said Hegel, is a poetically and rhetorically polished "mobile army of metaphors, metonyms and anthropomorphisms". As Gunnar himself put it: "telling the truth is consequently to be believed, to be an expert juggler of rhetorical tropes" (Olsson 2007: 97). There is a remarkable youtube video of Gunnar delivering a talk at Uppsala stadsteater in March 2008 (http://www.youtube.com/watch?v=tm6y82DJAK0). The sound track is barely audible. It matters not. It makes my point. The world need not just a new language: there must be a new rendering but also belief ("this issue of logic and rhetoric has been at the core of my entire life" (Olsson 2007: x).

I have a memory. Gunnar is serving as a respondent at Michigan – my recollection is that it was in response to a lecture by Edmund Leach, the great anthropologist. He strolled to the podium and said: "Professor Leach does not know what he is talking about". His point of course was that Leach's talk told us more about the language he was talking in than the subject he was talking about. More quizzical looks.

By 1973 the indispensable starting point for the new graduate student working with Gunnar was the logic of science, its internal structure and the struggle to maintain consistency and correspondence between the language of science and that which it purported to represent. How is truth to be preserved in the reasoning of scientific logic? The object-lesson – I did not fully grasp it at the time – was the limits of ordinary language, and the chasm between thought and action. It produced, as I saw it, a sort of paradox or a set of paradoxes (paradox incidentally was a running theme in our Geography 810 readings in 1974): on the one hand, a seeming disavowal of a deep American sense of the need to "do something" in and about the world (at a moment incidentally of profound crisis and change), and on the other, a turn to language, at a time of an explosion of interest in Marxism (the backdrop after all was the Althusserian and Habermasian moments – *Reading Capital* and much of the later Frankfurt School was appearing in English – and the publication in English of the *Grundrisse*). It was a close reading of Bertell Ollman's extraordinary book *Alienation*, published in 1972, which provided the insight into these paradoxes, and built a bridge to the land of dialectics (the alternative to fuzzy logics as I saw it). *Alienation* was a galvanizing text – anthropologist Bob Hefner who frequented Gunnar's seminar once said of the book that there was a dissertation on every page – because it spoke to the tenor of the times (*Antipode* after all was born in 1970) yet began with Pareto's remarkable insight into Marx's language. It led straight into the heart of relational thinking and dialectics and hence to Hegel and the late Wittgenstein (significantly Foucault was only a spectral presence in our discussions). I think it is fair to say that none of us – Gunnar included – quite knew what we were doing or where we were going.[9]

---

9   Inevitably I had my own moments of grave deep anxiety, nowhere more so than upon arriving in Nigeria in 1976 to start my dissertation plagued by military government, massive bureaucratic delays and a large measure of self-doubt. Not to worry Gunnar wrote: I was situated between "your colonial pasts and multi-national futures": I must above all

What I failed to grasp at first was the need to read the core texts as if they were deeply empirical, speaking directly to the human condition and the way we lived out lives (sometimes hard in reading *The Phenomenology of Mind* or the *Tractatus*). Or to put it differently: the fact that whatever his immediate concerns – whether Marx Ernst, Samuel Beckett or Georg Henrik von Wright – Gunnar's objects of scrutiny were consistently the same: the deep structure of power, why are we so disciplined, why do I believe what you say, how do I know the difference between you and me, what is constituted in purposeful action, how can I find my way in the world. It made for off-kilter readings of the canon. The commodity in *Volume I* and reflections on money in the *Grundrisse* led not to class struggle or state power but to the dollar and trust and inflation as broken obligations and to the magic of ontological transformations; a critique of capitalist planning led not to the charnel house of capital but to the question of why we are "not radical enough" and why "we are so compliant", so fearful.[10] He spoke about power but in neither Marxist nor Foucauldian senses; rather as a series of magical ontological transformations in which visible things are turned to invisible relations (another off kilter read, of Marx's commodity fetishism), the profane into the sacred, the sign into the signified. It was a project driven not by the brute realities of armies and states but by his repugnance – is this too strong a word, I think not – of the deadly seductions of power, of the noxious solicitations and admonitions to change the world, and above all by the call for purposive action among those who do not understand what they do and have no grasp of human tragedy. What was needed was a map of the "Territory of the Human", an "atlas of what it means to be human", a critique of "cartographical reason" (Olsson 2007: 9).

There were a number of beacons shining light on the road away from the inference problem. Bertell Ollman I have already mentioned, and the *Grundrisse* too. But as the translation problem and the equality sign came into sight we turned to George Steiner's *Babel*, Gregory Bateson's *Steps to an Ecology of Mind* (madness and the double-bind in particular), Marshall Sahlin's *Culture and Practical Reason* (he somehow seemed to write dialectically), and the linguist Peter Becker who, in a memorable seminar, showed us how Javanese shadow-puppetry had much to say about the question what does it mean to be human? But already the shift from social science to literature and literary criticism was underway. The limits of language and of translation could best be heeded by the siren call of Beckett, Mallarmé, and Joyce. By the time he arrived at NORDPLAN he could write this: "I am going even more desperately into the interface of the social sciences on the

---

"regain your courage and excitement….Vulgar pragmatism runs so strong that unless we show through our actions how we can overcome it we will all be crushed together" (letter dated 27 January 1977).

10   In a letter written to me in late 1978 Gunnar wrote: "But the Swedes are so afraid and so thoroughly alienated. If only I had the head and the bottom of Karl Marx I could write something now and here which would be as important in a few years as his work is at present. How long must one wait for the poets?" (17 December 1978).

one hand and literary criticism on the other....the pure social scientists though are just as dull here as anywhere else" (letter dated 6 January 1978). A reading list for his first cohort of NORDPLAN students (January 1978 'Vetenskap och planering') gave pride of place to Breton, Canetti, Cioran, Gombrowicz, Joyce, and Barthes.

More than anything else, the great moral of Ann Arbor seemed to be that the limits of ordinary language could only be met by learning to think and write in a way that the text was not about something but *is* the thing itself (echoes of Picasso here!). The goal was to achieve something like the most abstract of abstract painters: pure expression of relation (Olsson 2002: 260). Mondrian served as his muse. But Duchamp too. From his first exposure to the *Large Glass* in 1963 in Philadelphia, Duchamp's masterpiece had haunted him. What he saw in it – and in Ptolemy, Brunelleschi and Alberti much later – was a "clear cut case of the Geographic Inference Problem" (2007: 375): Duchamp was working backwards from the shadow to the shadowed. Inevitably the tenor and the form of his writing changed. The quality of the prose is remarkable: as distinctive and multi-layered a language as Joyce or Duchamp.

All of this – by which I mean the Ann Arbor years – seems to me, with the powers of hindsight, to have been a prelude to *Abysmal*. I think it took Franco Farrinelli – probably the only geographer from which Gunnar actually learned anything – to serve as the geographical bridge to the deep structure of being human: namely the map, the geographical holy of holies. By 2007, we finally have a geometry of power – the line, the point, perspective – and an account of cartographic reason. It was drafted in Uppsala, but completed in the northern California, in the dreamworld of Silicon Valley.

There was a larger canvas too, and it was of a hugely dysfunctional family. The Department of Geography at Michigan exhibited all of the attributes of almost any academic institution: a place in which, as Henry Kissinger noted long ago, bloodletting can occur because nothing is at stake. Michigan's recent history was of course inseparable from the quantitative revolution: two of its space cadets – Waldo Tobler and John Nystuen – were among its ranks. But it was as small, heterogeneous department suffering, inevitably, from the burdens of being little more than a herd of independent minds. Personality differences abounded, egos were insecure, most were compliant and uncritical. Personal animosities – as much as intellectual differences – came to dominate. There was little here to take joy in. Upon his departure he wrote me that "we left in the midst of wine and tears but with little feeling for the Department" (letter dated 12 May 1977). It was inevitable that the three individuals with something to say – Gunnar, Barney Nietschmann and Waldo Tobler – should all want to leave: two to the West Coast, and Gunnar back to Sweden. None of this was responsible for the closure of one of the most storied Departments of Geography in North America. This calamity lay in the collapse of the Detroit economy, the utter failure of leadership and governance from the Department itself, the petty-minded hostility of the university accountants and administrators who neither understood the discipline nor what a university represented, and the spinelessness of the academic senate who felt that uncomfortable sacrifices would prevent further rot. We – I mean we US-based geographers – are all still living with the consequences.

Mostly it is the conviviality. The museum of memory is of course incomplete and disorganized; there is mis-labelling and hidden boxes and corners, and unnamed rooms. What I recall most is that damn living room floor. And then much later: the walks in Värmland, the family farm, the Venice Biennale, dinners with Allan Pred in the Berkeley hills, the rambunctious birthday party at Thunbergsvägen, and a year together at Stanford. And the wine of course.

The departure from Ann Arbor was bittersweet. No regret over Geography perhaps, but the University had been a crucible of his creativity. He was leaving perhaps to flee death in exile but returning to an arena – planning – the critique of which had driven him deep into literature, art and the outer-margins of convention. Why return to *that?* What followed – a whirlwind of activity, posses of new students (planners quite unlike the students he had taught in Ann Arbor) and new intellectual communities in Prague, St. Petersburg, Krakov, Bologna and elsewhere – others can address better than I. From California – upon whose shores Gunnar and his students would occasionally wash up for their week long seminars – it seemed frantic, exhausting, and sometimes odd.

In the summer of 1977 all of this lay in a future without reference points, on a map without details. This is how it seemed when he wrote me from the middle of the Atlantic, aboard a very large ship:

> So now it is done…. But right at the moment we are in real limbo, somewhere in the middle of the Atlantic. Late tomorrow we will be landing at Cherbourg… and then to the hard work of Scandinavia. It is a tremendous challenge for all of us and it will be interesting to see how it works out. (Letter from Olsson, dated 12 May 1977.)

Sign off. Mission completed. Go home, Professor.

Bibliography

Barnes, T.J. 2010. Taking the pulse of the dead, *Progress in Human Geography* 34, 668–677.

Barthes, R. 1977. *Fragments d'un discours amoureux*. Paris: Seuil.

Bateson, G. 1972. *Steps to an Ecology of Mind*. New York: Ballantine.

Breton, A. 1960. *Nada*. New York: Grove Press.

Canetti, E. 1969. *Auto-da-fé*. New York: Avon.

Gombrowicz, W. 1968. *Ferdydurke*. New York: Grove.

Joyce, J. 1959. *Finnegan's Wake*. New York: Viking.

Marx, K. 1973. *Grundrisse*. Harmondsworth: Penguin.

Octavio, P. 1970. *Marcel Duchamp or the Castle of Purity*. London: Cape Goliard.

Ollman, B. 1971. *Alienation: Marx's Conception of Man in Capitalist Society*. Cambridge: Cambridge University Press.

Olsson, G. 1975. *Birds in Egg*. University of Michigan Publications, Department of Geography, Ann Arbor.

Olsson, G. 2002. Glimpses, in *Geographical Voices*. Edited by P. Gould and F. Pitts, Syracuse: Syracuse University Press, 237–268.

Olsson, G. 2007. *Abysmal: A Critique of Cartographic Reason*. Chicago: University of Chicago Press.

Sahlins, M. 1976. *Culture and Practical Reason*. Chicago: University of Chicago Press.

Steiner, G. 1975. *After Babel: Aspects of Language and Translation*. London: Oxford University Press.

Von Wright, G.H. 1971. *Explanation and Understanding*. Ithaca, NY: Cornell University Press.

Chapter 11

Environment and Planning D: Society and Space, 1987, volume 5, pages 249-262

The social space of silence

G Olsson
NORDPLAN, Box 1658, S-111 86 Stockholm, Sweden
Received 13 March 1987

Abstract. It is in the physical concreteness of the social space of silence that it cannot be abstracted. The reason is that every text is permeated by the notion of self-reference. The impossible challenge is nevertheless to think-and-act in such a way that there is no difference between the languages I am writing *in* and the phenomena I am writing *about*. Eventually this raises inescapable questions about the power of language and the language of power, about the truth of silence and the silence of truth, about mental prefixes and linguistic experiments, about you and me.

Everyone knows that I talk very little. But at certain times I was driven to talk by a force so compelling, I felt determined to transform the most simple details of life into so many insignificant words, that my voice, which was becoming the only space where I allowed her to live, forced her to emerge from her silence too, and gave her a sort of physical certainty, a physical solidity, which she would not have had otherwise.

(Maurice Blanchot: *Death Sentence* page 73)

*

The challenge is enormous. Not for the well ordered disciplines of the social sciences, but for its individual members exploring the limits of culture. Mallarmé was eons ahead:

NOTHING WILL HAVE TAKEN PLACE EXCEPT THE PLACE

EXEPTÉ

PEUT-ÊTRE

UNE CONSTELLATION

Toute Pensée émet un Coup de Dés
Hazerdous hazard

(Gunnar Olsson: "-/-" page 33)

*

"There's glory for you!"
"I don't know what you mean by 'glory'", Alice said.
Humpty Dumpty smiled contemptuously. "Of course you don't—till I tell you. I meant 'there's a nice knock-down argument for you'".
"But 'glory' doesn't mean 'a nice knock-down argument'", Alice objected.
"When *I* use a word", Humpty Dumpty said in a rather scornful tone, "it means just what I choose it to mean—neither more nor less".
"The question is", said Alice, "whether you *can* make words mean so many different things".
"The question is", said Humpty Dumpty, "which is to be master—that's all".

(Lewis Carroll: *The Annotated Alice* pp 268-269).

*   *   *

Questions spring from the constellation of quotes: WHAT IS THE DIFFERENCE BETWEEN YOU AND ME?

Rephrased and operationalized: Which difference makes a difference? What is an ontological transformation? How is touchable turned into untouchable, sound into silence, letter into meaning? Is power the power of meaning, meaning the meaning of power?

Is the power to produce meaning with the repres(s)entative editor, proudly appointed by society to guard and to legitimate the right to write? Is it with the creative writer, publicly relieving himself on sheets of paper? Is it with the responsive reader, privately experiencing anew what she already knew? Put differently: Is power in the authority of the author or in the auditing of the audience? Or is power at the same time everywhere and nowhere, always evasive yet in every glance, every touch, every mouth?

For answers, take your analytic prick, s'il vous plaît: Un oeuf à la Descartes in the elegant Bar de Saussure? Or some scrambled Humpties from the basement of Alice's Restaurant? The personal touch of the French or the common sense of Anglo–America? Remember, though, that the philosopher himself always had breakfast on eggs hatched for eight to ten days. Lonely babe in Babylone staring through his whoroscope.

"Damn it all! Can't have a paper without words, you know!" Which was Lord Peter Wimsey's way of saying that the language of power is in the power of language, the power of language in the language of power. No exit, for the phenomenon I wish to write *about* is at the same time the medium I am forced to write *within*. Once again caught in the familiar chains of self-reference, telling truths about lies and lies about truths. Confined within the prison house of communication, what I happen to say is not what *I* want to say but what the *saying* wants to say. All criticism is by necessity an exercise in metalanguage, all learning a struggle with paradox. Start talking with your cell mate, Epimenides!

It is in this paradoxical sense of foreclosure that language is both its own problem and its own solution; since I can not mention a word without using it, every negation is predicated on a form of affirmation. Perhaps this is why Penelope locked herself up with her tapestry woven at dawn and undone at dusk: activity for activity's sake; time for time; weaving not for the web but for the weaving of the web. The evasive principles of deconstruction had no doubt been in Ithaca, Greece, long before they reached Paris, Baltimore, New Haven, and Ithaca, NY. Shadows reflecting.

*

Simultaneously evasive and aggressive, most postmodernists tend to be explictly self-conscious. Every discourse is by them held to be heterogeneous, every meaning to embody a structure of undecidability, every text to be contextual. The only way to grasp and break out of such a reality is to perform radical experiments, for to *experiment* means literally to go outside limits, to refuse accepted categories, to fight codification. There is no alternative, for in categories lies the status quo of conservatism, in codes the totalitarianism of the norm(al).

It is as textual experiments that language approaches solutions to its own problems. The reason is that every utterance contains a crucial element of persuasion; there is no truth without opinion, no description without performance. Telling truths or lies is consequently not enough. Being convincing is equally necessary and that is regardless of whether the text in question is a theorem, a scientific paper, a novel, a poem, or anything else. The urge to express is linked with the urge to impress, the grasping of meaning with the evocation of meaning. Logic and rhetoric reach for one another. So do ethics and aesthetics.

As an integral part of the crisis of realism, there is fiction in every truth, truth in every fiction. And thus it is that Marcel Duchamp had no choice but to strip his bride bare, for new truths require new categories, new chattergories new forms of authority. But a comment on a text is itself a text. Etymology indeed suggests that to write a text is to weave a texture of words, to produce a ready-made tissue for blowing your nose or for whipping your ass. Is reading and writing a kind of anal erotics, a symbol of subject and object united? Who is to be master-baitor? That's all.

The analyst asks: Is the meaning of such critical texts in your reading from the outside or in my writing from the inside? Is reading a search for knowledge, writing a model of action? Is textual power in the inferential logic of factual reference or in the productive rhetoric of speech acts? Is the locus of power in the things denoted or in the words denoting? Where *is* the difference?

The questions prompt themselves. For it is in the theory and practice of current literature that language simultaneously reflects reality and constitutes it. If this is correct, then it must be explicitly recognized that every discourse has the potential both for manipulation and for evasion; every statement holds back and sets free at the same time. Does it follow that the power of words lies not in their users directly but in the social relations which knot them together? Is power in the abcdef-mindedness of rhythmic expressions? Is power a tautology anchored in conventional rules of inference? Are external and internal hiding behind the same veil of being? Is it there that ontologies are transformed?

Modern theories often imply that the power of words lies in other words. Signs are thought to embrace other signs. Signs copulate. Signs materialize. Word turns to body, body to word and eventually into a marching army of metaphors, metonomies, and anthropomorfeces. As a consequence, there has been a movement beyond the theoretical realms of logic and dialectics into the practice of rhetoric: possessing a language is for understanding and speaking alike.

Living with such a double possession is to experience how the present is always past, the past always present. In the here-and-now midst of the ruling metaphysics of presence, there is in fact a language of radical absence, for words never denote directly, always indirectly. A word is never the thing itself, merely a version thereof. It is nevertheless in their desire for presence that words take on some thinglike properties, partly as a function of the mythological tendency to reify denotation, partly as a reflection of phonology and the anagrammatic nature of letters. The alienation produced by these tendencies is not social or psychological but profoundly ontological. The conclusion stems from Jacques Derrida: Difference deferred is a difference that makes a difference. *Differance* is the term to dread a gain from.

Despite the current fashion of analytico – referential historiography and biography, it may therefore be writers like Nietzsche, Kafka, Musil, Blanchot, and Barthes that come closest to the evasive spirit of the present. It is within the fragmentary and aphoristic works of these authors that they allow the precision of ambiguity to rule over the vagaries of the historically specific. Employing such techniques, they manage to write both *in* and *of* power, to grasp ontological transformations in their process of transformation, to realize that tautology is a frozen state of reasoning and that concentration camps represent a degenerate form of power. Thus it is in the struggle with self-reference that both reader and writer are first caught in a maze of old cross-references and then asked to participate in the production of new meanings. Yet, the value of such texts is less in the hinted answers and more in the unknown questions. It is in fact in the tradition of deconstruction to aim for the unanswerable, not to affirm the positive but to reflect the negative; what is thereby destroyed is not meaning per se but the unequivocal domination of one meaning over all others.

Therefore: Insult the wor(l)d and it will reveal the taboos of the social space of silence! In everyday speech, language is idling. In avant garde literature, it works at a pitch, experimenting with the intelligabilities of today's text. It is in the crevice between convention and deviance that language becomes erotic.

*   *   *

One way to fight the totalitarianism of codification is always to begin anew. Another is to write in fragments. Since the lover's relation to the loved one is of an extreme solitude, these were also the approaches that Roland Barthes used in his moving book *A Lover's Discourse*. Like its own subject matter, this discourse does not set out to realize a small number of well-defined wishes. Instead it merely moves on, continuously and unpredictably new, guided by nothing but the randomness of alphabetese. Form and content are here brought together in a play of ambiguity, paradox, and irony. The hope is not to capture human relations in the nets of descriptive sociology, but to let them free in subversive action. SO IT IS A LOVER WHO SPEAKS AND WHO SAYS: Never bide your turn, never worry about the future. And so it comes that Barthes's quote from Tao can serve as its own verification: "He does not show himself and shines. He does not affirm himself and prevails. His work done, he does not attach himself to it, his work will remain" (Roland Barthes: *A Lover's Discourse* page 233).

Ready?

Can't stop them now. Here they come. Three words. Two empty spaces in between: I LOVE YOU.

Noun of love. Verb of love. Present tense. Action word doubly embraced. Subject and object linked together. First person reaching out for a second. Thus it is that when I say I LOVE YOU, I do not denote but perform. Whenever I repeat the banal words, I do not participate in an explanatory sermon but in the chanting of a hymn. I do not analyze but confess. My utterance lacks content, yet is full of meanings. As a tautological amen, it is a cry at the outer limits of language. It tells nothing but confirms everything. High above the crowd, Chagall's blue bride throws herself from one trapeze to another, floating in air, unquestioning, yet driven by an intense curiosity, reaching for the outstretched hands. No one understands, because the lover's discourse is so exceedingly solitary, itself a reflection of a loneliness that is not psychological but systemic.

Barthes's nontotalitarian epistemology has a formal counterpart in his use of pronouns and tenses. The he, she, it of the thingified third person are taken over by the I and you of the first and the second. The past was is replaced by the present is. As if to prove itself, love in fact turns into jealousy exactly at that moment when the persons of you and me are erased by the things of him or her; the ungraspable is destroyed exactly when cultural symbols are impoverished into material signifiers; torn by jealousy, the loved one remembers the simplicity of physical details and forgets the richness of total relations. What is grasped is the body not the spirit of the letter. What is lost is the insight that desire issues from lack not from presence; absence is transformed into an ordeal of abandonment.

Prior to that moment of alienation, however, I ask not with my mouth but with my eyes. You answer not with words but with glances. Love sees clearly. Love is not blind. Like Molly Bloom in the cool of the evening, I therefore ask you with my eyes to ask again. Yes my mountain flower. And I put my arms around you thereby proving that whenever in doubt, the body becomes the word's corrective.

The hell with the telephone! For "with my language I can do everything: even and especially *say nothing*. I can do everything with my language, *but not with my body*. What I hide by my language, my body utters. I can deliberately mold my message, not my voice. By my voice, whatever it says, the other will recognize 'that something is wrong with me'. I am a liar (by preterition), not an actor. My body is a stubborn child, my language is a very civilized adult ..." (Roland Barthes: *A Lover's Discourse* pages 43–44).

Like a stubborn child, love is not a conversationist. All it does is in effect to mumble to itself, far removed from the pursuit of compromise. Monologue is therefore the language form which best reflects the fact that the lover never resigns, never waits in line, never fights about the final word, never wrinkles his forehead, never shakes his fist. All he does is to laugh and cry. Not as a concerned sociologist or a powerful politician. But as the liberating artist he happens to be.

And thus. In the fantastic banalities of love I desire not *what* you are but *that* you are, not fetishized things but untouchable relations. Ecstacy and stillness come together in the moment that I love you so you love yourself. It is you I want, you   and in you myself. In the loving calm of your arms I become two in one, motherhood and sexuality united. And "her brother's body pressed so tenderly, so sweetly against her, that she felt she was resting within him as he in her; nothing in her stirred her now, even her splendid desire" (Robert Musil: *The Man Without Qualities*, as quoted in Roland Barthes: *A Lover's Discourse* page 224).

At that rare moment of perfect communication, the enchantment is turned into the bliss impossible to name. And whereof one cannot speak, thereof one must be silent. The unspeakable is the mystical, the mystical beyond the limits. It is from that empty space on the other side that I hear echoes of the banal chants. I LOVE YOU. "... und alles, was man weiss, nicht bloss, rauschen und brausen gehört hat, lässt sich in drei Worten sagen" (Ludwig Wittgenstein: *Tractatus Logico-Philosophicus*, dedication page).

*

In his final works Roland Barthes staged an utterance, not an analysis. The discourse is used not reduced, shown not said. A glimpse of nakedness. A fold in the velvet gown suggests that "language is a skin: I rub my language against the other. It is as if I had words instead of fingers, or fingers at the tip of my words. My language trembles with desire. The emotion derives from a double contact: on the one hand, a whole activity of discourse discreetly, indirectly focuses upon a single signified which is 'I desire you' and releases, nourishes, ramifies it to the point of explosion (language experiences orgasm upon touching itself); on the other hand, I enwrap the other in my words, I caress, brush against, talk up this contact, I extend myself to make the commentary to which I submit the relation endure" (Roland Barthes: *A Lover's Discourse* page 73).

*   *   *

Shocking discovery! Imagine. All your work made of words. Of words alone. "It is as though you had discovered that your wife were made of rubber: the bliss of all those years, the fears ... from sponge. It's worse than discovering your privates are plastic" (William H Gass: *Fiction and the Figures of Life* pp 27–28). Is that where power resides? In the plastic? In the rubber? In the physicality of the words themselves?

In deed. Touch the wordy world! Hold it in your hand! Caress it! Squeeze it! Move it to your tender fingertips! And it will come. From the depths of the unconscious well waves of meaningful images. It is in this type of relation that most writers stand to their words. No wonder, then, that language means so much. For to us, the late surrealists, language is like a lover, "not the language of love, but the love of language, not matter, but meaning, not what the tongue touches, but what it forms, not lips and nipples, but nouns and verbs" (William H Gass: *On Being Blue* page 11). The geographic inference problem of 1968 dressed up in the garb of 1984. Oh well! Ready maid.

But even though expressions are immediately physical, they have a nonphysical side as well. Marx called one of these sides use value and the other exchange value, Frege termed them reference and sense, Saussure signifier and signified, Austin brute and institutional facts, Derrida presence and absence. In general, preference is given to the first term in these pairs of difference. But to have power is to know how one face is turned into the other, to exhibit the visible and profit on the invisible, to whitewash with the transparent and color with the opaque. Power is always in the quest of security, in attempts to freeze the future.

It is nevertheless in the backrooms of the Bar de Saussure that commodities can be heard speaking directly to one another. After the tiniest sip of analysis, they begin to reveal their trade-secret: "Our use value may be a thing that interests men. It is no part of us as objects. What, however, does belong to us as objects is our value. Our natural intercourse as commodities proves it. In the eyes of each other, we are nothing but exchange values" (Karl Marx: *Capital* volume I, page 83). The same holds for expressions in general, including money and other promissory notes. In the process, rites of mythology are metamorphosed into laws of State, raw into cooked, cooked into cocked.

When this general power strategy is carried to extremes, the visible is over-emphasized and the invisible left in darkness. What is thereby repressed is the nonrepres(s)entable other. The effect is growing alienation of the type that Marx illustrated in his discussion of fetishism, and Lacan exemplified in his theory of castration. What counts are countable things. What is taboo are totemic relations. Foucault later drew attention to the same forces of reification through his remark that current ideology highlights the *scientia sexualis* and hides the *ars erotica*; there is a parallel in the development of his own writings from the history of madness to the history of sexuality. In the game of ontological transformations, the flipper is always a touch ahead.

Like the double-faced Janus, the flip-flopper knows how to join contradictions without going crazy. Perhaps it is therefore Janus, the janitor, who is the incarnation of modern power. Perhaps it is the priests of his congregation who know how to detect the difference between you and me. Perhaps it is they who separate the wheat from the chaff and who single out those traces of the past that are let through the Now-gate into the future.

But wait! Janus has a distant relative. Dionysus! Is it rather he who is the god of power? Proper question! For like the theory of expression itself, Dionysus is always caught in between, always dangling in the abyss between order and chaos. There is something to be re(a)d off his sheets. Is it the drops of blood from a sacrificial meal? Or just some wine spilled in an orgy? Neither, of course! Merely a piece of rhetoric to show how the intellectual's pen is his spear, the rubber his shield. Double entry, double protection. Castrated from the outset.

The differance? Repressing the presented or presenting the repressed? Why do some prefer the pleasures of ideology and others the ideologies of pleasure? But this is neither a Foucaldian pipe nor a Freudian cigar. This is this is this. What I write I write. What is a text.

Thus: "I have lost silence, and the regret I feel over that is immeasurable. I cannot describe the pain that invades a man once he has begun to speak. It is a motionless pain, that is itself pledged to muteness; because of it, the unbreathable is the element I breathe. I have shut myself up in a room, alone, there is no one in the house, almost no one outside, but this solitude has itself begun to speak, and I must in turn speak about this speaking solitude, not in derision, but because a greater solitude hovers above it, and above that solitude, another still greater, and each, taking the spoken word in order to smother it and silence it, instead echoes it to infinity, and infinity becomes an echo" (Maurice Blanchot: *Death Sentence* page 33).

It follows that tracing a text back to its experiential, referential or intentional roots is an infinite task, a rhythmic choir humming in an ecco chamber. There is no presentation without representation, no representation without presentation. When a text speaks about itself, it therefore speaks about the world, for all that can be said is in the saying; with Lacan, the unconscious is structured like a language, a symbolic reflection of imaginary differences that make a real difference. Whatever I do and whatever I don't, I am inevitably placed within sets of complex power relations, of meanings and distinctions. As a consequence, there is a politics of texts, of readers, writers, and expressions. Which is not to say, however, that all texts are political or that all life is art, merely to suggest that the term 'political' should be understood in its deeper meaning, "as describing the whole of human relations in their real, social structure, in their power of making the world; one must above all give an active role to the prefix *de-*; here it represents an operational movement, it permanently embodies a defaulting" (Roland Barthes: *Mythologies* page 143).

It is now well recognized that modern power arrangements are doubly dependent on the concept of difference, that is on the interplay of presence and absence. Society's need to normalize the deviant legitimizes the use of disciplinary techniques, while being different offers the individual a way of escaping them. Power and resistance thus require each other, just as a power of separation necessitates a separation of power. Through the operation of such a dialectic of opposition, even the most critical discourse becomes an integral part of what it aims to subvert; like its own language, power is consituted by the differences it seeks to overcome. Paradoxically self-reflecting, "power is tolerable only on condition that it mask a substantial part of itself" (Michel Foucault: *The History of Sexuality* volume 1, page 86).

What is masked is the insight that power is an exercise in ontological transformations. Once that veil is torn away even the tragedy of Marxism becomes intelligible. Thus it is in reading Hegel, Marx, Kierkegaard, and Nietzsche together that it becomes clear how history turns into hypothesis, hypothesis into thesis, thesis into norm, norm into counterfinality. Emerging as a critical issue is not the comparison of one utopia with another, but the social function of utopian thought itself. The reason is that "*utopias* afford consolation: although they have no real locality there is nevertheless a fantastic, untroubled region in which they are able to unfold; they open up cities with vast avenues, superbly planted gardens, countries where life is easy, even though the road to them is chimerical. *Heterotopias* are disturbing, probably because they secretly undermine language, because they make it impossible to name this *and* that, because they shatter or tangle common names, because they destroy 'syntax' in advance, and not only the syntax with which we construct sentences but also that less apparent syntax which causes words and things (next to and also opposite one another) to 'hold together'. This is why utopias permit fables and discourse: they run with the very grain of language and are part of the fundamental dimension of the *fabula*; heterotopias (such as those to be found so often in Borges) desiccate speech, stop words in their tracks, contest the very possibility of grammar at its source; they dissolve our myths and sterilize the lyricism of our sentences" (Michel Foucault: *The Order of Things* page xviii).

*

Stopping words in their tracks is to engage in ideology critique, to make silence speak. Ideology is in fact sometimes defined as the totality of that which goes without saying, as that taken-for-granted which is so transparent that its casts neither reflection nor shadow. But even though most secrets dwell in the whiteness of degree zero, some are in the words themselves. This holds especially for the case of prefixes.

What a word, that word! Nice and naked. PREFIX. Emperor without clothes. Form and content finally united. For to *pre-fix* is to fix in advance, to fasten and to castrate. As if to stress its own self-reference, the term itself contains a prefix; it does not refer to what it is, it is what it refers to.

The most important prefixes are *re-*, *pre-*, and *pro-*; in the *OED*, words with these beginnings take up an amazing total of 617 pages! Although not all of these constellations actually serve as prefixes, many of them lead directly to rewarding propositions about ideological preferences. Their function is to position meaning, less in time and more in figurative space. Even though *re-* holds a privileged position, the three are intimately related. As a consequence, they can not be analyzed one at a time but should be engaged in a free play with one another. There is a choreography of meaning, a dance of rhetorical reference.

This proposal of how to proceed reflects an epistemological premise. The presumption is that severing relations is to produce a picture of stale predilections rather than an image of living prospects. Recalling the proposals of Lacan, Barthes, and Foucault, splitting relations is in reality a procedure for masking: unrevealed tautologies are thereby called truths, not remedial prescriptions. While truth for Descartes is in the reassuring certitude of representation, for Nietzsche it is in the revolting ambiguity of the taken-for-granted. In practice, though, every preacher realizes that every premise contains a promise of predictable results, just as every promise contains a premise of profitable returns. In the process, products are reproduced, presentations represented, representations repressed. The repressed is indeed recaptured by the words themselves; the traumatic *re-* leads to the procreative *pre-*. All research searches back to the pregiven of its own initials: GO, professor, GO!

In this rerendering, progressive social science reemerges as a problematic program for the reinterpretation and subsequent representation of reality as it now presents itself. Likewise, representative democracy reappears as a procedure for reproducing remarkably repressentative proteges, ready to proclaim the regrettable prospects of the predominant ideology of presence. Why does proper reasoning preserve the truth of its premises in the predictions of its conclusions? Why does science retain its mythological belief in spiritual conception, regardless of the recognition that every representation is itself a presentation, every presentation itself a representation? Is the Virgin Mary the saint of reasoning? Is the Devil of rhetoric its preadamic seducer?

Weave Mary! Weave! Remember the members dismembered.

*

A text about you and me. A discourse on differance. A show of ontological transformations. A premediated refrain of pronouns and proverbs. A play on the power of silence and the silence of power. A touch of metaphors. A dance of ready-maids descending a staircase. An approach to limits. This is what this is.

And thus spoke Zarathustra: "I want to speak to the despisers of the body. I would not have them learn and teach differently, but merely have them say farewell to their own bodies—and thus become silent. But it is the child who says: 'Body and soul am I'. And why should one not speak like a child? [Yet] you say 'I' and you are proud of this word. But greater than this—although you will not believe in it—is your body which does not say 'I' but performs 'I'" (Friedrich Nietzsche: *Thus Spoke Zarathustra* my translation).

And yet. The smallest little child will soon discover that when the I is turned upside down, it still remains an I. When laid to rest, however, it becomes a dash— a mark of thought before a thought. But the same child will also learn that when a halo dot is placed above its head, the I looses its capital importance and becomes a small i. And a small i turned upside down is nothing less than an exclamation mark! The sign of infinity, though, is neither the dashing sleeping I nor the tilted relation /, but a lying 8, too tired of memories ever to stand erect again. In the meantime, the symbol -/- retains its power, for it signifies the practice of slash and burn. By no coincidence, it also happens that a slash is a slit made in a garment in order to expose to view a lining or undergarment of a different color.

Striptease as epistemology. Epistemology as striptease. Body relieved.

*   *   *

Accidentally, Werther's finger touches Charlotte's, their feet, under the table, happen to brush against each other. Werther might be engrossed by the meaning of these accidents; he might concentrate physically on these slight zones of contact and delight in this fragment of inert finger or foot, fetishistically, *without concern for the response* (like God—as the etymology of the word tells us—the fetish does not reply). But in fact Werther is not perverse, he is in love: he creates meaning, always and everywhere, out of nothing, and it is meaning which thrills him: he is in the crucible of meaning. Every contact, for the lover, raises the question of an answer: the skin is asked to reply.

(A squeeze of the hand—enormous documentation—a tiny gesture within the palm, a knee which doesn't move away, an arm extended, as if quite naturally, along the back of a sofa and against which the other's head gradually comes to rest—this is the paradisiac realm of subtle and clandestine signs: a kind of festival not of the senses but of meaning.)

(Roland Barthes: *A Lover's Discourse* page 67)

*

There are, indeed, things that cannot be put into words. They make themselves manifest. They are what is mystical.

......

My propositions serve as elucidations in the following way: anyone who understands me eventually recognizes them as nonsensical, when he has used them—as steps—to climb up beyond them. (He must, so to speak, throw away the ladder after he has climbed up it.)

He must transcend these propositions, and then he will see the world aright.

What we cannot speak about we must pass over in silence

(Ludwig Wittgenstein: *Tractatus Logico-Philosophicus* Props. 6.522, 6.54, and 7).

*

Preface

I SHOULD prefer this note to be read or, if skimmed, that it should even be forgotten; it teaches the skilful reader little which is placed beyond his penetration: but may disturb the simple one who has to apply his gaze to the first words of the poem in order that the following ones, placed as they are, may lead him to the last, the whole being without novelty except for the spacing out of the reading. The 'whites' indeed take on an importance, are striking at first sight … . The paper intervenes each time an image, of its own accord, ceases or withdraws, accepting the succession of others …

(Stéphane Mallarmé: "Un coup de dés").

Austin, John L, 1962 *How to Do Things with Words* Ed. James O Urmson (Oxford University Press, London)

Bair, Deirde, 1980 *Samuel Beckett: A Biography* (Picador, London)

Barthes, Roland, 1970 *Writing Degree Zero* translated by Annette Lavers and Colin Smith (Beacon Press, Boston, MA)

Barthes, Roland, 1972 *Mythologies* translated by Annette Lavers (Hill and Wang, New York)

Barthes, Roland, 1975 *The Pleasure of the Text* translated by Richard Miller (Hill and Wang, New York)

Barthes, Roland, 1977 *Roland Barthes by Roland Barthes* translated by Richard Howard (Hill and Wang, New York)

Barthes, Roland, 1978 *A Lover's Discourse: Fragments* translated by Richard Howard (Hill and Wang, New York) published in Swedish in 1983 as *Kärlekens samtal: Fragment* translated by Leif Janzon (Korpen, Göteborg)

Barthes, Roland, 1981 *Camera Lucida: Reflections on Photography* translated by Richard Howard (Hill and Wang, New York)

Bataille, Georges, 1979 *The Story of the Eye* translated by Joachim Neugroschel, and with essays by Susan Sontag and Roland Barthes (Penguin Books, Harmondsworth, Middx)

Beckett, Samuel, 1930 *Whoroscope* (Hours Press, Paris)

Bernstein, Richard J, 1971 *Praxis and Action* (University of Pennsylvania Press, Philadelphia, PA)

Blanchot, Maurice, 1978 *Death Sentence* translated by Lydia Davis (Station Hill Press, Barrytown, NY)

Blanchot, Maurice, 1982 *The Space of Literature* translated with an introduction by Ann Stock (University of Nebraska Press, Lincoln, NE)

Brown, Norman O, 1959 *Life Against Death: The Psychoanalytical Meaning of History* (Wesleyan University Press, Middletown, CT)

Brown, Norman O, 1966 *Love's Body* (Vintage Books, New York)

Brown, Norman O, 1974 *Closing Time* (Vintage Books, New York)

Canetti, Elias, 1982 *Kafka's Other Trial: The Letters to Felice* translated by Christopher Middleton (Penguin Books, Harmondsworth, Middx)

Carroll, Lewis, 1960 *The Annotated Alice* with an introduction and notes by Martin Gardner (Clarkson, N. Potter, New York)

Derrida, Jacques, 1974 *Glas* (Galilée, Paris)

Derrida, Jacques, 1978 *Of Grammatology* translated with an introduction by Gayatri Chakravorty Spivak (University of Chicago Press, Chicago, IL)

Derrida, Jacques, 1978 *Spurs: Nietzsche's Styles/Eperons: Les styles de Nietzsche* bilingual edition translated by Barbara Harlow (University of Chicago Press, Chicago, IL)

Derrida, Jacques, 1981 *Dissemination* translated with an introduction by Barbara Johnson (Athlone Press, London)

Derrida, Jacques, 1982 *Margins of Philosophy* translated with notes by Alan Bass (University of Chicago Press, Chicago, IL)

Dreyfus, Hubert L; Rabinow, Paul, 1982 *Michel Foucault: Beyond Structuralism and Hermeneutics* (University of Chicago Press, Chicago, IL)

Dummett, Michael, 1973 *Frege: Philosophy of Language* (Harper and Row, New York)

Dummett, Michael, 1978 *Truth and Other Enigmas* (Harvard University Press, Cambridge, MA)

Dumouchel, Paul; Dupuy, Jean-Pierre (Eds), 1983 *L'Auto-organisation de la physique au politique* Colloque de Cerisy (Seuil, Paris)

Evans, Gareth, 1982 *The Varieties of Reference* Ed. John McDowell (Clarendon Press, Oxford)

Foucault, Michel, 1965 *Madness and Civilization: A History of Insanity in the Age of Reason* translated by Richard Howard (Pantheon Books, New York)

Foucault, Michel, 1972 *The Order of Things: An Archeology of the Human Sciences* translated by Alan Sheridan (Vintage Books, New York)

Foucault, Michel, 1973 *The Birth of the Clinic: An Archeology of Medical Perception* translated by Alan Sheridan (Tavistock Publications, Andover, Hants)

Foucault, Michel, 1978 *The History of Sexuality* Volume 1, translated by Robert Hurley (Pantheon Books, New York)

Foucault, Michel, 1979 *Discipline and Punish: The Birth of the Prison* translated by Alan Sheridan (Penguin Books, Harmondsworth, Middx)

Foucault, Michel, 1980 *Power/Knowledge: Selected Interviews and Other Writings 1972–1977* translated and edited by Colin Gordon (Pantheon Books, New York)

Foucault, Michel, 1982 *This is Not a Pipe* translated with an introduction by James Harkness (University of California Press, Berkeley, CA)

Freud, Sigmund, 1950 *Totem and Taboo* (Routledge and Keagan Paul, Andover, Hants)

Gass, William H, 1968 *Willie Masters' Lonesome Wife* (Northwestern University Press, Evanston, IL)

Gass, William H, 1977 *On Being Blue: A Philosophical Inquiry* (Godine, Boston, MA)

Gass, William H, 1979 *The World Within the Word* (Nonpareil, Boston, MA)

Gass, William H, 1980 *Fiction and the Figures of Life* (Nonpareil, Boston, MA)

d'Harnoncourt, Anne; McShine, Kynaston (Eds), 1973 *Marcel Duchamp* (Museum of Modern Art, New York)

Hartman, Geoffrey, 1981 *Saving the Text: Literature/Derrida/Philosophy* (Johns Hopkins University Press, Baltimore, MD)

Heidegger, Martin, 1982 *Nietzsche, Volume 4: Nihilism* edited with analysis by David Farrell Krell (Harper and Row, New York)

Hofstadter, Douglas R, 1979 *Gödel, Escher, Bach: An Eternal Golden Braid. A metaphorical fugue on minds and machines in the spirit of Lewis Carroll* (Basic Books, New York)

Johnson, Eyvind, 1946 *Strändernas svall* (Bonniers, Stockholm)

Joyce, James, 1934 *Ulysses* (The Modern Library, New York)

Kaufmann, Walter, 1974 *Nietzsche: Philosopher, Psychologist, Antichrist* fourth edition (Princeton University Press, Princeton, NJ)

Kristeva, Julia, 1980 *Desire in Language: A Semiotic Approach to Literature and Art* translated with an introduction by Thomas Gora, Alice Jardin and Leon S Roudiez (Basil Blackwell, Oxford)

Kristeva, Julia, 1982 *Powers of Horror: An Essay on Abjection* translated by Leon S Roudiez (Columbia University Press, New York)

Lacan, Jacques, 1977 *Écrits: A Selection* translated by Alan Sheridan (W W Norton, New York)

Lacan, Jacques, 1977 *The Four Fundamental Concepts of Psychoanalysis* translated by Alan Sheridan (W W Norton, New York)

Lavers, Annette, 1982 *Roland Barthes: Structuralism and After* (Methuen, Andover, Hants)

Lemaire, Anika, 1977 *Jacques Lacan* translated by David Macey (Routledge and Kegan Paul, Andover, Hants)

Linsky, Leonard, 1967 *Referring* (Routledge and Kegan Paul, Andover, Hants)

Machado, Antonio, 1983 *Times Alone* selected poems translated by Robert Bly (Wesleyan University Press, Middletown, CT)

Mallarmé, Stéphane, 1965, "Un coup de dés", in *Mallarmé: The Poems* translated and introduced by Anthony Hartley (Penguin Books, Harmondsworth, Middx) pp 209–233

de Man, Paul, 1983 *Blindness and Insight: Essays in the Rhetoric of Contemporary Criticism* introduction by Wlad Godzich, second edition (Methuen, Andover, Hants)

Marx, Karl, 1967 *Capital* volume 1 (International, New York)

Marx, Karl, 1973 *Grundrisse* translated with a foreword by Martin Nicolaus (Penguin Books, Harmondsworth, Middx)

Musil, Robert, 1979 *The Man Without Qualities* translated by Eithne Wilkins and Ernst Kaiser (Picador, London)

Nietzsche, Friedrich, 1968 *The Will to Power* translated and edited with commentary by Walter Kaufmann (Vintage Books, New York)

Nietzsche, Friedrich, 1978 *Thus Spoke Zarathustra* translated by Walter Kaufmann (Penguin Books, Harmondsworth, Middx)

Olsson, Gunnar, 1980 *Birds in Egg/Eggs in Bird* (Pion, London)

Olsson, Gunnar, 1981, "On yearning for home: an epistemological view of ontological transformations", in *Humanistic Geography and Literature: Essays on the Experience of Place* Ed. Douglas C D Pocock (Croom Helm, Beckenham, Kent) pp 121–129

Olsson, Gunnar, 1982, "–/–", *Sub-Stance* 11 (2), 24–33; slightly different version also in *A Search for Common Ground* Eds Peter Gould and Gunnar Olsson (Pion, London) 1982, pp 223–231

Olsson, Gunnar, 1984, "The social space of silence" *You know where* September 11

Olsson, Gunnar, 1984, "Toward a sermon of modernity", in *Recollections of a Revolution: Geography as Spatial Science* Eds Mark Billinge, Derek Gregory and Ron Martin (Macmillan, London) pp 73–85

Parfit, Derek, 1984 *Reasons and Persons* (Clarendon Press, Oxford)

Paz, Octavio, 1978 *Marcel Duchamp: Appearance Stripped Bare* translated by Rachel Phillips and Donald Gardner (Viking Press, New York)

Philo, Chris, 1984, "Reflections on Gunnar Olsson's contribution to the discourse of contemporary human geography" *Environment and Planning D: Society and Space* 2 217–240

Roos, Anna M, 1939 *Sörgården* omarbetad (Bonniers, Stockholm)

Rorty, Richard, 1979 *Philosophy and the Mirror of Nature* (Princeton University Press, Princeton, NJ)

de Saussure, Ferdinand, 1959 *Course in General Linguistics* edited by C Bally, and translated by Wade Baskin (McGraw-Hill, New York)

Serres, Michel, 1982 *Hermes: Literature, Science, Philosophy* edited with an introduction by José V Harari and David F Bell and with a postface by Ilya Prigogine and Isabelle Stengers (The Johns Hopkins University Press, Baltimore, MD)

Sheridan, Alan, 1980 *Michel Foucault: The Will to Truth* (Tavistock Publications, Andover, Hants)

Stein, Gertrude, 1914 *Tender Buttons* (Claire Marie, New York)

Stein, Gertrude, 1933 *The Autobiography of Alice B. Toklas* (Harcourt, Brace, New York)

Stein, Gertrude, 1934 *The Making of Americans* (Harcourt, Brace, New York)

Steiner, George, 1975 *After Babel: Aspects of Language and Translation* (Oxford University Press, New York)

Stewart, Allegra, 1967 *Gertrude Stein and the Present* (Harvard University Press, Cambridge, MA)

Sturrock, John D (Ed.), 1979 *Structuralism and Since: From Lévi-Strauss to Derrida* (Oxford University Press, Oxford)

Ungar, Steven, 1983 *Roland Barthes: The Professor of Desire* (University of Nebraska Press, Lincoln, NE)

Wittgenstein, Ludwig, 1961 *Tractatus Logico – Philosophicus* translated by D F Pears and B F McGuinness with introduction by Bertrand Russell (Routledge and Kegan Paul, Andover, Hants)

Wittgenstein, Ludwig, 1968 *Philosphical Investigations* translated by G E M Anscombe (Macmillan, New York)

Last night, as I was sleeping, I dreamt—marvellous error!—that a spring was breaking out in my heart. I said: Along which secret aqueduct, oh water, are you coming to me, water of a new life that I have never drunk?

Last night, as I was sleeping, I dreamt—marvellous error!—that I hide a beehive here inside my heart. And the golden bees were making combs and sweet honey from my old failures.

Is my soul asleep? Have those beehives that labor at night stopped? And the water wheel of thought, is it dry, the cups empty, wheeling, carrying only shadows?

No, my soul is not asleep. It is awake. It neither sleeps nor dreams, but watches, its clear eyes open, far-off things, and listens at the shores of the great silence.

(Antonio Machado: *Times Alone* pp 42 – 45).

Chapter 12

Crazy Wisdom and Recovering the Human in Olsson's Method of Cartographic Critique

David Jansson

Indeed I have come to believe that our very survival depends upon improved abilities to be abstract enough; the most radical point is the point of insolvability.

Gunnar Olsson (1990: 109.)

Introduction

Gunnar Olsson's voyages into the uncharted frontiers of language have bedeviled, bewildered, and bewitched readers for nearly thirty years. His essays and books have posed serious challenges for the earnest reader, and not a few have given up in frustration. That Olsson's approach is provocative is clear, and in this chapter I want to address the following questions: Why is Olsson's method so perverse? Is there, in fact, a method to his madness? Why can't he just speak plainly, dammit?!

Most readers attempt to access the point of Olsson's work by looking for the meanings behind his words. While I do not wish to argue that this is the wrong approach, I want to suggest an alternative perspective that may help unlock some of the mysteries of Olsson's method. I will argue that it may be helpful to consider what Olsson shows, rather than what he means. More precisely, a potentially fruitful approach is to focus on the effects Olsson attempts to achieve more than the messages that he communicates. This is not at all to say that the messages are unimportant; indeed, Olsson's work is highly sophisticated because it works on many levels simultaneously. Most fundamentally, we might say that he seeks to smash the taboos of socialization that prevent human beings from understanding the workings of power and keep them obedient and pliable. In effect, he employs a method that attempts to challenge the thought processes that keep us "in line" so that we can recognize the otherwise hidden ontological transformations occurring all around us.

Olsson is certainly not the only scholar to confront the limitations of the written word to express radically new ideas. His close friend Allan Pred was one of those who experimented with different writing styles to try to capture nuances of social life that escape more traditional styles of representation. Many social theorists have also sought to challenge "the mental habits of linearity [that] persist in their hegemonic hold over our thinking" (Braidotti 2002: 1) through their choice of uncommon forms of expression. Feminist theorist Rosi Braidotti, for example, has explicitly explained her choice of writing style. In her own take on Olsson's fundamental question "What does it mean to be human?," Braidotti (2002: 2) argues that "the point is not to know who we are, but rather what, at last, we want to become, how to represent mutations, changes and transformations, rather than Being in its classical modes." She contends that our established modes of representation fail to adequately capture the processes of transition, hybridization, and nomadization that characterize the modern world. Thus:

> In order to do justice to these complexities I have opted for a style that may strike the academic reader as allusive or associative. It is a deliberate choice on my part, involving the risk of sounding less than coherent at times. It has to do with my concern for style not as a merely rhetorical device, but as a deeper concept …To attack linearity and binary thinking in a style that remains linear and binary itself would indeed be a contradiction in terms. This is why the poststructuralist generation has worked so hard to innovate the form and style, as well as the content, of their philosophy. This has been greeted by a mixed reception in the academic community. Assessed as 'bad poetry' at best, as an opaque and allusive muddle at worst, the quest for a new philosophical style that rejects the dualism of content and form has clashed with the mood currently dominant in scientific discourse. (Braidotti 2002: 8.)

Olsson himself is of course intimately familiar with the consequences of clashing with the dominant mood in scientific discourse. With reference to his own writing, Olsson might well agree with Braidotti's (2002: 9) hope that "what appears to be lost in terms of coherence can be compensated for by inspirational force and an energizing pull away from binary schemes, judgmental postures and the temptation of nostalgia."

While I am sympathetic to the need to find new modes of academic expression, and while I agree with Nicholas Blomley's (2008) desire for critical geographers to explore different languages for social criticism, I would also like to set the bar high in terms of the issue of when, and in what contexts, writers should resort to the more extreme unorthodox techniques. Much writing in social theory and cultural studies, for example, is impenetrable not because of the profundity of the author's ideas, which often can be easily expressed in plain language, but because of either an inability to write clearly or a conscious decision to complicate one's mode of expression in order to attach an air of erudition to one's work (Homi Bhabha comes to mind[1]). Furthermore, I do not agree with Braidotti when she argues against the utility of attacking linearity and binary thinking in a linear and binary writing style; while this approach may appear a contradiction in *terms*, it is not necessarily incapable of creating some of the desired *effects*. However, when it comes to the fundamental ontological questions that animate Olsson's work,

---

1 For a critique of Bhabha's often impenetrable prose, see Marrouchi (1998).

questions that take us into fields of uncommon abstraction, it is clear that our usual ways of expressing ideas are less well suited to the task. This is the task of understanding not only what it means to be human but also the nature of the very reality in which we find ourselves.

In adopting an unorthodox approach to the expression of an ontology that transgresses everyday understandings of reality, Olsson has company. I argue that we can learn much about how to approach the writings of Gunnar Olsson by considering the parallels of his work to another group of radicals seeking to correct our ontological understanding of existence. I am referring to the "crazy wisdom" or "holy madness" traditions of philosophical, religious, or spiritual exploration.[2] Some readers note that Olsson does not have much to say about the non-Western world and wonder how a consideration of other intellectual and philosophical traditions might inform his analysis (e.g. De Weerdt 2009, Philo 2008). This chapter contributes something to such an inquiry, though in doing so, it sins against the academic convention that locates mystical practice as off limits, on the wrong side of the boundary. But what could be more appropriate in an essay on the MO of GO? In this chapter I compare the methods of a controversial group of spiritual explorers with those of a certain dabbler in the black arts of cartographic reason for the purposes of suggesting a potentially fruitful way to understand Olsson's work. So now to the practitioners of holy madness.

Holy madness and the crazy wisdom tradition

When considering the different manifestations of the holy madness tradition, it is striking how often one encounters elements and themes that permeate the work of Gunnar Olsson. In this section we review some of the major social roles embodied by this tradition.

An early archetype of the holy madness tradition is that of the trickster. Particularly associated with the North American Indians, the trickster serves to amuse and instruct.[3] Stories of the trickster's antics are intended to communicate something about the whimsical, arbitrary, and dangerous forces represented by Nature—an aspect of being that a civilized consciousness prefers to deny. For Jung (1968: 260), the trickster "is a faithful reflection of an absolutely undifferentiated human consciousness, corresponding to a psyche that has hardly left the animal level." The trickster embodies the anticultural forces that threaten human society, which "are kept at bay by the countless institutions that compose the skeleton of

---

2   For reviews of holy (or divine) madness and crazy wisdom approaches, see Feuerstein (1992, 2006), McDaniel (1989), Trungpa (2001), Kinsley (1974), Kakar (2009), Hyers (1996), and Nisker (2001, 2008).

3   For more on the trickster, see Hyers (1996, 1969), Campbell (1968), and Jung (1968, 255–272).

culture: the rites, myths, dogmas, scientific theories, interpersonal arrangements, personal beliefs, and so on" (Feuerstein 2006: 4).

> The majority of trickster myths present a hero of sorts ... who wanders from one adventure and misadventure to another. This itinerant hero has neither a clear place in the "scheme of things" nor a clearly defined social identity. In his nomadic meanderings he is involved in a miscellaneous series of episodes that have little logical or dramatic connection, except that they are usually concerned with survival and with getting in and out of tight spots ... Among the Algonquins, the trickster falls into a hollow tree and only manages to escape by throwing himself out, piece by piece, from a small hole in the side of the tree. In Winnebago tales, the trickster gets his arm caught in a tree fork while trying to stop the tree from squeaking in the wind, and as a result some passing wolves help themselves to a free meal of raccoon which the trickster has just trapped and prepared. (Hyers 1996: 177–8.)

The trickster also moves along the borderland of sexuality, that often taboo-laced social sphere that connects us most intimately with Nature. Thus "it is no surprise to find sex organs, sexual desire, and efforts to organize sexual behavior as themes in trickster mythology" (Hyers 1996: 178).

Related to the trickster is the figure of the religious or ritual clown.[4] The clown occupies a space between categories, moving along the borderland "between idiocy and saintliness" (Feuerstein 2006: 8), and this is precisely where the clown draws his or her power. Conrad Hyers (1996: 144) refers to the clown as "the lord of ambiguity and relativity:"

> The clown is lord of that no-man's-land between contending forces, moving back and forth along all those human lines drawn (not without arbitrariness) between law and order, social and antisocial, reason and irrationality, friend and foe, fashionable and unfashionable, important and unimportant. The clown is now on one side, now on the other, and ultimately both and neither.

As they tread this line between reality and unreality, clowns entertain and disrupt simultaneously. While the word "clown" may signal to Westerners an image related to the modern circus, the clown is in fact part of a lineage that plays a more subversive role than its modern counterpart might suggest.

The archetype of the jester is also relevant here. The jester was that rare person who could laugh at the monarch and get away with it. Indeed, it was the jester's job to poke fun at the station of the ruler, to question the King or Queen's unquestionable power. "The jester's function is humorously to profane the categories and hierarchies with which we would capture the ultimate truth about things, domesticate it, and add it to the electronic data bank." The jester

---

4   For more on the clown, see Zucker (1969) and Hyers (1996, chapter 7).

refuses to take with absolute seriousness the mores and metrics of the court; the jester finds that the moat imprisons the king as much as it protects him. "Hence, the neat patterns of rationality and value and order with which we organize and solidify our experience are confused and garbled. Sense is turned into nonsense, order into disarray, the unquestionable into the doubtful. The jester does not fit into, indeed refuses to fit into, the established conventions and hallowed structures of this or that human sphere."[5]

The trickster, clown and jester can thus be considered public figures who play an important social role. Through their amusing antics, they highlight the inherent contradictions and absurdities characterizing social life in the communities in which they operate. Trickster, clown, jester—they all represent an archetype that "stands outside ordinary consciousness and beyond confines of social conventions, sacred taboos, and rational enclosures" (Hyers 1974: 172). This figure thus represents a pure form of *freedom*. A freedom which entails an existence beyond or against the law, even transcending the law. The trickster, clown, and jester, "by occupying an ambiguous space between the holy and the unholy, good and evil, wisdom and ignorance, reason and nonsense, are particularly suited to this task of pointing beyond all such distinctions" (Hyers 1974: 172). Nothing less than a foundational challenge to the laws of rationality and identity, in fact to all law. These holy fools are foolish because they refuse to make the distinctions required by society, refuses to participate in the illusion of the same and the not-same. Taboos violated, enclosures erased, law transcended—thus the clown and the fool point the way to a radical freedom, one that appears nowhere (and now-here) on the maps that guide our socialization. In fact, one could argue that these characters truly embody the "excluded middle" that has engrossed Olsson since *Birds in Egg*; they occupy that abysmal void of ambiguity and indeterminacy. Thus they perhaps belong to the same category as the avant-garde cartographer, mischievously erasing the lines of society's maps and redrawing them in bold and potentially liberating ways.

In this way, these figures rehearse the unorthodox teachings of the rogue guru, the practitioner of crazy wisdom or holy madness. Taken together, they all "represent the *axis mundi*, the world axis, and are thus radically "concentric" … Yet, from the viewpoint of conventional society, [they] appear to be "eccentric"— out of focus, out of line, out of their rational mind;" they all walk "the fine line between transcendence and immanence, between sacredness and darkest profanity" (Feuerstein 2006: 8, 9). A fundamental difference, however, between the trickster/ clown/jester and rogue guru is that the former operate more publicly and at a more collective or societal level. The rogue guru's methodology is, on the other hand, more personal and esoteric (though exceptions exist). The practitioner of holy madness typically seeks to cultivate a guru-disciple relationship, a relationship

---

5   Both quotes are from Hyers (1996: 129). Appropriately enough, given the subject of this chapter, Hyers (1996: 117) notes that "Even university professors could get part-time positions entertaining as palace fools!"

which in many cases is described as *the* way to spiritual enlightenment, to a fundamental reevaluation of the nature of reality.[6]

For perhaps as long as there have been human beings, there has existed a small group of individuals that sees its role in society as facilitating the development of human consciousness. Some of these individuals become the high priests of various religious traditions, while others stand outside the religious establishment and preach a more idiosyncratic way to the goal of spiritual enlightenment. As Roger Walsh (2006: ix) writes:

> There is considerable unanimity among the world's religions, and especially among the contemplative traditions, that we have overestimated our usual state of mind, yet greatly underestimated our potential. These traditions, which together form the perennial philosophy, perennial wisdom, or perennial psychology, consider our usual awareness to be only semiconscious dreams, *maya*, or a consensus trance. Yet these same traditions claim that we are capable of escaping from this trance and of thereby realizing what has been variously called enlightenment, liberation, salvation, *moksha*, or awakening.

The Sanskrit word maya can be roughly translated as "illusion," and is built upon the parts ma ("not") and ya ("that")—so that the daily world that is associated with maya is really "not that," or, not the Real. The "perennial philosophy" developed as a way to help individuals see beyond the illusion to the deeper truth of existence—that separation, the mind-body duality, is not our "true" condition. Around this insight an entire apparatus of religious institutions and mystical traditions has been built. Establishment religion has been seen as a way to encourage moral, ethical and spiritual development in the masses, while the more esoteric and mystical strains of these religions have provided techniques for inner development for those individuals who may be more receptive to the deeper ontological truths that are mostly communicated symbolically by the mainstream religions.

Sometimes these mystical practices are advanced by what we might call "rogue gurus," teachers who transgress the more conventional boundaries of the mystical traditions. As Walsh (2006: xi) notes, "some religious practitioners and masters have clearly seemed bizarre by conventional standards. Some appear to have deliberately flouted convention, provoked authorities, and offended their listeners. Indeed, some of them have appeared so bizarre as to have been labeled by such names as holy fools, crazy-wisdom teachers, or god intoxicants." These rogue gurus inhabit the apex of a tradition that extends from the more narrowly-defined archetypes of the trickster, clown, and jester, to the pinnacle of spiritual achievement allegedly occupied by sages who see it as their duty to initiate their more receptive followers into the mystical secrets that lead to higher levels of

---

6   As a result of the cultivation of an intimate relationship between guru and disciple, this form of spiritual exploration demands a high degree of maturity from both parties and, when such maturity is lacking, is susceptible to considerable abuses.

consciousness. Their "madness" lies in their method of teaching, of communicating the fundamental truths of existence:

> Holy madness, or crazy wisdom, is a radical style of teaching or demonstrating spiritual values ... What [these approaches] have in common is an adept... who typically instructs others in ways that are designed to startle or shock the conventional mind .. From the conventional point of view, the crazy-wise teachers are eccentrics who use their eccentricity to communicate an alternative vision to that which governs ordinary life. They are masters of inversion, proficient breakers of taboos, and lovers of surprise, contradiction, and ambiguity. (Feuerstein 2006: 3.)

Such "crazy" methods are understood by the guru to be necessary to trigger spiritual development in the guru's followers.

In the East, the crazy wisdom tradition had prominent practitioners in India, Tibet, China and Japan. In the school of Chinese Buddhism known as Ch'an (Zen in Japanese), some masters used a variety of "surprise" methods to induce a state of enlightenment in their followers—"including sudden shouting, physical beatings, paradoxical verbal responses, and riddles" (Feuerstein 1992: 49). Practitioners of the "sudden school" could be found in China and Japan, flouting "abstract speculation, pale logic, icons, idols, doctrines, scriptures, gods, and other religious paraphernelia. Lin-chi admonished his disciples not to think of the Buddha as the ultimate Reality but to regard him as the peephole in the latrine. The eighth-century master Tan-hsia burned a statue of the Buddha to keep himself warm, and the adept Te-shan spoke of the Buddha as a 'piece of shit'" (Feuerstein 1992: 51).

Such scatological language is also a common feature of the crazy wisdom approach in the West, as these gurus thumb their noses at conventional norms and rationality. Unconventional means and outrageous behavior are standard fare for the practitioners of crazy wisdom/holy madness, and these tactics challenge generally accepted ideas about the proper behavior of religious or spiritual leaders. In fact, they often choose to communicate in a way that many find unfathomable. Chögyam Trungpa (2001:169), himself identified as a practitioner of crazy wisdom, has written that "the crazy-wisdom guru does not speak or teach on the ordinary level, but rather, he or she creates a symbol, or means. A symbol in this case is not like something that stands for something else, but it is something that presents the living quality of life and creates a message out of it."

What message might the crazy-wisdom guru communicate? As Georg Feuerstein (1992: 14) sees it, the goal of such practitioners is to "maximize the conditions that would allow them to demonstrate and to school themselves in the art of what the great German mystic Meister Eckehart calls *gelâzenheit*, 'letting go.'" We are to let go of our attachment to duality, to the illusions that surround us in our daily lives, to our socialized containment, to the powers that spellbind us and keep us locked into an existence that denies our true state. In the context of modern spiritual traditions, these powers are symbolized by the ego that has invested its desires in the very world image that spiritual practice is intended to shatter. Thus the ego, which cannot be separated from the socialization process, is the primary obstacle that must be overcome, and it is a major target of the crazy-wisdom gurus. It is the role of such figures to "tear off all our cultural blinders and rational pretensions so that we may see reality unmasked. Speaking for Nature and the unconscious, both characters turn the conventional universe upside down and inside out. The trickster's impulse to spread "strife" is analogous to the guru's desire to disrupt the disciple's automaticities (called "culture") and to induce in him or her a state of discontent—a crisis in consciousness" (Feuerstein 1992: 5). This crisis in consciousness is a precondition of "enlightenment"—the limits of one's understanding must be breached before a new awareness can emerge. Holy madness practices are one way to induce such crises in the service of enlightenment.

In some cases, genuine existential crises can lead to a rather unexpected experience: laughter. While Olsson flags the importance of the radical point of insolvability, he does not mention that this point has a sound. That sound is laughter. Indeed, "the most characteristic symbol of this tradition of "sudden" enlightenment is laughter ... This is the laughter of liberation, the laughter of the being who has recovered authenticity, or innate buddhahood" (Feuerstein 1992: 52). Laughter, then, is a sign of freedom.

> At every level of manifestation, humour spells freedom in some sense and to some degree. Humour means freedom. This is one of its most distinctive characteristics and virtues. Here, however, the freedom to laugh which moves within the conflicts and doubts and tensions of life—the freedom, therefore, which is still relative to bondage and ignorance—becomes the freedom to laugh on the other side (the inside) of enlightenment. He who is no longer in bondage to desire, or to the self, or the law, he who is no longer torn apart by alienation and anxiety, and who is no longer defined and determined primarily by seriousness, can now laugh with the laughter of little children and great sages. (Hyers 1972: 168.)

In a real sense, genuine humor and laughter are no less than signs of our deepest humanity liberated. In the context of philosophical (or spiritual) practice, "laughter is a sign not of gleefulness but of victory over the shadow side of the human psyche" (Feuerstein 1992: 53). It is tempting to think of Gunnar Olsson's hearty laugh as evidence of an internal state of freedom.

The hol(e)y madness of Gunnar Olsson

We can think of the crazy wisdom of Olsson's project from the perspective of the reception of his work and of the form and substance of his argument. With regard to the former, it is interesting to note that two Swedish reviewers of *Abysmal* have argued that the style of the book reminds them of "litteratur skriven i en esoterisk tradition."[7]

Det vill säga, texter som inte bara påstår att det finns ett svar på de stora livsfrågorna utan också gör anspråk på att sitta inne med detta svar (ett svar som dessutom ofta härstammar från någon mer eller mindre bortglömd eller hemlig urkund och lika ofta kretsar kring sifferkombinationer och geometriska symboler, till exempel pyramider) och som är författade på ett sådant sätt att de kan begripas endast av en mycket liten exklusiv skara av redan invigda.

[That is to say, texts that do not just state that there is an answer to life's major questions but also make a claim to fundamentally grasp this answer (an answer that moreover often originates in some more or less forgotten or secret document and just as often revolves around number combinations and geometric symbols, for example pyramids) and that are written in a way that can be understood by only a very small, exclusive crowd of the already initiated.] (Eriksson and Nordlund 2008, my translation.)

While Olsson in *Abysmal* deals mostly with "documents" (including works of art) that are hardly secret or forgotten, this reaction to the book is nevertheless instructive. The esoteric traditions were developed in order to give a small number of more "advanced" individuals the opportunity for inner development that the established religions, calibrated more for the masses, could not offer. This is indeed a kind of "elitism," in the sense that the esoteric traditions of the major religions were accessible only to an elite few who were willing and able to make the effort (e.g. study, self-discipline, and meditation) required to access the deeper truths that religion basically mythologized. To pursue the esoteric path means to see the mythologies of religion as metaphors that dress up these truths in various kinds of clothing and to realize that one must transcend these metaphors in order to experience the true nature of reality. The holy madness path is one way that individuals have used to access what they consider to be the truth of existence.

Other readers have hinted at the basic relevance of approaching Olsson from the holy madness perspective, even if they have not made this connection explicitly. For Tom Conley (2008: 104), Olsson takes on the persona of a "wizened magus" along the lines of the magician Alcandre in Corneille's play *Illusion comique*, who "writes to enthuse and inspire." Jörn Seemann (2008: 133) describes Olsson's "writing and personality" as: "important, timely, utopian, intellectually progressive, subtle, passionate, powerful, open-ended, ambiguous, theatrical, scatological, and, occasionally, smutty"—words that could easily apply to any number of crazy wisdom adepts. Furthermore, Seemann (2008: 133) notes that some readers "complain about Olsson's smug, verging on arrogant style and his narcissist and sometimes sexist attitudes that turn his writing into

---

7 "...literature written in an esoteric tradition."

a solipsist monologue and an easy target for gender-conscious citizens and more puritan readers." Such comments fairly describe the taboo-breaking, self-centered personal style of many contemporary Western gurus, as well as the reactions they often inspire.

In fact, self-centeredness seems to be a hallmark of (especially the contemporary) rogue gurus, as they cultivate an aura of omniscience or godliness among their followers; the guru-disciple relationship is an intensely personal one. A guru's followers put their trust in the guru and follow his (and sometimes her) teachings (often experienced as "orders" by their followers) devotedly, until such time as they may lose faith in the guru. The teaching is thus difficult to separate from the teacher. In this context, it is interesting to note J.-D.C. Dewsbury's (2007: 391) reaction to *Abysmal*: "It is peculiarly autobiographical throughout. I am convinced by the man, and therefore by his compass." However, I should point out a crucial difference here: while the guru is convincing largely because of an assumed otherworldliness (i.e., holiness) and achievement of a higher level of consciousness, one gets the sense from Dewsbury's review that it is in fact Olsson's *humanity* that he finds convincing, his way of showing what it means to be human. In both cases, though, it is the personal credibility of the teacher/author that is the issue.

Marcus Doel (2003: 154) develops this idea in some detail. He notes, for example, that "there is no escaping the fact that Olsson personalizes everything … Everywhere Olsson impresses his 'I' upon us." But this personalized monologue is directed outward: thus Olsson's writing is "highly personal and extremely intimate. His texts are replete with I's, me's, and you's. By turning away from the mirage of objectivity towards the trap of reflexivity (from one game of seduction to another), there is little doubt that it is Olsson himself who is addressing us from the Land of Thought-and-Action … and not each of us in general, but rather each of us in particular .. He expects to find you totally alone." Gurus are sometimes said to have the ability, while giving a public talk, to make eye-to-eye contact with everyone in the room, no matter how many people are present (e.g. Lowe 1996). Direct communication. As if no one else is there. The sense of aloneness, in fact, comes up in some of the accounts that former students have given of their gurus, when the intense connection they experienced in the presence of the guru revealed their (metaphorical) nakedness in front of this person of power (Feuerstein 1992, 2006). And what is the point of this direct communication? Here again, Doel (2003: 157) is helpful: "the more that personalization is impressed upon us, the more one should be impressed by co-relation." This personalization is (in part) about unveiling the true nature of the relationship between You and Me, about that space between individuals and between the individual and society.

Doel (2003: 142) also recognizes a degree of "madness" in Olsson, "not least because [he] wishes to disclose the madness of Reason while affirming the creative force of what is estranged from Reason … If madness—and laughter— is to figure in Olsson's work, it will be as a figure that disturbs the smooth running of representation (*conformity*), enumeration (*calculation*), and capital (*apportionment*); a figure that disturbs both the taken-for-granted and critical carving up of social space and co-relation." So madness—and laughter—are key to the deconstruction of the world we believe in, a mental map that erases the true nature of human existence and obstructs human freedom.

One of the purposes of Olsson's passionate, powerful, ambiguous, theatrical and smutty writing style is, as I see it, to upset the apple cart of everyday (socialized) consciousness (or as Olsson refers to it, the taken-for-granted) and push the boundaries of the known and the methods through which we express the known. Olsson's work is a critique of ontological transformations, and through his writing strategy Olsson hopes to inspire a new way of seeing the world and in fact a new way of being in the world.

At the same time, he seems to understand that his method will not be grasped by some readers, as in the following passage where he considers the reaction of the reader:

> Some will see my activity as an exercise in falsification of institutional facts. Others will call it construction by negation. A few will agree that wrestling with paradoxes is the only way to learn. But most will merely move their lips, parrot the words, and shrug their shoulders. They are the ones who shall never understand, for they will never ask before they know the answer. (Olsson 1980: 61b.)

Understandably, comments such as these come across as arrogant to some readers. We might also consider these sentiments a frank admission that not everyone is ready, willing and able to do the hard work of critical abstract thinking. In *Abysmal*, however, Olsson (2007: 197) hints at a way in which we might understand what he is getting at:

> To prepare the canvas for the reception of the unknown the artist therefore kills it with layers of gesso, a practice that is highly reminiscent of the socialization process through which you and I are made so obedient and so predictable. The parallel is obvious, for while some readers will immediately grasp the unheard messages of the present book, others will neither understand nor remember. In through one ear, out through the other. No traces left in-between.

We must appreciate the force of socialization here. Our perceptions are shaped through socialization, and it is a primary goal of power to eliminate the ability to conceive of alternatives. "Instead of building a world for the constant groping of autonomous man, we are confining ourselves within spatial and social prisons which will serve only to increase our sense of loss and futility. Instead of preserving ambiguity, we impose certainty" (Olsson 1980: 204b). "We" impose certainty at the behest of the powers that construct cartographical reason, that organize our mental maps in such a way to imprison us in our own categories and block all the exits. If such efforts of power are successful in reifying social relations, the number of individuals who would be receptive to critiques of cartographic reason are minimized, but when

> we become self-conscious about them, we are free to ignore their commands, for social relations are internal relations which themselves are in the words of myth. They get their power from being solidified in institutions, where they are kept alive by sacrificial offerings and guarded by their own inmates. When we take their veil as granted, we cannot see it. When we realize that we are weaving it, we dare to lift it. (Olsson 1980: 150b)

Thus whether Olsson's comments about the reception of his work are best characterized as honesty or arrogance is largely in the I/eye of the beholder and is intimately related to the process of socialization.

To stay with this trajectory for a moment, the generation of contradictory responses is in fact a hallmark of the rogue guru. While followers may perceive divinity in the guru, others may see a vulgar criminal (Feuerstein 1992). In Olsson's case, some consider his writing strategy to be clear and informative. Chris Philo (2009: 206) finds *Abysmal* "challenging and yet perhaps surprisingly accessible at every turn," and for John Pickles (2007: 396), while "the manuscript is long, the writing is clear, even spare, and always crisply analytical." Others are less charitable. Seemann (2008: 133) refers to readers who find that Olsson's "selective language makes [his texts] almost impenetrable, unapproachable, and extremely difficult to understand. Olsson's peculiar way of writing and his frequent references to 'dense' works of modern philosophy result in an uncomfortable and frustrating feeling of simply not having a clue to what he is writing about." Likewise, based on a belief that "language should be intelligible to many readers as well as to the author," Hugh Prince (1980: 295) prefers not to "follow Gunnar Olsson into his private world of ambiguity, where he wanders in search of creativity." I would suggest that such diverging reactions are particularly symptomatic of the kind of transformational work in which Olsson (like the crazy-wisdom gurus) is engaged.

Olsson himself does not make any direct links between his project and those of the holy madness crowd, aside from a suggestive comment arguing that, with regard to requests academics may receive from government bodies to conduct particular social science studies, "the responsibility of independent intellectuals is not to stand with hat in hand, but to be jesters, sometimes performing and sometimes not. But at whose mercy is that jester who cuts too closely to the truth he is supposed to suggest but never tell?" (Olsson 1980: 31e). While Olsson does not develop this metaphor further, his writings suggest various affinities with the crazy wisdom tradition.

As we follow Olsson on his journey we are presented with amusing anecdotes, oblique references, a mish-mash of philosophy, art, literature (etc.), the cryptic statements, so pregnant with and devoid of meaning at the same time. What, for example, could he possibly mean by: "Dealing with modal logic is like eating a doughnut from the inside. It is to chew away at fluffy possibility and to digest it into precise and agreed-upon knowledge" (Olsson 1991: 43)? Or, consider the "silence" of the blank pages in between some of the chapters of *Lines of Power*, itself a fairly "mad" gesture—unconventional, ambiguous, drawing the reader into a closer relationship to the author, as the former experiences the breaking of the taboos of writing convention. As Olsson (1991: 3) writes, "Impossible to say, easy to show." If the "social space of silence" is impossible to communicate through words, it is (perhaps) easy to show on the blank page. All very Zen-like, signaling the limits of language and the importance of performance, of the doing, the place of bodily expression; but this experimentation with language and form is not done simply for the sake of experimentation. Philo (1984: 227), for example, considers Olsson's project as directed towards "some sort of 'personal emancipation.'" This is understandable given such comments as: "I was forced to write this seventeenth chapter thereby marking yet another stage on the road of liberation" (Olsson 1980: 30e). But his is not exclusively a personal project. In fact, Olsson's ontology could be described with reason as in harmony with the feminist slogan "the personal is political." In addition to transforming individual awareness, Olsson clearly sees a need to transform social institutions—he is adamant in his critiques of social science, planning, and state capitalism, for example, and even if he may appear to avoid grappling with the messy real-world issues of the day (Philo 1994), his analysis is quite useful for not just the deconstruction of these institutions but also the reconstruction of a more human world.

There is, as noted above, an undeniable element of "personal emancipation" involved in Olsson's project. What might this personal emancipation be? Olsson (1991: 54, emphasis added) informs us that "*the* major task of a critical social science," as he sees it, is the decomposition of the invisible "ethical glue that simultaneously keeps individual and society together and apart." Olsson seeks to challenge the logics of social ethics and interpersonal relations, to unsettle their foundations and to expose the workings of power which, at their worst, deform our humanity. Socialization is a formidable enemy, because "every actor is so entrenched in his role that he takes the shadow play to be reality and reality to be the play. It is indeed an integral part of all internal relations (and thereby of all ideologies and all mythologies) that we obey their commands without hearing them, and without knowing where they come from" (Olsson 1991: 20). This is precisely why "to tell the truth can sometimes be an evil, to lie can sometimes be a duty" (Olsson 1991: 30). In which case we can consider Olsson's oeuvre to be a big fat lie. At the same time, Olsson (1991: 20) despairs that breaking the shadowy spell completely "is impossible, for under the mask I shed there is always another. And the next veil I always fail to notice because it is one with my own sight; I see the mote in my neighbor's eye, but not the beam in my own." The poet e.e. cummings (1950: 65) appears to be more optimistic when he writes "now the ears of my ears awake/now the eyes of my eyes are opened." But *whose* eyes and ears are signaled by the "my"?

Olsson is not quite a pessimist, though. He suggests a way to use his work: "Anyone who understands me eventually recognizes these stripteasing questions as nonsensical, when he has used them—as steps—to climb up beyond them" (Olsson 1991: 98). These "stripteasing questions" recall the approach of Zen teachers who pose nonsensical riddles ("koans") as a way to trigger a higher form of consciousness among their students. Furthermore, Olsson's comment recalls Wittgenstein:

> My propositions serve as elucidations in the following way: anyone who understands me eventually recognizes them as nonsensical, when he has used them – as steps – to climb up beyond them. (He must, so to speak, throw away the ladder after he has climbed up it.)
>
> He must transcend these propositions, and then he will see the world aright.
>
> (TLP § 6.54) (Curry 2000: 95, quoting Wittgenstein)

To "see the world aright" is to transcend the veils of power. However: "Power is another word for the relation between the I and the non-I. It follows that the process of liberation can never end, for its driving force is in the emancipation and creation of the self" (Olsson 1991: 102). Thus liberation is not a destination, but it is the way (like the Tao, we might say).

Olsson (2009, personal communication) considers his work to be "art about social science, not social science about art." In order to appreciate and even understand his work, then, one must approach it with an appropriate *attitude*. This requires an openness to the work as something that one will not necessarily understand in the traditional sense. Philo's (1994) admission that he cannot claim to understand what Olsson means at various points in his argument is illustrative: Philo shows that one can accept a certain ambiguity or even discomfort in the reading experience and still gain from the work. In the same sense that a guru might exhort his followers to open their hearts to him, a reader of Olsson is challenged to perform the highly un-academic maneuver of opening her or his heart to him, and thus it may be that for a reader to benefit from reading Olsson, the right attitude with which one approaches his work is required. An attempt to understand the meaning of every sentence of one of Olsson's publications is likely to frustrate even the most dedicated reader and to be counterproductive.[8] Instead, much like when taking in a work of art (or the teachings of a guru), one benefits from letting the work "do its thing." In a somewhat different context, Olsson (2007: 109–10) writes that the "key question is not what a given sentence means but what it does, especially how it does whatever it does." We might say that Olsson's work is an attempt to demonstrate what it means to be human. The key question that follows is: do we really *want* to know what it means to be human?

In demonstrating one aspect of what it means to be human, Gunnar Olsson seeks to preserve his personal integrity, and his writing style is a very deep expression

---

8   Especially since many of the seemingly obscure allusions Olsson employs are chosen not first with the reader in mind but as a way for Olsson himself to keep track of where he is in his line of thinking (Olsson 2009, personal communication).

of who he is as a person. His "solipsism" is in a way a stubborn commitment to honoring his own humanity, one might even say "spirituality," and his writing style is a way to say to the reader "you are going to meet me on my own terms;" by truly expressing his I he allows for there to be a You. And who is this I? It is a social scientist who seeks to develop a new language for his science that maximizes the possibilities for human freedom. For when our social science and planning techniques are grounded in categorization and linearization, we "come close to creating a society for human beings who themselves are peculiarly categorical and linear ... Instead of creating a world for becoming, we are creating thingified man; by treating the relations between people as if they obediently followed the multiplication table, we are ridding ourselves of that challenging ambiguity, which alone makes life worthwhile (Olsson 1980: 359–60)." A world without ambiguity has no place for the excluded middle, no role for the trickster, clown, or jester. We might also say that such a world is in fact in desperate need of these figures.

And of artists and poets.

> Am I now about to suggest that if we want our social scientific findings to be translated into action, then we should throw away our equations and instead begin to speak as poets? Am I suggesting that we should cease being manipulative social engineers and instead begin to behave as the sensitive explorers we actually are? Perhaps I am! Perhaps I am not!
>
> But—at the moment—perhaps I really am! The reason is that we seem to live in an age in which the precarious balance between the classical conservatism of certainty and the romantic creativity of ambiguity is being seriously threatened... The implication is of course that you and I should be made to behave according to the strictures of scientific methodology rather than the other way around. (Olsson 1980: 10e)

But the way of the poet is not necessarily welcomed in our scientific age, characterized by an increasing focus on the quantification of academic "productivity" and a tooth-and-nail competition for research funding. Olsson (1980: 17e) is quite aware that while he sees before him a road that many artists have traveled before,

> few social scientists have ever dreamed of entering it. I deplore the fear as I recognize the cause. For just as the Bourgeoisie, the Church, and the Revolutions all have expelled their poets, so has the Academy. And for the same reasons of domination in the name of mass communication and ritual communion! ... 'Power to the People!' will remain a futile cry until preceded by 'Power to the Imagination!' Rather than suppress the individual in the name of the group, we must do exactly the opposite.

Thus it is not enough to simply attempt to show through his writing one example of what it means to be human if one ignores the institutional context within which we, as humans, live. Thus Olsson goes beyond the guru's focus on changing the ontological perspective of the individual to draw attention to the power and problems of socialization. In this way one might say that he is doing both the private work of the guru and the public work of the trickster, clown, and jester.

Conclusion

> If any one among you thinks that he is wise in this age, let him become a fool that he may become wise. (1 Cor. 3:18)

> To most, my comments thus far will appear far out. (Olsson 1980: 13e)

Far out, in deed. For many readers, Olsson's comments will appear not just far out, but *too* far out, beyond the borders of the acceptable and recognizable. In part, this is due to his unconventional writing style. It is thus interesting to consider what it is that makes Olsson's work hard to understand. The reader will quickly notice that his language is very simple, and quite concrete—it is certainly justifiable that the artist in Olsson likens his writing to sculpture, rather than to, for example, painting (Olsson 2009, personal communication), as his words are truly embodied. One becomes familiar with the salty beard, the touch of grandchildren, the force of sexual desire, the visceral sensation of memory. While all of this is expressed in clear and concrete language, conventional meaning may remain elusive because of Olsson's desire to push our thinking to a more abstract level.

A considerable degree of abstraction is necessary in order to be able to discuss the "the abyss between visible and invisible" (Olsson 2007: 205). Spiritual adepts, especially those influenced by Buddhism, speak of the "emptiness" that is "behind" all we see and feel. Different kinds of "holes" (perhaps), but in both cases the need for abstraction is a fundamental prerequisite for true understanding. It is by acquainting ourselves with the holes in our social relations, the ambiguity of the excluded middle, and the abyss between the visible and invisible that we can steal a glimpse of freedom. And what is Gunnar Olsson doing with his writing performances other than showing us what it means to be free, and ultimately what it means to be human. For what does it mean to be human? To live! To love! To laugh! To cry! To enjoy! To fornicate! To be confused! To be frustrated! The question is whether we are able to hear/here this guru's crazy-wise message through the haze of socialization.

And just why is it that this Swede feels a need to explore what it means to be human? In perhaps no other country is the population in such doubt about its own humanity. In a book published in 2006 two Swedish authors wondered in the title *Är Svensken Människa?* [Is the Swede Human?]. In 2009 the humorist and linguist Fredrik Lindström provided the reassuring answer in his wildly successful one-man show called *Svenskar är också människor*—Swedes are people, too. How do we explain such existential confusion in a modern, wealthy society? For Olsson the culprit is none other than the social engineering of the Swedish welfare state. Modernity in Sweden has been built upon an extreme standardization that has no place for diversity, for difference, for *deviance*. It is a cliché of identity studies that without difference there is no identity, and Olsson would no doubt argue that without difference there is no *humanity*.

Swedish has two main words for "different": "olika" and "annorlunda." The former refers to difference in a neutral sense, as in different colors to choose from when buying a dress. The latter is more judgmental, with a connotation closer to "strange" or "deviant." Thus one can be different (olika) without being Different (annorlunda). The Swedish authorities have very little administrative tolerance for the annorlunda, and the same could reasonably be said for Swedish society in general, and even Swedish academia. It is this social and professional context that has inevitably shaped Olsson's perspective; indeed, it is perhaps even true that can cannot *fully* understand Olsson's critique, in its substance and style, without having lived in Sweden for a significant period of time. Swedish society (though not *only* Swedish society) desperately needs to embrace the spirit of the annorlunda embodied by the geographies of Gunnar Olsson. However, faced with a hyperlegitimate welfare-capitalist state on the one hand, and an academia where the law is the law and rules are rules (except when the powers that be wish to find ways around them, in which case everyone knows the spells that make the laws and rules conveniently disappear), the likelihood of any substantial change in Swedish society in the near term remains minimal. In the meantime, Olsson seeks refuge in the excluded middle and invites us to join him there, or at least stop by for a visit. It can be at times a lonely place, though the warm laughter and good wine makes it more than worth the trip. In any case, the views from there are amazing.

Bibliography

Blomley, N. 2008. The spaces of critical geography. *Progress in Human Geography*, 32(2), 285–93.

Braidotti, R. 2002. *Metamorphoses: Towards a Materialist Theory of Becoming*. Cambridge: Polity Press.

Campbell, J. 1968. *The Hero with a Thousand Faces*. Princeton: Princeton University Press.

Conley, T. 2008. Review of *Abysmal. Imago Mundi*, 60(1), 104–105.

cummings, e.e. 1950. *XAIPE: Seventy-One Poems*. New York: Oxford University Press.

Curry, M.R. 2000. Wittgenstein and the fabric of everyday life, in *Thinking Space*, edited by M. Crang and N. Thrift. London: Routledge, 89–113.

De Weerdt, H. 2009. Book review. *World History Connected*, 6(2). [Online]. Available at: http://worldhistoryconnected.press.illinois.edu/6.2/br_de_ weerdt.html [accessed: 17 November 2010].

Dewsbury, J.-D.C. 2007. Abysmal: in the "Bar de Saussure, the Nietzschean hangout where not a member is sober." *Geografiska Annaler*, 89B(4), 389–93.

Doel, M.A. 2003. Gunnar Olsson's transformers: the art and politics of rendering the co-relation of society and space in monochrome and Technicolor. *Antipode*, 35(1), 140–67.

Eriksson, M. and Nordlund, C. 2008. Tankens kartografi. *Upsala Nya Tidning*, 6 April, B14.

Feuerstein, G. 1992. *Holy Madness: The Shock Tactics and Radical Teachings of Crazy-Wise Adepts, Holy Fools, and Rascal Gurus*. New York: Arkana.

Feuerstein, G. 2006. *Holy Madness: Spirituality, Crazy-Wise Teachers, and Enlightenment*. Revised and expanded edition. Prescott: Hohm Press.

Hyers, M.C. 1969. The dialectic of the sacred and the comic, in *Holy Laughter: Essays on Religion in the Comic Perspective*, edited by M.C. Hyers. New York: The Seabury Press, 208–40.

Hyers, M.C. 1974. *Zen and the Comic Spirit*. London: Ryder.

Hyers, C. 1996. *The Spirituality of Comedy: Comic Heroism in a Tragic World*. New Brunswick: Transaction Publishers.

Jung, C.G. 1968. *The Archetypes and the Collective Unconscious*. Second Edition, in *The Collected Works of C.G. Jung*, Vol. 9, Part 1 (Translated by R.F.C. Hull). Princeton: Princeton University Press.

Kakar, S. 2009. *Mad and Divine: Spirit and Psyche in the Modern World*. Chicago: University of Chicago Press.

Kinsley, D. 1974. "Through the looking glass": divine madness in the Hindu religious tradition. *History of Religions*, 13(4), 270–305.

Lowe, S. 1996. The strange case of Franklin Jones. [Online]. Available at: http:// www.flameout.org/flameout/gurus/dafreejohn_franklin.html [accessed: 17 November 2010].

Marrouchi, M. 1998. Counternarratives, recoveries, refusals. *boundary 2*, 25(2), 205–57.

McDaniel, J. 1989. *The Madness of the Saints: Ecstatic Religion in Bengal*. Chicago: University of Chicago Press.

Nisker, W. 2001. *The Essential Crazy Wisdom*. Berkeley: Ten Speed Press.

Nisker, W. 2008. *Crazy Wisdom Save the World Again!* Berkeley: Stone Bridge Press.

Olsson, G. 1980. *Birds in Egg/Eggs in Bird*. London: Pion.

Olsson, G. 1990. Lines of power. *Nordisk Samhällsgeografisk Tidskrift*, 11, 106–16.

Olsson, G. 1991. *Lines of Power/Limits of Language*. Minneapolis: University of Minnesota Press.

Olsson, G. 2007. *Abysmal: A Critique of Cartographic Reason*. Chicago: University of Chicago Press.

Philo, C. 1984. Reflections on Gunnar Olsson's contribution to the discourse of contemporary human geography. *Environment and Planning D: Society and Space*, 2, 217–40.

Philo, C. 1994. Escaping Flatland: a book review essay inspired by Gunnar Olsson's *Lines of Power/Limits of Language*. *Environment and Planning D: Society and Space*, 12, 229–52.

Philo, C. 2009. Review of *Abysmal*. *Annals of the Association of American Geographers*, 99(1), 205–209.

Pickles, J. 2007. Radical thought-in-action: Gunnar Olsson's critique of cartographic reason. *Geografiska Annaler*, 89B(4), 394–97.

Prince, H. 1980. Review of *Humanistic Geography: Prospects and Problems*. *Annals of the Association of American Geographers*, 70(2), 294–96.

Scott, A.J. 1976. Review of *Birds in Egg*. *Annals of the Association of American Geographers*, 66(4), 633–36.

Seemann, J. 2008. Review of *Abysmal*. *Geographical Review*, 98(1), 133–35.

Trungpa, C. 2001. *Crazy Wisdom*. Boston: Shambhala.

Walsh, R. 1992. Foreword, in *Holy Madness: The Shock Tactics and Radical Teachings of Crazy-Wise Adepts, Holy Fools, and Rascal Gurus*, by G. Feuerstein. New York: Arkana, xv–xvii.

Zucker, W.M. 1969. The clown as the Lord of Disorder, in *Holy Laughter: Essays on Religion in the Comic Perspective*, edited by M.C. Hyers. New York: The Seabury Press, 75–88.

Chapter 13

Environment and Planning D: Society and Space, 1993, volume 11, pages 279–294

Chiasm of thought-and-action[†]

G Olsson
Nordiska institutet för samhällsplanering, Box 1658, S-111 86 Stockholm, Sweden
Received 19 November 1992; in revised form 7 January 1993

Abstract. How do I know the difference between you and me and how do we share our beliefs in the same? How are we made so obedient and so predictable? As a minimalist approach to these questions I imagine human thought-and-action as a double helix. It is assumed firstly, that man is a semiotic animal, a species whose individuals are kept together and apart by their use of signs; secondly, that every sign within itself combines elements of drastically different ontologies. This invisible world is then captured in a three-dimensional coordinate system whose axes are those of identity, difference, and intentionality. While the resulting map is anchored in fix-points of silence, the real world of socialization and understanding is always in flux. The paper closes with a pastiche on Carl von Linné's *Flora Suecica*; in the current world of thought-and-action, signifier and signified are assigned the same ordering functions as stamina and pistil once were in the world of plants. How do I draw the invisible lines of the taken-for-granted? How do I project a dematerialized point onto a transparent plane?

There is a double helix in the social sciences too, a chiasm of thought-and-action, an epistemological braiding of ontological antinomies, an imagination of what it is to be human. Life is form, form the modality of life.

I prefer the term 'imagination' to that of 'theory', for in my experience every theory carries within it a trace of commission, an (un)conscious adjustment to a particular interest, an attitude that it is not enough to understand the world, but that·I must change it as well. Built into the theoretical is in fact a hostility to abstractness, a temptation to thingification, a utilitarian shift from social science to social engineering. 'Can' turns to 'ought', opportunity to obligation, possibility to necessity. Court painters nevertheless lead the dangerous lives they deserve, for their masters hate ingratiation as much as humility, integrity as much as laughter. Social realism is bad art for the same reasons that social engineering is bad ethics, less because knowledge is power, more because power is knowledge. Ass-lickers become ball-biters, castrated eunuchs.

My own imagination has emerged gradually, in stages without breaks. Thus there are clear affinities between my current concerns and the etchings of *Birds in Egg/Eggs in Bird*, the watercolors of *Antipasti*, and the oils of *Lines of Power/Limits of Language*. How do I know the difference between you and me and how do we share our beliefs in the same? To which extent is it I who speak through language and language that speaks through me? How are we made so obedient and so predictable?

★

† An imagination first sketched at the symposium "Limits of Representation" held in Villa Clarke, Bagni di Lucca, June 25–29, 1991. I am grateful to all participants in that event, but especially to Franco Farinelli, Ole Michael Jensen, Allan Pred, and Dagmar Reichert. Kent Karlsson drew my attention to Claude Mellan's self-referential version of Veronica's kerchief; *FORMATUR UNICUS UNA non alter* THE ONE FORMING THE ONE no other.

If only I could know others, then others could know me. Impossible, for just as the 'now-here' of time and space is a linguistic shifter, so is the 'I' of personal identity. Once caught in the cultural net of à priori categories, the I does not know whether it is itself or one of its too many duplicates. Since the image in the mirror simultaneously re-flects reality and in-forms the ego, a man cannot get rid of the relation to himself any more than he can get rid of himself. The logical, the moral, the aesthetic are analogous, for all are self-referential. It is in deed in the proper name of Kant itself that his critiques must be taken to their limits. But a limit can never be understood, for understanding is itself a limit.

Such is the desire of my imagination: a minimalist rendering of how we define ourselves, an acknowledgement that linguistic signs are forms *of* art shaped by forms *in* art. Like Piet Mondrian and Samuel Beckett I therefore dream of a work in which "the expression of things gives way to the pure expression of relation", where my writing is "not *about* something, [but] *is* that something itself". Preliminary sketches have appeared before, first under the heading "Malevic sfigurato", then as the two chapters "Squaring" and "Malevich Torpedoed", most recently in the tourist brochure "Invisible Maps". If desire is mimetic, how do I draw the likeness of Nothingness?

★   ★   ★   ★   ★

Common to my various sketches are two assumptions. One is that man is a semiotic animal, a species whose individuals are kept together and apart by their use of signs; the other that every sign within itself combines elements of drastically different ontologies. The latter stem from the legacy of Descartes, even though their specific names vary with the contexts in which they occur; sometimes they are called use value and exchange value, sometimes sense and reference, mind and matter, presence and absence, signifier and signified. Oversimplified, one part of the antinomic pairs is in the physicality or corporeality of the sign, the other in the intentionality of its cultural meaning. The former is open to the five senses, the latter to the sixth. And yet it is important to recall not only that every thought occurs to a flesh but also that there can be no art without matter; poetry is written with words, not with ideas, paintings are painted with paint, not with concepts.

Partly for the sake of the French, mainly for that of analysis, it is necessary to distinguish between signifier, S, and signified, s. Talk about things and relations belong to different language games. It must nevertheless be remembered that every sign within itself contains both ingredients at the same time. *Mens sana in corpore sano.*

Like the eye of anatomy, also the I of the sign can both see and be seen. In neither case, however, is what and how I see independent of the particular world I am a part of. It is in this double sense (of the distinctive correspondence between outside and inside, inside and outside) that the carnal being is the prototype of Being-in-the-world. I hear myself not through my ears but through my tongue. I see myself not through my eyes but through my touch. I eat what I am. Therefore, whenever in doubt, always trust your body, for body is biological matter and cultural mind intertwined.

My body is neither thing nor idea, but the measure of a thing. And thus it is that even when a woman talks, I can never fully understand her. Lips are lips, kisses are kisses. The marble of the sculpture is not the marble of the quarry, the gold on the finger not the gold in the bank. Yet you know what it means to miss New Orleans. A rose is a rose is a symbol of love.

★

Just as the body carries the initiation scars of circumcision, pierced ears, and tattooed chests, so does the mind. Thus, the normal psyche is marked by its ability to draw and to symbolize real distinctions, by its unconscious memories of the imaginary. Understanding the difference between you and me is in practice to live a split life without going crazy. In contrast, the psychotic neither makes nor shares distinctions, for in the Land of Psychosis there are no initiation rites, no marks, nothing social. The loss of not knowing the loss of inclusion is an echoless scream in a mountainless valley. In the words of Jean-Jacques Rousseau's *Heloise*: "Wanting to be what we are not, we come to believe ourselves something other than what we are, and this is how we get mad."

It follows that any scar is better than no scar, for without exclusion and inclusion nothing exists, not even nothing. 'And that, Herr Goldschmidt, that is why you must submit, when I now execute the will of my Führer. Bitte, pull down your pants! Prove with your cock who you are, just as the beriddled Oedipus once did with his swollen foot. What does your name mean, Herr Goldschmidt? Was it not your Lord who declared that the cut in the flesh of your foreskin is a sign of the covenant between Him and you?'

Alas for the seed of men. Identity as sameness and difference brought together. Parricide and incest as transgressions of the sacred boundary between man and beast. The tragic hero as violator of the limits he himself has established, victim of his own words, law maker as law breaker. How come that we are so obedient and predictable that when someone points it out we feel uncomfortable? Titian's *Flaying of Marsyas* is the most gruesome of paintings, the Holocaust the most unmentionable of acts.

★

And so it is that every sign carries an invisible mark of distinction, itself visualized in the line of

$$\frac{S}{s} \quad \text{and} \quad \frac{s}{S}.$$

I would even go so far as to suggest that every sign can be condensed into the dash of the fraction line. This is not to say, however, that my minimalism is a reductionism, only that there is a fundamental narcissism in all vision, hence in all thought-and-action. Put differently, even though S and s are obsessed by a desire to be the same, that desire can never be satisfied. It follows that the semiotic animal is thoroughly paradoxical, for it can be what it is only by being what it is not.

One reason is that meaning does not reveal itself in the identities intended, but in the differences achieved. Another is that immediately I write

$$\frac{S}{s} \quad \text{or} \quad \frac{s}{S},$$

then the signified ceases to be a pure signified and turns into a signifier. In my own conception, the fraction line serves as a symbol of the real castration that bears the transmission of culture. Perhaps the bar is the trace of the real, for it is exactly here that the five senses touch the sixth. Desire is not the desired, desire is desire. Desire is not the meaning of the meaning of the upper-case or lower-case s, desire is the invisible −. The semiotic animal wants nothing. All it asks is to be believed, all it needs is response. Desire is the desire of having one's desire recognized.

Not *Desidero ergo sum*, but *Desiderare ergo sum*. Not that "*I* desire therefore I am", but that "*Desire* therefore I am". Human beings lend signification to everything, especially and foremost to their thoughts-and-actions. Seducer seduced. Tracers tracing traces tracing tracers. Fox-hounds. Sour grapes.

★   ★   ★

Put together, the two assumptions of man as a semiotic creature and the sign as a fraction lead to a conception of thought-and-action as a play of ontological transformations. Whenever he thinks-and-acts, man is a juggler of Cartesian categories, one torch in his hand, the other in the air. The ruling paradigm remains that of the *Genesis*: "Let there be!—And there was."

In performing its speech acts, the Word becomes flesh and dwells among us, full of grace and truth and honey-sucking bees. Socialization writ large, creativity as the art of making the present absent, the absent present. And yet it is in the brutality of facts that reality shows itself less in copies of the external, more in gestures of the unconscious. Thus, I once again catch a glimpse of that Hopper woman in the airport lounge, blue dress, one leg on top of the other. The angle of her knee returns my glance, thereby confirming that I exist. And then in the coolness of the evening I thought well as well her as another and then I asked her with my eyes to ask again yes and would I yes to say yes my mountain flower. The keys to. Given!

To exercise power is in this perspective to perform the double trick of turning things into relations, relations into things, inner into outer, outer into inner. And as a way of illuminating its own function, the word 'image' carries connotations not only of 'phantom' but also of 'statue', not only of fantasy but also of *Homo erectus*—Phallus of the phallus.

It cannot be said more clearly: the world of thought-and-action is bat-like. Viewed from one direction it looks like a bird, from another like a mouse. No wonder that power is so poorly understood, for already the *Leviticus* classified the bat among the unclean. To be unclean is by definition to be a member of the class of the non-classifiable, to be neither this nor that, neither fish nor fowl. To be in the taboo is to be in a limit. In-between is an abyss, the bottomless pit of primal chaos. The American savages were taught the Spanish grammar, for without grammar the Conquistadores could never conquer.

★

For the conventional theorist, ambiguity presents an insurmountable obstacle. Unclean is the hare because it chews the cud but does not part the hoof, unclean the swine because it parts the hoof but does not chew the cud, unscientific the study whose subject matter hops capriciously about. Without order no knowledge, without knowledge no power.

In my own imagination, the ambiguous order of thought-and-action can be condensed into a dematerialized point. This origo of abstractness is itself a version of Wassily Kandinsky's geometric point. In its minimalism it belongs to language and signifies silence. In its stillness it belongs to contemplation and signifies change. Graphically:

★

To bring the silence of these lines into common parlance, they must be properly baptized. I therefore now name the vertical axis "IDENTITY" and the horizontal "DIFFERENCE". Both are nevertheless silent lines, the former white and cold, the latter black and warm. In the origo is the muteness of the taken-for-granted, hence the orthogonality of the coordinate net. It follows that before a line can break into speech it must be tilted; for Piet Mondrian—prime investigator of silence—the diagonal was so full of deceit that when Theo van Doesburg began to paint it, their friendship was over. While the classicism of modernism accepts the frame, the baroque of postmodernism tries to break it.

At the end-points of the two axes lie extreme forms of silence. These extremes can themselves be given specific names, and it is on your acceptance of them that the credibility of my imagination eventually hinges. In a sense, they represent special cases of the sign and thereby the fix-points of thought. More specifically:

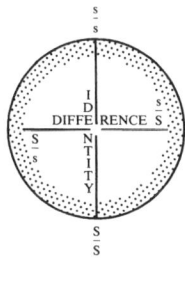

★

It is obvious from this drawing that I imagine thought simultaneously as the telescope of my rifle and the deer of my hunt. At the empty center of the cross lies the dematerialized point with which I aim at the world; at the periphery is the ring that represents the limit of language, hence of thought. The end-points of the horizontal line touch this limit from the inside, those of the vertical mark it from the outside. Moving the rifle up and down, I capture the world in its metaphoric likeness; moving it sideways, I trace the displacements of the metonymic. To capture is to metaphorize, to trace is to metonymize.

Roman Jacobson's analyses of aphasia come readily to mind. But so does also the image of Diana's bath, important both to Titian and to Jacques Lacan. You remember the story: Diana has been hunting the whole day. She is warm and when she comes upon a stream of water she strips bare and wades into it. But behind a bush hides Actaeon in his spying caught by Diana's eye. Trouble in the making, for possessed she will no longer be the unpossessed virgin that drives gods and mortals mad, unseen she will not be THE woman Actaeon imagines. Perhaps that is why the voyeur had to be turned into a stag, pursued and eventually killed by his fifty hounds. Huntress hunted, desire as lack, lack as desire. To desire is in essence to metonymize, to realize that to desire is not to have an object but to be a subject. Yet there can be no desire without embodiment, no sucking without breasts. To be human is to be mammalian.

Thus there is an erotic in every truth, for 'truth' is itself a mode of socialization. And as if to prove its own point, this particular essay is threatening less because of its doctrines, more because of its passions. See what I mean. Sucker!

Exaggerating? Of course! How else could I understand? How else could I convince? Not that she shouldn't, not that she wouldn't, and I know it's not that she couldn't. It's simply because she's the laziest gal in town.

★

It is in this same vein of understanding and convincing that I associate the ring of my telescope with the final paragraph of Ludwig Wittgenstein's *Tractatus*. The reason is that after Plato understanding is geometric and communication phallogocentric; here nobody enters who does not know his geometry. It follows first that of what one cannot see, thereof one cannot speak, and then that "whereof one cannot speak, thereof one must be silent". In actuality, I see not with my I's eye but with my pupil, not with my pupil but with my brain, not with my brain but with my mind. The world is a fantasy, an imaginary construction of the mind.

Thus it is with my mind's eye that I can sense how the two end-points of the identity axis reach toward the outside limit of language. Properly speaking, these two points cannot be named, for they are devoid of difference. At one extreme is the

$$\frac{s}{s},$$

which is the imagined sign of pure spirit, at the other the

$$\frac{S}{S},$$

the imagined sign of pure matter. The former is in the ether that clouds the top of Olympus, the latter in the rocks that strew the shores of Scylla and Carybdis.

The allusions to mythology prompt themselves, for at least since antiquity man has tried to define who he is. This has led him into fighting a kind of two-front war, where one enemy is that of pure spirituality, the other that of pure physicality. Put differently, one boundary dispute concerns the difference between man and god, the other the distinction between man and beast. In the former No-Man's Land threatens the hubris of the superhuman, in the latter the degradation of the subhuman. In between is the imaginary of the human itself, for only man knows how to imagine. What turns Oedipus Tyrannos into a tragic figure is that he oscillates between being the equal of gods and being the equal of nothing at all. His sin is that he is a slayer of distinctions, father and son in one, husband of his mother, child of his wife. Identity is never indifferent, for certainty rules in a world of ambiguity. Cloven foot, rabbi run.

The front lines move back and forth. To illustrate, the Greeks prior to the fifth century did not even possess a word for "human will", for to them a man could not act on his own, only as an instrument of the gods. In addition, gods were condemned to eternal life, which is why no Greek ever wanted to be one of them. But then, two and a half millennia later, Friedrich Nietzsche could define the human as the will to power. God is dead, Man is god! Proud rulers are no longer descendants sent from Heaven, but representatives of the Electorate; *l'État c'est moi*, the Whip of self-reference.

In the meantime, the other front line has fluctuated too. While for the Greeks women and slaves were not really human, current issues in ethics involve such matters as the rights of animals and fertilized eggs. This is in essence what Environmentalism is about: the limit between the beastly and the human, the cloven hoofs and the silken paws.

Whether the total territory of the imaginary has increased or decreased is a moot question. Some would argue that man has invaded the Land of Gods, others that idiots have peopled the Land of Man. Perhaps it is the gods of television that today govern mankind, perhaps the engineers of thought that command atoms and genes. *Das also war des Pudels Kern*, the inevitability of the Faustian pact, the conflict between the Word of the Old Testament and the Deed of the New. Beware of your soul! Not everything is for sale.

On our way from dust to dust, the questions remain the same: Who am I and which is the difference between you and me? In contrast, the answers keep changing, for answers are always contextual. In the utopian No-where of now-here they point to the political ideologies of the twentieth century, to paradox and predicament, to intentionality caught by the tail of its own tale.

Such is the current price of being obedient, for such is the speech of silence: *Beyond* the limits of language. Babble's wall from the outside.

★

The signs at the end-points of the identity axis are silent, because their nominator and denominator are the same, their fraction equal to one. The crossing of the bar has gone unnoticed, the circumcision leaving not a scar of proof. The end-points of the difference axis are silent as well, but here it is because the signifier and the signified are too different to be connected. This is typical of what I elsewhere have called the "crisis of the sign", the experience that there are no words for what I really feel, no political representatives of who I really am.

Also the signs of difference take two extreme forms, viz.

$$\frac{S}{s},$$

and

$$\frac{s}{S}.$$

The first denotes an expression in search of its meaning, the second an intention in search of its expression. The former is the sign à la Jacques Lacan, the latter à la Ferdinand de Saussure. The arts of surrealism and postmodernism are in the first mood, the politics of nazism, communism, and social democracy in the second.

The survival rates of these ideologies seem inversely proportional to their degree of inhumanity. For that reason, I am eternally grateful to the Fate that in 1935 let me be born in Sweden and not in Germany or the Soviet Union. But this happy circumstance must not keep me from understanding that everything has a price, also the politics of the welfare state. How do I insult a power which is so powerful that it is faceless? How do I learn about difference, when difference is defined away? How do I topple a regime which has no statues erected in its honour? How do I find my way in a jungle of paragraphs? In short: How can I live in a culture which is so proud of its penis that it is unaware of its Phallus? Why is it so hard to detect the difference between the *Nom-du-Père* and the *Non-du-Père*?

Perhaps there is in these questions yet another connection with the rhetorical tropes of metaphor and metonymy. Thus whereas metaphor is what is said and shown, metonymy is what is heard and seen. What I give is a metaphor, what you receive is a metonymy. Metaphor is the spark, metonymy the explosive. The metaphor of metaphors is the anchor, the metaphor of metonymies the arrow. By saying how things are, I show who I am. I speak less for the purpose of informing, more for speaking the speaking.

It is in this double sense of holding fast and letting loose that both knowledge and power are structured as a language. While the conscious focuses on the metaphoric of the sign's nominator, the unconscious hides in its denominator. The locus of power is nevertheless in the invisibility of the fraction line, in the simultaneous separation and joining of the S and the s, the s and the S. Emperor on the balcony, child in the gutter. Spiders in the bedrooms of the Kakanian castles.

Such is the silence of speech: *Within* the limits of language. Babble's wall from the inside.

★   ★   ★

Thus is my imagined vision of the plane of thought: a cross centered on a dematerialized point and surrounded by a halo of silence. Where is the action?

The action is in another dimension, captured by a third axis. This line, which I hereby name "INTENTIONALITY" or, more properly, "DESIRE", passes through the dematerialized point at a right angle. Graphically:

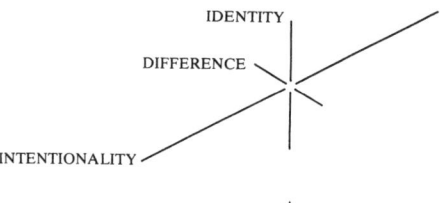

★

Also the end-points of the intentionality axis can be named, and also here the variation is in terms of thinghood. At one extreme, desire is totally thingified, at the other completely spiritualized. In the former case, the signifier is most properly written as $\underline{S}$, in the latter as $\bar{s}$. My wording indicates that I see action less from the ethical viewpoint of intentions and more from the psychoanalytic standpoint of the unconscious; while intentions serve as colonizing projections into the future, the unconscious is the memory of the forgotten. In certain circumstances nothing is more noticeable than a lack of purpose.

What I denote by $\bar{s}$ comes close to the concept of the sublime, that which stems from the *sub limis*, that which is uplifted from under the threshold. To feel the sublime is to experience how imagination approaches its own limits, for sublimation does not represent the thing lost but actually recreates it. In painting, the sublime is that which remains invisible; in chemistry, sublimation is the conversion of a substance from the solid to the vapour state without its being liquid in between.

In contrast, the $\underline{S}$ comes closer to the repressed. In Freudian terms, the individual who is perfectly repressed is also perfectly socialized. In Lacan's world, on the other hand, repression is less an issue of forgetting desire and more of speaking the truth. Since truth by definition is unspeakable, repression constitutes the paradoxical 'representation' of the ultimate silence of death. René Girard's theory of mimetic desire and the killing of the double comes to mind as well, for it is when desire momentarily reduces itself to the scapegoat of need that it effectively guarantees its own perpetuation. It is in this sense that desire reveals itself through its own negation; the dialectics of desire is a dialectics of signification, of making a thing present by its absence.

In normal situations, both sublimation and repression are part of the defense mechanisms without which anybody would go crazy. Carried to their ultimate consequences, however, they reach toward silence and thereby madness. To distinguish these cases, I denote the SUBLIME and the REPRESSED of madness with the letters s and S, and the sublime and repressed of normalcy with the barred notation $\bar{s}$ and $\underline{S}$. The former touch the limits of language from the outside, the latter from the inside.

Finally, it should be noted explicitly that the axis of intentionality cuts through the plane of thought at the dematerialized point of the equal sign. It is in this same point that action appears in its most purified form, farthest away from the extremes of both $\bar{s}$ and $\underline{S}$. Here, as in the aufgehoben state of the man without qualities, nothing stirs in her now, not even her splendid desire. It is through this point that all thought-and-action must pass, for it is in this point that everything turns to its opposite.

★

When turned around the intentionality axis, the coordinate system forms a three-dimensional volume shaped like an American football. It is this suspended figure that in my imagination offers the most abstract picture of thought-and-action:

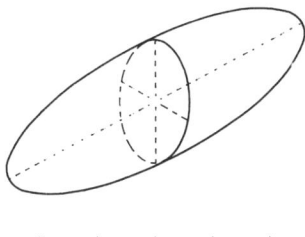

★   ★   ★   ★   ★

Thus is the form of my imagination: black marks projected onto a screen of white paper. The Name-of-the-Father performing its double functions of law and order, legislation and punishment.

And just as Carl von Linné could build his classificatory system on observations of stamina and pistils, so I now think of the concepts of signifier and signified as serving similar functions in the realm of thought and language. And just as the boiling and freezing of water led Anders Celsius to the design of his thermometer, so I dream of the various forms of silence as fix-points in a set of maps from the Land of Action. The territory to be explored is like America between Cristóbel Colón and Amerigo Vespucci; unknown to the white Spaniards, familiar to the red Indians. Stanley, stay home! Mother Matrix rules.

In actuality, the signs of thought-and-action seldom occur in the clearcut forms of the proposed structure. Instead they show themselves at various stages of imbalance; sometimes the signifier is more prominent, sometimes the signified, sometimes the sublime, sometimes the repressed. The most immediate task is therefore to discover, name, and order a set of illustrative cases. It must nevertheless be borne in mind that it is through abstraction that an object becomes more real than the real; what a revolutionary discovery it was when Paul Cézanne suddenly realized that he no longer painted landscapes but literally pictures, not mountains and houses but triangles and rectangles, not content but form. Yet there is an inherent conflict between the attempts to elevate art and the desire to be grounded in reality, between the aesthetic truths of the point, line, and plane on the one hand, and the persuasive power of representation on the other. Nothing is more powerful than the power of the example.

★   ★   ★

With these comments on abstraction and rhetoric constantly in mind, I now return to the picture of the American football. I cannot really explicate how and why, but when I look at this peculiar image with my eyes closed, I see that the shell of the oval consists of a set of tightly packed threads, sometimes shaped as a double helix, sometimes as a Moebius band. Graphically:

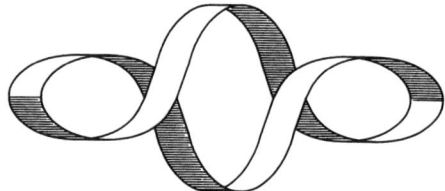

From the center of this figure rules the dematerialized point, the origo of imagination. Along the innumerable threads which connect the eight fix-points of silence lie the transition forms of human conduct, always in flux yet always kept in place by its dialectical counterparts of individual and society. These forces of checks and balances can themselves be represented by a set of lines, all running through the dematerialized point at an angle. I have set this angle equal to 45 degrees, partly because Kandinsky insisted that the diagonal is the most talkative of lines, mainly because I thereby introduce asymmetry (hence movement) into the otherwise symmetrical figures. Whether the resulting drawings remind me more of the biological scriptures of DNA or of the alternating currents in an electric generator, I do not know.

At any rate, the dialectical forces can be depicted as follows:[1]

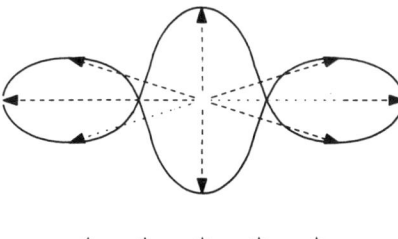

★　　★　　★　　★　　★

As illustrations and elaborations, I now proceed to a brief presentation of two sets of cases. The first approaches the speech of silence and thereby some crucial forms of socialization: By which means are we made so obedient and so predictable? The second captures the silence of speech and thereby some well-known modes of understanding: Through which categories do I learn about difference and by which languages do we establish the same?

In my imagination, the double helix of socialization is drawn as

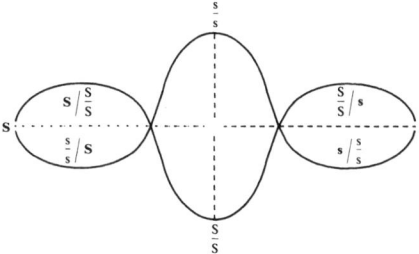

while that of understanding looks like

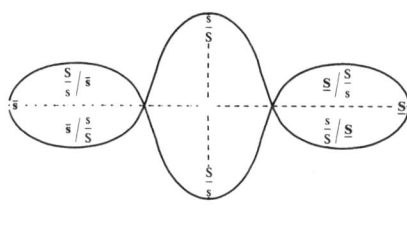

★

(1) It should be reemphasized that the lines of the dialectical forces run through the dematerialized point at angles of 45 degrees. The discrepancy between the text and the drawing is therefore not real but apparent, an unavoidable result of projecting a three-dimensional image onto a two-dimensional plane. Yet I shall be the first to testify that lodged inside a rotating football it is easy to get dizzy.

What now remains to be discovered is a set of empirical examples which can be attached to the noted transition forms. In a sense, the two schemes as hitherto presented are mute, for they are *langue* without *parole*. To get them to speak, I must therefore link the generic names of the drawings with a set of illustrative cases, not unlike what Bertrand Russell did with his theory of proper names and definite descriptions, and not unlike what Carl von Linné did in his *Flora Suecica*. The difficulties are nevertheless tremendous, for in the realm of thought-and-action the meaning of a name is not given by the object it denotes but by the context in which it occurs.

Here almost everything is hidden. What I shall pass over in the space of a sentence, others have covered in yards of books. Yet I cannot resist the temptation. Can't stop me now! Out it comes, a small but representative cut from the imagined *Flora* of thought-and-action. First some specimens collected in the deep forests of socialization. Then some colorful growths from the open fields of understanding.

★

S   s

the *REPRESSED*, desire of the completely obedient.

$$S \left/ \dfrac{S}{S} \right.$$

the transition form of the *fetish*, a material object of magical powers. Arrest of the metonymic, expression of the belief that lifeless things have a spirit, that a stone is more than a stone, a bear tooth more than a bear tooth, a shoe more than a shoe. And yet one should never ignore the power of Bertrand Russell's remark that "a robust sense of reality is very necessary in framing a correct analysis of propositions about unicorns, golden mountains, round squares and other pseudo-objects." The practice of operationalization belongs to the ideology of materialism and speaks the rhetoric of the concrete. I know for a fact what a fact is. *Factum-verum*. Man the forger forging the uncreated conscience of his race. Phallus of the phallus.

$$\dfrac{S}{S}$$

silence of *STONES*, word without meaning. Petrified matter.

the *SUBLIME*, without a trace swept under the threshold.

$$s \left/ \dfrac{s}{s} \right.$$

the transition form of the *icon*, an attempt to picture the non-picturable. Liberation of the metaphoric. A holy image, a link between God and Man, a symbol of incarnation. The light of an icon comes from everywhere, for it is not the viewer who looks at an image but the image that looks at a viewer. Perspective is reversed, time cancelled out. The icon has no frame, yet it lies at the heart of Western culture. The Iconoclastic Controversy is itself a consequence of Christianity's double roots, one in Judaism with its prohibition against graven images, the other in Hellenism with its habit of representing the gods in statues. *Homo erectus*. Trust as a matter of obedience. Belief as rememberance of members dismembered.

$$\dfrac{s}{s}$$

silence of *SPIRITS*, meaning without words. Evaporating mind.

$$\frac{S}{S}\bigg/ s \qquad\qquad \frac{s}{s}\bigg/ S$$

the transition form of the holy *communion*, the ritual sharing of beliefs. Performance of ontological transformations in which wine turns to blood turns to love. Incarnation in reverse. Submission under the social pressures of being normal, of not being alone with the non-expressable. The refusniks are in deed frightening: the hermit, the autist, the anorectic. How do I know difference and how do we share the same? *Corpus domini nostri Jesu Christi custodiat animam tuam in vitam aeternam*	the transition form of the state *prison*, legalized means of correction. Discipline and punish, institutionalization of the idea that the body is the mind's corrective. Operationalization as thingification, the power of the eye and the index finger. Yet there is an unthinkable difference between a body in chains and a mind without desire, between classes of objects and unique individuals. *Cogito ergo sum* yields first to *Desidero ergo sum*, then to *Desiderare ergo sum*. Therefore I sentence you to

s   S

CONVERTED. Eternal life. Ether of Mount Olympus.	*CONVICTED.* Monstrous death. Rocks of Scylla and Carybdis.

Graphically:

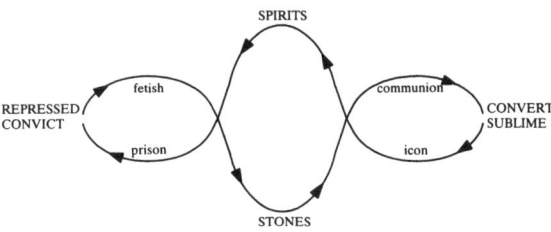

★

s̄   S̱

the *sublime*, recreation of the invisible, uncovering the marvellous.	the *repressed*. Violence of the social. Hiding the scar of the horrible.

$$\bar{s}\bigg/ \frac{s}{S} \qquad\qquad \underline{S}\bigg/ \frac{S}{s}$$

the transition form of *inwardness*, the mode of understanding which is in the poetry of Søren Kierkegaard and Stéphane Mallarmé. The fact that existence is unspeakable does not mean that it is nothing, for what cannot be	the transition form of *outwardness*, the mode of understanding which is in the science of Karl Marx and the politics of the twentieth century. It is not enough to understand the world, the point is to change it. Yet, not even

said can sometimes be shown. Within the prisonhouse of language, every thought gives off a throw of dice, not a predetermined verdict but perhaps a constellation. There is an echo in Mallarmé's cry for his dead son: "It is you that I want, you and in you myself." Man is not a stone but a living being with the courage of giving an answer that is not an answer. 'Predicament' is the name of the game, 'anxiety' the price of the unnameable. All that is solid melts into air.

Marx himself was a pure materialist, for it was he who stated that "even though relations between commodities initially may appear as relations between things, they are in fact relations between people." What distinguishes the worst of architects from the best of bees is that the former raises his structure in the mind before he builds it in wax. And yet, 'paradox' is the analyst's major enemy, 'tragedy' the inevitable outcome. All that is vapourish solidifies into matter.

$$\frac{s}{S}$$

the silence of *religion*, the art of I and Thou.

$$\frac{S}{s}$$

the silence of *politics*, the practice of delegation.

$$\frac{s}{S}\bigg/ \underline{S}$$

The transition form of the *symbolic*. Fly in the ceiling of Solomon's Temple, Chagall's blue bride over mythical roof-tops, bull's eye.

$$\frac{S}{s}\bigg/ \bar{s}$$

the transition form of the *real*. Malevich's white square on white, plane of the taken-for-granted, Mose's tablets.

$$\underline{S} \qquad\qquad \bar{s}$$

the repressed of the double. Killing of Girard's *scapegoat*. Scream in the desert. Death of an individual. Birth of the social.

the sublime of Veronica's *kerchief*, twist of an eye, tone of a voice. Birth of an individual, death of the social.

Graphically:

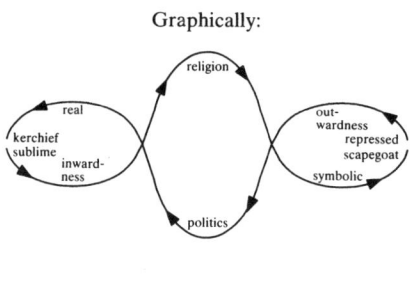

     ★   ★   ★   ★   ★

Mystical figures, creatures of imagination. Cure and poison in the same breath. Memory of Pharmakon, dream of dreams. Seeing that the naked eye is blind.

The *Anemona Nemorosa* of the Linnean system is characterized by its own sexuality. Although I tried, I could never figure it out. What I do know, however, is that the same flower weaves the white carpet over which fairies dance in spring-time. And in my own country school every ten-year old was certain that *Prostafis Luctarilla* comes with a special flagrance and a bombastic bang.

A promise is a promise. But in the drunkenness of the Bar de Saussure I have been told that desire cannot be named. Yet I know how to recall them: Anna, Mona, Emma, Rosa. All in my herbarium.

★ ★ ★ ★ ★

Auerbach E, 1953 *Mimesis: The Representation of Reality in Western Literature* (German original, 1946; Princeton University Press, Princeton, NJ)
Brown N O, 1974 *Closing Time* (Vintage Books, New York)
Cassirer E, 1953–57 *The Philosophy of Symbolic Forms* (German originals, 1923–29; Yale University Press, New Haven, CT)
Derrida J, 1987 *The Truth in Painting* (French original, 1978; Chicago Press, Chicago, IL)
de Saussure F, 1983 *Course in General Linguistics* (French notes, 1907–11; Duckworth, London)
Focillon H, 1989 *The Life of Forms in Art* (French original, 1934; Zone Books, New York)
Freud S, 1950 *Totem and Taboo* (German original, 1913; Routledge and Kegan Paul, London)
Girard R, 1987 *Things Hidden Since the Foundation of the World* (French original, 1978; Stanford University Press, Stanford, CA)
Hegel G W F, 1977 *Phenomenology of Spirit* (German original, 1807; Clarendon Press, Oxford)
Joyce J, 1934 *Ulysses* originally published 1922 (New Library, New York)
Joyce J, 1958 *Finnegans Wake* originally published 1939 (Viking Press, New York)
Kandinsky W, 1977 *Concerning the Spiritual in Art* (German original, 1914; Dover, New York)
Kandinsky W, 1979 *Point and Line to Plane* (German original, 1926; Dover, New York)
Kant I, 1949 *Critique of Practical Reason* (German original, 1788; University of Chicago Press, Chicago, IL)
Kant I, 1951 *Critique of Judgment* (German original, 1790; Harper, New York)
Kant I, 1966 *Critique of Pure Reason* (German original, 1781; Doubleday, New York)
Kierkegaard S, 1959 *Either/Or* (Danish original, 1843; University Press, Princeton, NJ)
Kristeva J, 1987 *Tales of Love* (French original, 1983; Columbia University Press, New York)
Lacan J, 1977 *Écrits: A Selection* (French original, 1966; Tavistock Publications, Andover, Hants)
Lachterman D R, 1989 *The Ethics of Geometry: A Genealogy of Modernity* (Routledge, New York)
Lacoue-Labarthe P, 1989 *Typography: Mimesis, Philosophy, Politics* (French originals, 1975–86; Harvard University Press, Cambridge, MA)
Malevich K, 1968 *Essays on Art* (Russian originals, 1915–33; Borgen, Copenhagen)
Mallarmé S, 1977 *The Poems* (French original of "Un coup de dés", 1897; Penguin Books, Harmondsworth, Middx)
Mallarmé S, 1983 *A Tomb for Anatole* (French original, 1961, written in the 1870s; North Point Press, San Francisco, CA)
Marx K, 1967 *Capital* (German original, 1867–94; International Publishers, New York)
Merleau-Ponty M, 1968 *The Visible and the Invisible* (French original, 1964; Northwestern University Press, Evanston, IL)
Musil R, 1979 *The Man Without Qualities* (German original, 1930–42; Picador, Basingstoke, Hants)
Nietzsche F, 1968 *The Will to Power* (German originals, 1883–88; Vintage Books, New York)
Olsson G, 1980 *Birds in Egg/Eggs in Bird* (Pion, London)
Olsson G, 1990 *Antipasti* (Korpen, Göteborg)
Olsson G, 1991 *Lines of Power/Limits of Language* (University of Minnesota Press, Minneapolis, MN)
Olsson G, 1991, "Malevic sfigurato" *Slam* **3** 7–10
Olsson G, 1991, "Invisible maps: a prospectus" *Geografiska Annaler* **73B** 85–91
Panofsky E, 1991 *Perspective as Symbolic Form* (German original, 1927; Zone Books, New York)
Spencer Brown G, 1969 *Laws of Form* (George Allen and Unwin, London)
Stella F, 1986 *Working Space* (Harvard University Press, Cambridge, MA)
Vernant J-P, Vidal-Naquet P, 1990 *Myth and Tragedy in Ancient Greece* (French originals, 1972–86; Zone Books, New York)
Watson J D, 1968 *The Double Helix* (Atheneum, New York)
Whitehead A N, Russell B, 1910–13 *Principia Mathematica* (Cambridge University Press, Cambridge)
Wittgenstein L, 1953 *Philosophische Untersuchungen/Philosophical Investigations* (Basil Blackwell, Oxford)
Wittgenstein L, 1961 *Tractatus Logico-Philosophicus* (Bilingual original, 1922; Routledge and Kegan Paul, London)

© 1993 a Pion publication printed in Great Britain

Chapter 14

Chiasm
Rubric

Jette Hansen-Møller

Nineteen years ago Gunnar Olsson published "A chiasm of thought-and-action" (Olsson 1993).[1] It was a minimalistic rendering which illustrated the processes of thought and action shaped as the double helix of a Moebius band and compared to an American football. In the meantime, this image was substituted by a pyramid (Jensson 2000; Olsson, 2007). Never mind. The desires were similar: to produce a text in which the expression of things gave way to pure expression of relations; to write a text not *about* something, but *being* that something in itself. A metaphor was thrown; others invited to juggle *de* Saussure *a* Lacan with Olsson. To me this opportunity is better than most to employ the identities and differences of Charles S. Peirce's[2] on the same stage as Olsson's to get a better understanding of the choreography of the game. Let's witness the clubs swapping.

In front of the curtain

> The celebrated Mr. O.
> performed his feat as article in magazine.
> …
> Having been some days in preparation a splendid time was guaranteed for all.[3]

In ancient Egypt jugglers were highly appreciated as can be seen in Figure 14.1. After the fall of the Roman Empire they were accused of practicing witchcraft, and in the 19th century they were restricted to present their art between what was considered the "real" performances (Wikipedia, 2008a). So, instead of "topping the bill" like Mr. K in Beatles' quoted song text above, Olsson's inspiring fountain

---

1   In the following references from this text are written in Italics followed by a numeral in parenthesis as for example *temptation of thingification* (279). The number refers to the page in which the text is found.

2   In the following references to Peirce are written as (CP 2.233) CP referring to his Collected Papers, the number indicating the section number in Past Masters CD-ROM Databases edited by Hartshorne, Weiss and Burks, 1994.

3   Inspired from Lennon/McCartney, 1967.

of rings in rings was presented in front of the curtain while stage workers replaced set pieces behind it; in this case that of Lagopoulous' with that of Jensen's Red River Valley (Lagopoulous 1993; Jensen 1993).

Figure 14.1   Representation of wall painting from c. 1994 – 1781 BC. found in the 15th tomb in the Beni Hassan area, Egypt[4]

After the performance the applause was light and random, and Olsson was almost accused of being a gleeman (Gren 1994; Doel 2003; Sparke 1994). What neglect. For other reasons this author hesitated for more than two decades to try to squeeze the metaphors presented: one, a fear of the *temptation of thingification* (279) whereby they would lose their attraction; another, lack of an appropriate instrument by which the text could be comprehended, a text which, "resists a single, unambiguous reading."[5]

Now the first step has been taken to share what I witnessed in the anatomic theatre from a seat between the American semiotician Charles S. Peirce and the French psychoanalyst Jacques Lacan. In the following my understanding will be presented first as a waltzing with rings in rings. Then I will steal some of Olsson's balls through a side swapping and transfer them to my own juggling pattern in order to gain an in dept understanding of their constructions, content and relations. After an inspection of the everyday objects presented by the gentleman juggler I conclude with a presentation of Olsson's tabooed.

Music!

---

4   The painting appears to depict jugglers. Being represented in an ancient Egypt tomb it signifies something which the diseased wished to bring to the next world and suggests a religious significance of some kind. For example round things were used to represent solar objects, birth and death (Wikipedia 2008a). Note that the jugglers are women.

5   The most correct remark of Sparke's in his critique of Olsson's text (Sparke 1994).

Waltzing with rings within rings

Horns flourish. In enters the wizard from Uppsala. Hair cut, ear pierced, eyes flashing. Go Olsson, GO!

Standing on the shoulders of Other's, the *juggler of Cartesian categories* (282) starts switching his rings of rings from wrist to wrist: "What Jacques (La)can, I can do better." *See what I mean. Sucker!* (284). Pure re-Joyce.

As Olsson so often taught, the opening of a text should be a continuation of its final assertion. In that light the title *Chiasm of thought-and-action* could be interpreted as a continuation of *FORMATUR UNICUS UNA non alter* (223) appearing under the representation of Mellan's Etching of Veronica's kerchief but that would be wrong. The reason is that this etching is located, not at the end of the article but as the first of the references. As such it serves instead as Olsson's reference to himself. The explanation is that Mellan's etching is made of one spiralling line beginning and ending in the nose of Christ' (Olsson 1994a). Thereby the juggling pattern of the article, the relation between the text of the etching and the title of the article, plus the structure of Olsson's presentations of understanding and socialization are all chiastic structures. From a literary point of view concepts are placed symmetrically in chiastic structures to emphasize their importance and perhaps for aesthetic and mnemonic reasons (Wikipedia, 2008b). The term chiasm also refers to the cross of the optic nerves in the brain. To certain genotypes, like Siamese cats and white tigers, this wiring is disrupted why they have to compensate by squinting (Wikipedia 2008c). So has this reader of Olsson's Chiasm to follow the double spaced poetic ending of his *Flora of thought-and-action* (290).

The Latin term "una" used by Mellan in the subtitle of his etching refers to the spiralling line by which it was made. Being feminine this definite article made Olsson recall the iconoclastic battle for and against the use of graven images (Olsson 1994a). Particularly women were active in the defence of the right to represent the holy, and the controversy was also settled by one, the Byzantine Empress Irene in 787.

The somewhat concealed hints to women Olsson followed up immediately after by the question *How do I know the difference between you and me and how do we share our beliefs in the same* (279)? This observation I take as an invitation to try to *share* Olsson's *belief* and participate in his juggling. But how can we share something if all I can do if *in doubt ... [is] to trust my body* (280)?

Although he has two arms and two legs, a torso and a skull as I have, my body is radically different from Olsson's. In contrast to him I have a uterus; my sense of balance is disturbed by a bewildered otolith and my right hip bone replaced by one of titanium. Perhaps these differences explain why I prefer the Chiasm to the later developed Pyramid (Jensson 2000). Another reason is that the shape of the first is not set but a net, and that its structure, in contrast to that of the pyramidal glass coffin anchored in granite, is susceptible to the surroundings. As such I perceive the Chiasm as more of an inmate's experience of thought-and-action than the crystal palace which rather demonstrate the power of cartographical reasoning "plugging the gap of an [-other?] abyss": the relationships between the powers of science, art and religion (Olsson 2007: 415-437. Abyss being a notion also used by Lacan 1989: 166).

Despite these differences between bodies and preferences, I do agree with Olsson that *man is a semiotic animal, a species whose individuals are kept together and apart by their use of signs* (280). In my draft:

> What I see is nothing but my own concepts[6].
> What I hear is nothing but my own imaginations.
> What I feel is nothing but my own limits.
> What I taste is nothing but my own digestion
> What I smell is nothing but my own distance differed.
>
> Yet there are differences between
>
> envisioning, looking and watching;
> overhearing, listening and bugging;
> embodying, encompassing and embracing;
> salivating, consuming and dining;
> scenting, sniffing and nosing;

In other words *[T]here is a fundamental narcissism in all vision, hence in all thought-and-action where the five senses touch the sixth* (281).

To define the end and beginning of a ring is impossible. Did the birds exist before the eggs or the eggs before the birds? In the last meaning of the Chiasm-text Olsson wrote… *desire cannot be named. Yet I know how to recall them: Anna, Mona, Emma, Rosa. All in my herbarium* (293). By these statements he should not be categorized as a sexist because Anna, Mona, Emma, Rosa do not refer to women desired but to the Latin terms of plants used by Linnaeus in his endeavour to structure their relations. Similarly Olsson employed 'S' and 's' in (293) Chiasm of thought-and-action (279) in his quest to investigate understanding and socialization. No end no beginning. All that exists is circulation around a Lacanian emptiness.

At the proscenium of Society and Space centrifugal and centripetal forces are balanced by the rhythm of props switching from one hand to the other. The horizontal force employed to maintain the circulation Olsson categorised *DIFFERENCE* (283). Vertically the urge to transcend is hindered by gravity no matter the strength put into the throw by the juggler. Olsson baptised this force *IDENTITY* (283), see Figure 14.2. Together these axes define the plane of thought. In addition a third force is involved. It demonstrates what juggling is all about: neither an urge to be recognised as a unique soloist by the audience in front, nor to maintain the right to cash a paycheque every month, but a desire to keep the props in the air. This latter force *cuts through the former plane at the dematerialized point of the equal sign.*Here *action appears in its most purified form* and *nothing*

---

6   Inspired by "What we see is nothing but ourselves looking", Harnoncourt, A.D.H. and K. McShine (eds) *Marcel Duchamp*. Museum of Modern Art, New York, 1973 (quoted in Olsson 1980: 42e).

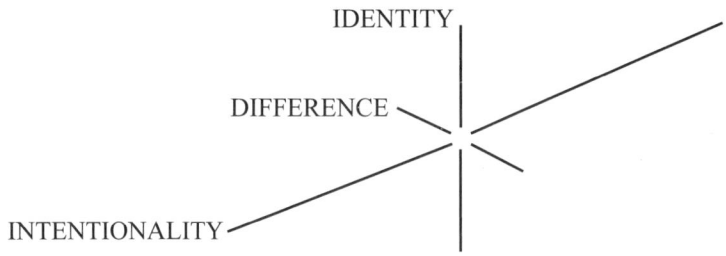

Figure 14.2 The axis in Olsson's Chiasms (Olsson 1993: 286)

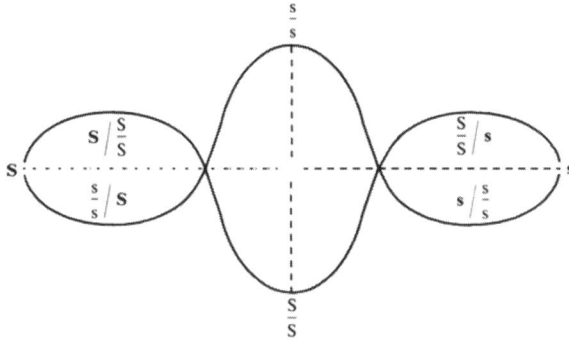

Figure 14.3 The double helix of socialization (Olsson 1993: 289)

stirs... for it is in this point that everything turns to its opposite... (287). Olsson referred to the diagonal force as that of *INTENTIONALITY or DESIRE* (286).

Based on the plane and the force which penetrates the chiasm, two similarly structured double helixes are developed by abduction rather than deduction or induction. The first, represented in Figure 14.3, Olsson called the double helix of socialization. The other, forwarded in Figure 14.4, he named the double helix of understanding.

The weaving together of the patterns was represented by *black marks projected onto a screen of white paper* (287) like ordinary juggling patterns (Figure 14.5).

Olsson described this pattern as *a shell of an oval* which consisted *of a set of tightly packed threads* (288) sometimes referred to as a double helix, sometimes a Moebius band and pet named the American football (See also Lacan 1991: 156).

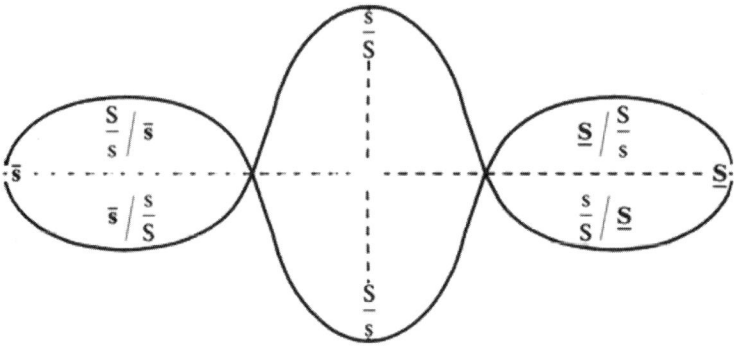

Figure 14.4   The double helix of understanding (Olsson 1993: 289)

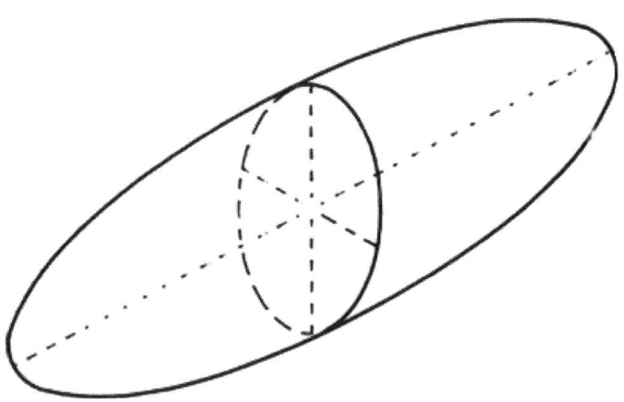

Figure 14.5   Olsson's American football (Olsson 1993: 287)

Too many directions to choose from now. Better pull down the "necklace of rings in rings" to prevent them from dropping (Lacan 1989: 153. See also Lacan's (1975: 107–123) borromean knots).

Did you catch a glimpse of an 'I' in the act; Olsson's, mine, or perhaps your own (Olsson 1991a)?

Side swapping and stealing balls

There is a double helix in the social sciences too (279).

A new beginning, but why the "too"? Neither the title of Olsson's paper nor its abstract can help us understand what he referred to by "too" had it not been for the signified of the concept "double helix" itself. Within Molecular Biology the double helix is the structure of DNA identified by James D. Watson and Francis Crick. A double helix is composed by two congruent helices with the same axis, differing by a translation along the axis, half way or not, within geometry. As to Plato's Academy *nobody enters* Olsson's cartography *who does not know his geometry* (284).

There is a triad in my[7] interpretation too; neither because it was inspired by Olsson's three axes[8], nor because Peirce preferred to present his thinking as bifurcations but for three other reasons. First, because the latter defined a sign as being "something which stands for something to somebody in some respect or capacity" (CP 2.228). Second, because Peirce incorporated three modalities in his categories in contrast to Kant. And third, because I found a rectangular representation of Peirce's semiotics by Nöth (2000), (see Figure 14.6) useful in my quest to develop a framework by which the similarities, differences and relations between humanity and nature could be investigated.

	Interpretant	Object- Relation	Representation
Firstness	Rheme	Icon	Qualisign
Secondness	Dicisign	Index	Sinsign
Thirdness	Argument	Symbol	Legisign

Figure 14.6 C.S. Peirce's nine sign-classes according to Nöth (2000: 66) horizontally reversed around the centre column by the author

---

7   When I in the following refer to my own renderings these are to Hansen-Møller, 2004, 2006 and 2009.

8   As a curiosum it can be mentioned that Olsson (1991b) two years before the publication of the Chiasm-article wrote that he had come to suspect that the habit of thinking in threes had reached its limits Yet, he continued to do so in the Chiasm-article. For example he referred to his previous works as *the etchings of Birds in Eggs/Eggs in Birds, the watercolors of Antipasti, and the oils of Lines of Power/Limits of language* (279) and listed three preliminary sketches of the present text under the headings: "Malevich sfigurato" (1991c); two chapters in "Lines of power/Limits of language" (1991a) "Squaring" and "Malevich Torpedoed"; and finally "Invisible Maps: a prospectus" (1991b). More importantly he forwarded the three axes of *IDENTITY, DIFFERENCE* and *INTENTIONALITY* in the Chiasm-article.

Peirce found that the Interpretant, Object-Relation and Representamen all expressed themselves as monadic potentialities of Firstness, dyadic actualities of Secondness and, triadic habits generating patterns called Thirdness. I translated these categories and phenomena to the rubric, the Natursyns model, forwarded in Figure 14.7.

Phenomena \ Modalities	Humanity	Landscape	Nature
Potentiality Firstness	Sensibility	Habitat	"Nature"
Actuality Secondness	Cognition	Area	Environment
Continuity Thirdness	Argument	Symbol	Habit

Figure 14.7   Natursyns-model; a conceptual framework by which meanings of nature can be investigated, explained and compared (Hansen-Møller 2009)

Of course my intention to develop the above rubric can be declared a *commission* and an *(un)conscious adjustment to a particular interest* (279). Yet, I have constantly tried to avoid to *turn 'can' to 'ought, opportunity to obligation or possibility to social engineering* (279). Anyhow, *I can no longer resist the temptation* (290) to make a side swap and steal some of Olsson's balls to try to integrate them into my own pattern. (Did anyone think penis envy?) What will take place is *a translation from one "language-parole" to another* (Olsson 1991a: 95). The name of the game is semiosis. It is a praxis we play, not because we have learned a set of rules, but because we cannot resist to generalize and associate. Above all it is the habit of understanding i.e., the law of the mind (Santaella, 2001). The outcome will be a map. But as "telling the truth is to claim that something is something else and be believed when you do it" (Olsson 1991a: 167), and because a map "at the same time [is] a picture and a story: a *de*-scription of a set of points and a *pre*-scription of how we find our way among them... a concrete illustration of an abstract argument" (Olsson 1994a: 216) I feel obliged also to describe the pattern to which the balls are transferred.

From Peirce's so-called phaneroscopy (phenomenology) it can be derived that he found the "thinking of nature" and the "nature of thinking" analogue (Nöth 2001). In addition he understood "matter to be effete mind, mind frozen into regular routine" (CP 6.277). These considerations made me interpret humanity and nature as endpoints on a continuum of which landscape was but one possible relation. In the Natursyns model the Interpretant was translated to Humanity although Peirce did not limit his concept to comprise human subjects. The Object-relation was translated to Landscape and the Representamen to Nature. In the translation I gained from Peirce's pragmatic inclusion of the relation to an object in his semiotics in contrast to what Saussure did in his semiology. Some might therefore claim that I make a mistake in the following when I relocate the Signifiers and Signifieds of Saussure-Lacan-Olsson in my Peirce-Nöth-inspired framework. However, a translation is never an exact copy of the original and usually it is from misinterpretations that we learn the most.

To Peirce semiosis was a unidirectional evolutionary, iterative process which mind and matter had in common. It ran vertically from monadic, qualitative potentialities over dyadic, actual quantities to triadic statements or conclusions. To make the names of these modalities more telling than Peirce's concepts Firstness, Secondness and Thirdness I labelled them Potentiality, Actuality and Continuity respectively.

As the overall structures of the actual juggling-patterns now are sketched we are ready to begin swapping.

To Olsson the taken-for-granted was the origo where the IDENTITY and DIFFERENCE axis crossed. Within semiotics I see it as the general law influenced by all the modalities of the Interpretant. Therefore, an analysis of for example a text must begin with an unraveling of the signifying chain from the taken-for-granted and backwards. In the following the Chiasm-text will be examined by this procedure. However, to follow the transfer of balls from one context to another some acquaintance with Olsson's definitions of them seems necessary.

Olsson's primary signs, the **S** and **s**, he snatched from Lacan who borrowed them from Saussure. By the latter, **S** was baptized *signifier*, and **s**, *signified*. In his psychoanalytical interpretation of Saussure's semiology Lacan defined **S** as the in-between between the signifiers of speech, i.e. the subject (Hyldgaard, 1998: 42). Olsson denoted *the SUBLIME and the REPRESSED of madness with the letters s and S* which *touch the limits of language from the outside* (287) on his intentionality-axis in the double helix of socialization. On the analogous axis in the double helix of understanding, yet located at the inside of language, he denoted the endpoints $\underline{S}$ and $\bar{s}$ described as *the REPRESSED and SUBLIME of normalcy* (287) respectively.

In his linguistic approach to language Saussure understood the sign as the fraction

$$\frac{s}{S}$$

even if he never presented it as such himself.[9] To Olsson Saussure's fraction denoted *an intention in search of its expression* (285), *the silence of religions, the art of I and Thou* (292).

In contrast to Saussure, Lacan (1989: 163) employed the algorithm

---

9   Yet, according to Lacan (1989:149) the notation can be found in Saussure's *Course de Linguistique Générale* published by a group of his disciples.

$$\frac{S}{s}$$

which Olsson described as *an expression in search of its meaning* (285), *the silence of politics, the practice of delegation* (292).

Supplied with these definitions the analysis of Olsson's text and the transfer of balls can begin. Olsson illustrated his conclusion in a double spaced poem-like passage which Peirce probably would have categorized as a Symbol. To Peirce a Symbol was "a sign whose special significance or fitness to represent just what it did lied in nothing but the fact of there being a habit, disposition or other effective general rule that made it be interpreted as such" (CP 4.447). Accordingly he classified all conventional signs as Symbols. That is the reason why I kept the concept Symbol as stand-in for all types of representations of Landscapes in my rubric (Figure 14.7).

Olsson used the notation

$$S \Big/ \frac{S}{S}$$

to indicate a similar function, *the transition form of the fetish, a material object of magical powers* (290).

Peirce's Symbols were dependent of an interpreter in order to act as signs but did not lose their character as signs if the object they referred to did not exist. The interpretant Peirce called Argument, the object he baptized Legisign. I kept the concept Argument in my framework but exchanged Legisign with the term Habit in correspondence with Peirce's definition of the Legisign as laws usually established by men. In his phaneroscopy Peirce described the conclusion of semiosis as "mind frozen to regular routine" (CP 2.246; 6.277).

This reminds me of Olsson's algorithm

$$\frac{S}{S}$$

which he described as the *silence of stones, words without meaning. Petrified matter* (290) in his double helix of socialization.

Among the taken-for-granted of Olsson's was that man was a semiotic animal whose individuals were kept together and apart by their use of signs. Geometry and language were his guiding laws of representation and within the latter he understood the sign as a fraction inspired by Saussure and Lacan. For example he claimed that *poetry is written with words, not ideas* as *paintings are painted with paint, not with concepts* (280). He also trusted his *own body* whenever in doubt (280).

He argued that it was the combination of the facts that man was as a semiotic animal and that the sign was a fraction that made thought-and-action a play of ontological transformations and *man a juggler of Cartesian categories* (280). Additionally, he explained that it was the same relation which made the semiotic animal so *paradoxical* that it could only *be what it is by being what it is not* because *meaning does not reveal itself in identities intended but in differences achieved* (281). Finally he reasoned that we should trust our own bodies because *body is biological matter and cultural meaning intertwined, neither a thing nor an idea but the measure of things* (280); a measure which *lent signification to everything, especially thought-and-action*. By that it becomes clear why he often stated that he did *never fully understand when a woman talks* (280). The above quotes I suppose Peirce would have characterized as Arguments (CP Endnotes, book 5), a term which reminds me of Olsson's ball, **S**, the *repressed Violence of the social* (290).

Now we have portrayed the Continuity in Olsson's juggling pattern at the level of Thirdness and are ready to take one step up to the level of Secondness and investigate what Olsson understood as Actualities.

When a ball lands in my hand my body and mind are impressed by its energy. The moment I pass it on I assert its existence. As such it can be understood, first as a Peircean Sinsign "*sin* meaning being only once as in single or simple, an actual but not yet interpreted sign or event" (CP 2.245). In respect to Nature I substituted the term Sinsign with the concept Environment. Secondly, when the ball is re-cognized, it has affected my experience or Cognition and is turned into what Peirce categorized a Dicisign (CP 2.251). The resulting object-relation Peirce named Index (CP 2.283). In my interpretation any recognized part of an Environment likewise is turned into an Area.

Olsson's Environment is comprehensive. It includes concepts and names of persons from philosophy, the humanities and the natural sciences as well as the arts of literature and painting. They are all assigned a meaning and position in Olsson's juggling pattern through which he illustrated that a signified can only operate if it is already present in the subject and has been passed over to the level of the signified, i.e. to Peirce's Dicisign or Lacan's algorithm (Lacan, 1989: 155).

Olsson, born, educated and married in Sweden taught ten years in Michigan, USA and *heard himself not through his ears but through his tongue*. He *saw* him*self not through* his *eyes, but through* his *touch*. (280), *not with* his *I's eye but with* his *pupil, not with* his *pupil but with* his *brain, not with* his *brain but with* his *mind* (284) and *ate what* he *was* (280). By saying how things are he showed who he was. In other words, Olsson's Index or *world was a fantasy, an imaginary construction of the mind* (284) shaped like the double helixes of understanding and socialization in this context.

To Peirce an Index was a "sign determined by virtue of being in a real relation to its object in such a way that it served to identify that object and assure us of its existence or presence to the extent that the Index would lose its character as sign if its object was removed, but not lose that character if there were no interpretant" (CP 4.447; 8.335; 2.283). As such the double helixes would have no character without socialization or understanding being parts of Olsson's Environment whereas they would maintain their meaning whether anybody read the Chiasm-article or not.

At the core is the *dialectics of desire* that is, the *dialectics of... making a thing present by its absence* (287) illustrated so convincingly by Butades' daughter[10] when she drew a contour line on the wall around the shadow of her beloved the night before he was sent off to war. Representation and mapmaking in practice, more influenced by the object denoted than by the meaning connoted. "Mirror, mirror upon the wall, who is the fairest fair of all?" (Grimm, 1898). Freud termed the process Vershiebung ("Displacement" according to Lacan 1989: 160). Lacan baptized the passing-over the Imaginary (Lacan 1992), a phenomenon he found common to all human beings as well as to sticklebacks (Lacan 1988: 137) and doves (Lacan 1989: 3–4).[11]

Peirce stated that Indicies stressed differences. Likewise Olsson underlined that *every sign within itself combines elements of drastically different ontologies* (280). *Oversimplified, one part of the antinomic pairs is in the physicality or corporeality of the sign, the S, the other in the intentionality of its cultural meaning, the s,... And yet it is important to recall not only that every thought occurs to a flesh but also that there can be no art without matter* (280).

Now we are prepared to take the final step up to the level of Firstness which comprises "that of which we cannot speak" (Wittgenstein 1993). Perhaps it is analogous to what Lacan labeled the Real and therefore particularly difficult to re-present (Lacan 1992). At the core is the endless repetition to pursue and present the sublime.

Peirce invented the concepts Qualisign and Rheme for the Representamen and Interpretant of the sign at this level and baptized their relationship Icon.

To Peirce the term Qualisign represented an embryonic sign (CP 2.244). For this aspect of Nature I used the word "Nature" the quotation marks indicating the qualitative potentialities of the not yet embodied aspects of matter and mind for that matter. "Nature" and Qualisign resemble Olsson's $\bar{s}$, *the sub limis*, ... (287) in its modality. It is that which Olsson, like other scientists, strives to discover and express throughout the article; discover by asking one question after the other: *How do I draw the invisible lines of the taken-for-granted* (279)? *How do I project a dematerialized point into a transparent plane* (279)? *If desire is mimetic, how do I draw the likeness of Nothingness* (280)? *How do I insult power which is so powerful that it is faceless* (286)? *How do I learn about difference, when difference is defied away* (286)? *How topple a regime which has no statues erected in its honour* (286)? and express in diagrams.

Perpendicularly to the Qualisign Peirce located Rheme, which signified the qualitative potentiality of the Interpretant in his semiotics and unconscious feelings in his phaneroscopy (CP 2.250). Such feelings neither involved analysis, comparison, nor any process whatsoever, nor consisted in any act according

---

10   By this example Stoichita (1997: 35–36) described one of the ways in which Alberti imagined the invention of representation. Another was the myth about Narcissus.

11   The Estonian semiotician Jacob von Uexküll (2001) indentified a similar phenomenon in all living beings and called it a *Funktionskreis*, functional circle.

to Peirce. In my draught Sensibility is used as a synonym for human flair for sensation and imagination. Perhaps "Sensability" had been a more correct term if it existed. To this field I suggest Olsson's ball

$$\frac{s}{\overline{s}},$$

the silence of SPIRITS, meaning without words. Evaporating mind (290) transferred.

The relationship between Rheme and Qualisign Peirce baptized Icon, a term which I substituted by the biological concept Habitat; an outsiders reference to an insider's perception of being here-and-now. Peirce described an Icon as a sign which referred to the object it denoted merely by virtue of characteristics of its own, and which it possessed just the same, whether any such object actually existed or not. Further, it had no dynamic connection with the object it represented; it simply happened that its qualities resembled those of that object, and excited analogous sensations in the mind, for which it was a likeness (CP 2.299). Metaphors and diagrams served Peirce as examples of Icons. The metaphor was also a key notion to Olsson's transition form from $\frac{s}{\overline{s}}$ to $\frac{s}{s}$. He described it as *an attempt to picture the non-picturable. Liberation of the metaphoric. A holy image... a symbol of incarnation of which the light comes from everywhere, for it is not the viewer who looks at an image but the image that looks at the viewer* (290). Olsson presented it as

$$s\left/\frac{s}{s}\right.,$$

At the core is sublimation. According to Olsson *To feel the sublime* referred to: a) *the potentials of the five human senses open to the physicality or corporeality of the sign (280)*, i.e. like the relationship between Rheme and Qualisign and Sensibility and "Nature"; and comprised b) *to experience how imagination approaches its own limits, for sublimation does not represent the thing lost but actually recreates it* (287); as well as c) *an attempt to picture the non-picturable* (290). *In painting, the sublime is that which remains invisible* (287) Olsson stated. But, as Magritte, who more than most painters granted us metaphors, remarked: "Only the visible can be concealed, the invisible hides nothing" (Magritte in Foucault 1983: 57).

From my point of view Olsson's key metaphor for the sublime is *the dash of the fraction line* (281), i.e. the *bar-de-Saussure*. Thereby and by all the additional metaphors spread out in the article Olsson demonstrates his desire as well as his ability not only to write a text *about* something, but a text *being* that something in itself. An analysis of these metaphors must wait till the final section of this article as it is now time to present Olsson's balls in their new context, see Figure 14.8, a context which they have influenced themselves.

	MIND	FEMININITY	FLESH
REAL	$\dfrac{s}{s}$	$\dfrac{s}{s}\Big/ s$	$\bar{s}$
IMAGINARY	$\dfrac{s}{\bar{s}}$	$\dfrac{s}{\bar{s}}\Big/\dfrac{\bar{s}}{s}$	$\dfrac{s}{s}$
SYMBOLIC	$\underline{s}$	$s\Big/\dfrac{s}{\bar{s}}$	$\dfrac{s}{s}$

Figure 14.8 Olsson's (1993) balls transferred to the juggling pattern a la Hansen-Møller

Inspired by the above and particularly Lacan I will now change the labels of the modalities of the vertical axis: Potentiality, Actuality and Continuity with the Real, the Imaginary and the Symbolic. The categories of phenomena on the horizontal axis are simultaneously changed from Humanity and Nature to Mind and Flesh, of which Femininity substitutes the object-relation, Landscape. Perhaps I could have recycled the term Humanity and used it instead of Femininity, as Humanity originally was chosen instead of Man in order not to appear chauvinistic. However, as I have no experience of what it means to be composed by a double pair of x-chromosomes, of being categorized as male by society or symbolized as son, husband or father I prefer the term Femininity in this context. Mind, Femininity and Flesh are considered structurally similar but their stuffing differ. As I have shown already Olsson's axis of *DIFFERENCE* resembles the horizontal relation of the Imaginary. In my juggling pattern that of *IDENTITY* is represented as the vertical column Femininity. The new rubric comprises two diagonals. The one relating the lower left field of $\underline{S}$ with the upper right field of $\bar{s}$, what Olsson described as *the REPRESSED and SUBLIME of normalcy* (287) and which represents what makes us so *obedient and predictable* (281); *that only semiotic animals can be* (Olsson 2007:11). In contrast, the perpendicular diagonal which links the field in the lower right corner, $\frac{s}{\bar{s}}$, with that in the upper right one, $\frac{\underline{s}}{s}$, embodies what makes the semiotic animals so disobedient and unpredictable as only she can be, exactly because of her ability to represent and thereby also imagine the obscure.

Juggling with everyday objects

To me the above interpretation can be recognized in the video installation *Surrender*, Figure 14.9, produced by the American artist Bill Viola in 2001 as part of a series called *The Passions* (Ebbesen 2005). *Surrender* consisted of two rectangular flat-panel plasma screens mounted to a wall the one above the other. On the screens, two persons mirror one another from the navel up, the one wearing a blue t-shirt, the other a red one. Depending on your time of arrival in front of the screens you would for example see two blurred, moving images, one red and one blue, slowly solidifying into two persons who rise from what turns out to be water. In upright positions, still dripping with water eyes closed, their faces express all signs of grief, then slowly starting to bend forward against one another into the water and while blurring they merge. Red turns to blue, blue to red, the whole sequence repeats itself infinitely. *Now-here, no-where.*

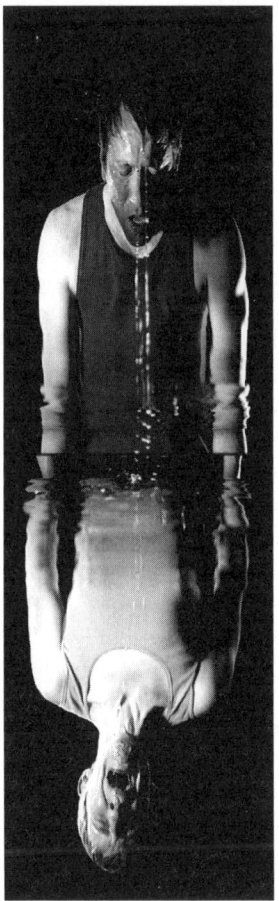

Figure 14.9   Still from the video installation Surrender 2001 by Bill Viola[12]

Viola made his recordings in real time, but presented them in extreme slow-motion in a quest to reveal transformations. To me the installation offers more than a representation of a signifier aiming at its signified and a signified in search for its expression. The images were positioned vertically, not horizontally.

12   Printed with permission from Bill Viola Studio LLC

Besides, they did not just represent bodies surfacing from water and diving into it again. Instead, the images were mirror images *in* water of the emergence and disappearance of the persons, the lower parts of their bodies remaining hidden to the spectator. In addition, as I learned from the exhibition catalogue afterwards, one of the images represented a women the other a man (Ebbesen 2005). Sliding under the surface I now experience the urge to be recognised by the Other as the same and the realization of the impossibility thereof; a metaphor of the human condition engraved by language at the *Bar de Saussure*. As such *Surrender* not only illustrates the switch from the horizontal axis of *DIFFERENCE* to the vertical of *IDENTITY*. It also challenges the relation between the unconscious and the take-for-granted about the sexes: first, by presenting representations of the colour codes normally attached to the sexes reversed, the woman wearing a blue T-shirt, the man a red one; second, by hiding the most important bodily differences of the persons under the "surface" and, third, thereby suggesting the differences to be discursive. In sum, the installation concurrently *attempt*[ed] *to picture the non-picturable liberating the metaphoric* and was *a material object of magical powers arresting the metonymic* (290). The reason is that … *since the image in the mirror simultaneously re-flects reality and in-forms the ego, a* [wo]*man cannot get rid of the relation to him*[her]*self any more than* [s]*he can get rid of him*[her]*self* (280). Whether this interpretation is right or wrong no one can confirm. However, the most important places where Viola's work subsist is "not in the museum gallery, or in the screening room, or on television, and not even on the video screen itself, but in the mind of the viewer" (Krogh 2005). Up and down, down and up, right to left, left to right such is the jugglery macabre of the semiotic animal.

Stamina and pistils, telescope and riffle, arrow and anchor were thrown in the air in the most demanding of the gentleman jugglers performances: that of juggling with everyday objects. And in the concluding fountain, Olsson launched an American football, a Moebius band, a DNA string and the currents of an electric generator (288). Thing upon things, image upon images. "One word for another" (Lacan 1989: 164). What a "dazzling tissue of metaphors" (Lacan 1989: 156–157) a *metaphor* being *what is said and shown, metonymy what is heard and seen. What I give is a metaphor, what you receive is a metonymy. Metaphor is the spark, metonymy the explosive* (286).

To Lacan a "metaphor was a symptom, as desire was metonymy" (Lacan 1989: 175). However, not any conjunction of two signifiers was sufficient to produce the poetic spark of a metaphor. What it took were two signifiers: "one which takes the place of the other while the other remains present through its (metonymic) connexion with the rest of the signifying chain" (Lacan 1989: 157). To Olsson the Chiasm served this purpose as its "flesh and function" were taken as "the signifying elements" (Lacan 1989: 175) running through the transformations of a double helix, an American football, a Moebius band, a DNA string, the currents of an electric generator and the line in the etching of Mellan. Together these signifiers simultaneously inhibited and promoted Olsson's life and served as the phantasms covering up the essentials of his statement.

Therefore: back to the beginning.

The title of the article, *Chiasm of thought-and-action,* focused our attention, first to the middle preposition "of." From that we were inclined to expect a comparison between a chiasm on the one hand side and thought-and-action on the other, the hyphens between 'thought', 'and', plus 'action' indicating that thought and action were taken to be one. This promise was both met and not.

The article had a chiastic structure and presented and compared geometric helices. However, these representations presented understanding and socialization more explicit than thought and action. In that respect it could be argued that the title of the article had been more informative had it been called "Chiasm of understanding-and-socialization" or "Chiasm of understanding-and-action." Yet, I suspect Olsson had more reasons to choose what he did. First, it was of importance to him to avoid the *temptation to thingification* which he found built into every theory and the *attitude that it is not enough to understand the world, but that I must change it as well* (279). Because of that he probably preferred to use the *term 'imagination' to that of 'theory'* (279). Second, when linked to 'action' by the hyphens the term 'thought' is probably not immediately exchangeable with 'understanding'. Instead, 'thought-and-action' perhaps refers to INTENTIONALITY *or more properly "DESIRE"* (286) which was Olsson's own answer to the question: *Where is the action?* (286). In other words, another argument underlining that what the article was *about* and what it illustrated *was* a chiasm of *Desiderare ergo sum* (282). However, and luckily to us, Olsson did not completely avoid thingification but followed Wittgenstein's example when the latter stopped philosophizing and made drawings of architectural constructions instead.

Some find drawings more abstract than words and therefore harder to interpret, others that they are easier to discuss and take apart. In any case, and to this author in particular, they initiated the use of the eye and the index finger, or what Olsson has claimed to be the "metaphors for grasping the distantness inherent in all subject formation" (Olsson 1991a: 48). So, in that and against his declared intention, Olsson changed at least one part of the world: mine.

When Olsson had presented his coordinate system and its axes he served an American football and claimed it to be *the most abstract picture of thought-and-action* (287). I found it a somewhat awkward image to choose by a declared solipsist. But perhaps this choice involved a particular clue. A football simultaneously symbolizes a social game with its own taken-for-granted laws and arguments and incarnates a mimetic desire to relate to others. Moreover, this image did not just fall from the *clouds* (284) or *the Ether of Mount Olympus* (291). Our attention was already directed at least to 'America' and to 'games', although within languages, where Olsson used the Spanish Conquistadores' domestication of the American savages by grammar as an example of socialization (282).

When Olsson finally let the football out he at once called upon *the Name-of-the-Father* and *its double function of law and order, legislation and punishment* (287). Among these fathers Carl von Linné's name and his stamina and pistils were first. Then Anders Celsius and his identification of the boiling and freezing points of water were presented. The stamina and pistils and the boiling and freezing points Olsson forwarded as metaphors for his dream to identify *various forms of silence as fix-points in a set of maps from the Land of Action. The territory to be explored... like America between Cristóbal Colón, and Amerigo Vespucci.* (288). For the benefit of non-geographers it is found relevant to reveal that Cristóbel Colón and Amerigo Vespucci are signifiers not of places but of persons. Colón is the Spanish version of the name Columbus and Vespucci the surname of the explorer of the east coast of South America whose first name, Amerigo, was used to symbolize the new world from 1612 and on. By these comparisons Olsson first unveiled the importance of time in favor of place in his demonstration of desire. Secondly, he demonstrated his awareness of the obstacles he raised for himself when he compared his own accomplishment to those of Linné's and Celcius'. Perhaps the middle part of the article, the mappings, in that respect resemble the burial hymn to little Haiwatha in the middle of Dvorak's New-world-symphony? At least to me it seems as Olsson, by his calling upon the Names-of-the-fathers, did as Antigone when the choir did not understand her reason for breaking Creon's law (Hansen-Møller 1995: 99–101). He consciously entered "le point infranchissable *ou celui* de la Chose" (Lacan 1986: 155). This point Lacan understood as the mediating function between the real and the symbolic. It was always represented by an emptiness, because it could not be represented by anything else or more precisely, because it could only be represented by something else as it referred to an absence. To be in that abyss is probably not to be *on* a limit, rather to be *the* limit, as Olsson described it elsewhere (Olsson 1991a: 132).

The person who finally made Olsson rest at peace was Cézanne when the latter *realized* that he painted *not content but form* (288). However, when or if I 'realize' something I have already left the situation and consider what happened. Therefore, Cézanne did probably not paint content *but* form but rather content *and* form simultaneously like Olsson actually did when he symbolized his visions in words and drawings. Yet, what Cézanne actually lost, according to Olsson's explanation elsewhere, was "perspective, that single point of vision that hitherto ha[s]d stabilized what people saw" (Olsson 1991a: 139). And Olsson continued: "In this act of deconstruction, he rediscovered the art of hieroglyphic writing, where it is so clear that the cultural bounds of meaning are contained within the bodily limits of form." I presume that here we witnessed the incident where the right part of Olsson's brain double-crossed his left. The reason is that when I read "the cultural bounds of meaning are contained within the bodily limits of form" I immediately recognized the link between $\underline{S}$ and $\frac{S}{s}$, i.e. the symbolic, but when I look at Cézanne's oeuvres I rather envision the challenge of the link between the potentials of the sixth sense of the real, $\frac{S}{s}$, and the taken-for-granted continuity of the symbolic, $\frac{S}{s}$. Besides, Olsson probably employed the word 'realized' in both articles because it corresponded with the introductory part of the sentence which he phrased as follows: *It must nevertheless be borne in mind that it is through abstraction that an object becomes more real than the real* (288). Cézanne's desire then, more than Linné's, Celsius', Columbus' or Vespucci's, was the desire of Olsson's Other.

Had I been more attentive or better trained I would probably have discovered this detail earlier as already on the second page of the Chiasm-article Olsson stated that he shared the dreams of *Piet Mondrian and Samuel Beckett* to produce *a minimalist rendering of how we define ourselves, an acknowledgement that linguistic signs are forms* of *art shaped by forms* in *art* (280). Like in Viola's video the point is what slides under the surface of the surface.

Geographer on the balcony,
    artist in the basement.
Nothing new in that.
Desire re-cognized (281).
End of show.
Curtain!

Rubric
Chiasm

Bibliography

Dinsen, A.M. 2004. Natursemiotic og logic ifølge Peirce, in *Semiotiske undersøgelser*, edited by A.M. Dinesen and T. Thellefsen, København: Hans Reitzels.

Doel, M.A. 2003. Gunnar Olsson's transformers: The art and politics of rendering the co-relation of society and space in monochrome and Technicolor. *Antipode*, 35(1), 140–167.

Ebbesen, L.V. 2005. Surrender, in *Bill Viola. Visions*, edited by J.E. Sørensen, Århus: Aros.

Foucault, M. 1983. *This is not a pipe*. Berkeley: University of California Press.

Gren, M. 1994. *Earth writing – exploring representation and social geography in-between meaning/matter*. Gothenburg: University of Gothenburg, Department of Geography, Series B, no. 85.

Grimm, J. and W. 1898. Little Snow White, in *Grimm's Fairy Tales*, edited by Grim J. and W. London: Ernest Nister.

Hansen-Møller, J. 1995. *Den skjulte diagonal. En lanskabsfortælling i ord og billeder.* København: Christian Ejlers.

Hansen-Møller, J. 2004. Landskab: Habitat/Område/Symbol. En model til analyse af meninger med landskab, in *Mening med landskab. En antologi om natursyn*, edited by Hansen-Møller J. København: Forlaget Museum Tusculanum.

Hansen-Møller, J. 2006. The Meaning of Landscape – a diagram for analyzing the relationship between culture and nature based on C. H. Peirce's semiotics.

Place and Location, Studies in Environmental Aesthetics and Semiotics, Vol. 5, 85–108.

Hansen-Møller, J. 2009. Natursyns model: a conceptual framework and method for analysing and comparing views of nature. *Landscape and Urban planning*, Vol. 89, Issues 3–4, 65–74.

Hyldgaard, K. 1998. *Fantasien til afmagten. Syv kapitler om Lacan og filosofien*. København: Museum Tusculanums Forlag.

Jensen, M. 1993. Red River Valley: geo-graphical studies in the landscape of language. *Environment and Planning D: Society and Space*, 11, 295–301.

Jensson, G. 2000. *MAPPA MUNDI UNIVERSALIS a commentary on the power of geographical reason*. Uppsala: Uppsala International Contemporary Art Biennial Eventa 5, Section II – Obog Per apsera ad astra, Uppsala Cathedral, September 4–10.

Krogh, A. 2005. Bill Viola, Visions, in *Bill Viola. Visions*, edited by J.E. Sørensen, Århus: Aros.

Lacan, J. 1975. *Livre XX, Encore 1972–1973*. Paris : Seuil.

Lacan, J. 1986. *Le Séminaire, livre VII: L'éthique de la psychanalyse*. Paris: Seuil.

Lacan, J. 1989. *Ecrits: A Selection*. London: Routledge.

Lacan, J. 1991. *The Four Fundamental Concepts of Psychoanalysis*. London: Penguin Books.

Lacan, J. 1992. Det symbolske, det imaginære og det reelle, 8. juli 1953. *Almen Semiotik*, 5, 8–24.

Lagopoulous, A.P. 1993. Postmodernism, geography, and the social semiotics of space. *Environment and Planning D: Society and Space*, 11, 255–278.

Lennon/McCartney.1967. Being for the Benefit of Mr. Kite! in The Beatles, *Sgt. Pepper's Lonely Hearts Club Band*. Parlaphone, UK.

Magritte, H. 1983. Letter to Foucault May 23 1966, in Foucault, M. *This is not a pipe*. Berkeley: University of California Press.

Miller, J.A. (ed.) 1988. *The Seminar of Jacques Lacan, Book I: Freud's Papers on Technique 1953–1954*. Cambridge: Cambridge University Press.

Nöth, W. 2000. *Handbuch der Semiotik*, 2. edition, Stuttgart: J. B. Metzler.

Nöth, W. 2001. Ecosemiotics and the semiotics of nature. *Sign Systems Studies*, 29(1), 71–82.

Olsson, G. 1980. *Birds in Egg/Eggs in Bird*. London: Pion Limited.

Olsson, G. 1990. *Antipasti*. Körpen: Göteborg.

Olsson, G. 1991a. *Lines of Power/Limits of Language*. Minneapolis: University of Minnesota Press.

Olsson, G. 1991b. Invisible maps – a prospectus. *Geografiska Annaler*, 73B(1), 85–92.

Olsson, G. 1991c. Malevich sfigurato. *Slam*, 3: 7–10.

Olsson, G. 1993. Chiasm of thought-and-action. *Environment and Planning D: Society and Space*, 11, 279–294.

Olsson, G. 1994a. Job and the case of the herbarium. *Environment and Planning D: Society and Space,* 12, 221–225.

Olsson, G. 1994b. Heretic Cartography. *Eucumene, A Journal of Environment, Culture, Meaning*, 1(3), 215–234.

Olsson, G. 1998. Towards a Critique of Cartographical Reason. *Ethics, Place and Environment*, 1(2), 145–155.

Olsson, G. 2007. *Abysmal: A Critique of Cartographical Reason*. Chicago: University of Chicago Press.

Olsson, G. 2011. Min Kant han är tre kanter, en trekant är min kant, in *Sociologik – tio essäer om socialitet och tänkande*, edited by C. Abrahamsson, F. Palm and S. Wide, Santérus förlag: Stockholm.

Peirce, C.S. 1994. *Collected Papers 1931–1958, I–VIII*. Edited by C. Hartshorne, P. Weiss and A.W. Burks. Charlottesville: Past Masters CD-Rom Databases.

Santaella, L.B. 2001. Matter as effete mind: Peirce's synechistic ideas on the semiotic threshold. *Sign Systems Studies*, 29(1), 47–62.

Sparke, M. 1994. Escaping the herbarium: a critique of Gunnar Olsson's "Chiasm of thought-and-action." *Environment and Planning D: Society and Space*, 12, 207–220.

Stoichita, V.I. 1997. *A short History of the Shadow*. London: Reaktion Books.

Sørensen, J.E. 2005. *Bill Viola. Visions*. Århus: Aros.

Uexküll, J. v. 2001. An introduction to Umwelt. *Semiotica*, 134(1–4), 107–110.

Wittgenstein, L. 1993. *Tractatus Logico-Philosophicus*. København: Gyldendal, Moderne tænkere.

WIKIPEDIA < http://en.wikipedia.org/wiki/Juggling> downloaded 11/04/2008a.

WIKIPEDIA<http://en.wikipedia.org/wiki/Chiastic_structure#The_ABC.E2.80. A6CBA_chiastic_structure> downloaded 08/07/2008b.

WIKIPEDIA Guillery R.W. and J.H. Kaas. Genetic Abnormality of the Visual Pathways in a "White" Tiger. *Science*. 180(92):Jun. 1973: 1287–89 and Guillery, R.W. Visual pathways in albinos. *Scientific American*. 230(5): May 1974: 44–54. <http://www.sciencemag.org/wiki/Optic_chiasm> downloaded 17/04/2008c

WIKIPEDIA, WATSON & CRICK, *Nature* 171, 1953: 737 <http://en.wikipedia. org/wiki/Double_helix#cite_note-2> downloaded 08/08/2008d

WIKIPEDIA <http://demonstrations.wolfram.com/DoubleHelix/> downloaded 04/09/2008e.

Part D
MAPS

Chapter 15

Projection of Desire / Desire of Projection[1]

Gunnar Olsson

In this performance I shall pay yet another visit to my own world of thought-and-action, a world which to me has many affinities with the overall theme of the present volume as a whole. It will quickly become clear why I take that world to be inherently interesting.

To understand why, just pay close attention to that last word itself. <u>Interesting</u>! Have a taste of that expression, take it in your mouth, press it with your tongue against the inside of your own body, and it will reveal its secret through its own etymology. For the word 'interest' comes from Latin 'inter esse', two words which literally mean 'in-between-being'. To say it clearly: to be inter-esting is to be in the razor-sharp limit of ontological categories.

These alternative categories of being carry different aliases in different contexts. But they all stem from two families only. One is called MIND, the other MATTER; as with other Mafia families the one is impossible without the other, the other impossible without the one. It is the local feud of these two families that this performance is about; a feud full of desire, a desire that never can be satisfied, a series of monologues in which neither side will get the last word. What the struggle is about is the eternal question of what it means to be human.

* * *

For as long as we know – perhaps since the foundation of the world – man has tried to define who he is. In that struggle with definitions he has been forced to confront two types of silence, one lodged in the pure spirituality, the other in the pure physicality. At one border is consequently that spirit which is so extremely spiritual that it lacks all forms of expression. Even its name is taboo; like the name of the god of the Old Testament it must not, and cannot, be mentioned. As recalled, the Hebrew JHWH means 'I am who I am', a tautology, by definition true but not informative. At the other border is the stone or rock which is so extremely physical that no matter how much I try, it gives away no meaning, no associations.

---

1   Non-revised notes for a lecture given in the series "Räumliches Denken", ETH, Zürich, February 1, 1996. Translated and published as "Die Projektion des Begehrens/ Das Begehren der Projektion", in *Räumliches Denken*, Hsrg von Dagmar Reichert. Zürich: Vdf, 1996.

And in that refusal it can sometimes itself become a silent god; reification turns to deification, deification to reification.

To be the semiotic animal that the human being inevitably is, is to be involved in a sort of two-front war, whose purpose is to determine the boundaries toward the utterly alien. More specifically, one of the boundary disputes is about the difference between man and the gods. The other concerns the distinction between man and the beasts.

As an early illustration, just recall the story of King Oedipus. Who is that creature? Is he the equal of the gods or is he nothing at all? Is he the king or is he an incestuous murderer? His great sin is in fact that he is a slayer of distinctions, father and son in one, husband of his mother, child of his wife, brother of his children. An inter-esting figure if one every was, for if anyone lives in the land of in-between, then it is certainly he; Oedipus as a paradigmatic inhabitant of the ontological realm that is one with its name: INTER ESSE.

Mapping this imaginary territory involves the drawing of two lines, the settling of two boundaries. Both of these lines are receding horizons of silence, two versions of Ludwig Wittgenstein's proposition number 7: "whereof one cannot speak, thereof one must be silent". One of the lines marks the boundary to the gods; it is meaning without expression. The other marks the boundary to the stones; it is expression without meaning. The resulting map looks like this:

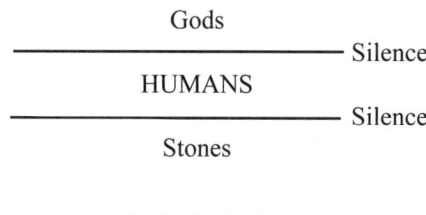

Gods

——————————————————— Silence

HUMANS

——————————————————— Silence

Stones

*   *   *   *   *

History is of course replete with attempts to reach a final solution, to end once and for all the eternal struggle between the gods and the stones, between the flesh and the spirit. The examples are legion, and most of them are more closely related to aesthetics than to ethics. For by what would the true artist be obsessed, if not by the impossible dream of ending the ontological family feud; of merging expression and impression, impression and expression; of a desire to make visible the invisible, to say the unsayable, to transcend the limits of language. Put differently, the arts, mythology, religion and politics all grapple with similar problems of representation.

But perhaps the most remarkable of all remarkable attempts to end the ontological struggle is in the social invention known as Jesus Christ. In making this suggestion, I am thinking less of the historical figure, whose life and death is retold in the Gospels of the New Testament, and more of how that same creature came to be defined in the various ecumenical councils, especially in the Creed of Niceae from the year 325.

In that powerfilled document – in that peace treaty from years of hard negotiation – it is (with later amendments and emphases added) decreed as follows:

I believe in one God the Father almighty; Maker of heaven and earth, and of all things visible and invisible.

From that beginning, the Confession then moves on to the second article. It is there that we find what I take to be the most unambiguous projection of desire ever to be offered. In part the definition is as follows:

And (I believe) in one Lord Jesus Christ, the only begotten Son of God, begotten of the Father before all worlds, Light of Light, very God of very God, begotten, not made, being of one substance with the Father; by whom all things were made; who for us men and for our salvation, came down from heaven, and was incarnate by the Holy Ghost of the Virgin Mary, and was made man....

And then, finally, the third article according to which

I believe in the Holy Ghost, the Lord and the Giver of Life; who proceedeth from the Father and the Son; who with the Father and the Son together is worshipped and glorified....

*

The translation from the words of the Confession to the lines of the map is straight-forward. It looks as follows:

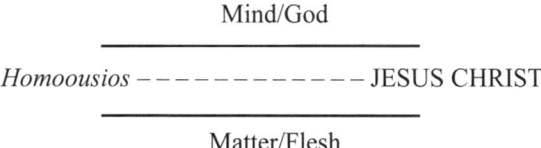

Mind/God

Homoousios – – – – – – – – – – – JESUS CHRIST

Matter/Flesh

Now it should be noted explicitly that the baptizing of the inter-esting line as 'JESUS CHRIST' hinges on the Greek word *homoousios*, which can be translated into English as 'of one substance'. Put differently, the definition is such that Jesus Christ at the same time, and without separation, is both god and flesh, both matter and mind. A paradoxical figure in constant limbo, a theological entity located in the thin transition zone between ontological categories, a lonely inhabitant in the no-man's land of INTER ESSE. In that position he is by necessity beyond common understanding, reachable only through an individual, Kierkegaardian, leap of faith.

And yet. What cannot be said, can sometimes be shown. For even though you and I never can be one with the thin-thin line in-between-being, we may nevertheless asymptotically approach it. The ideal time for so doing falls between Easter and Pentecost, those fifty days when Jesus himself is floating around, when it is difficult to determine exactly where he is, indeed whether he is at all. Is he dead or alive, is he ascended into heaven or is he buried in the ground? Is he real or imaginary? Does he dwell in the world of decaying matter or in the world of the Holy Spirit? A paradigmatic case of genuine ontological uncertainty, if such as case can ever be imagined. Like a Giacometti sculpture, the Jesus of Niceae appears as the epitome of the sublime: tentative without being vague.

But the rebirth – the Renaissance – is near, for Pentecost or Whitsunday commemorates the descendence of the Holy Spirit on the disciples. And for this very reason, Whitsunday is the most appropriate day to be baptized, for just as Christ is then reborn into a new existence, so the convert, through the sacrament of baptism, is entering into a new life. Here, as elsewhere, naming is the name of the game; with the new name as his pass-word, the believer is allowed to cross from one category to another. And the color of the garment is white, a color that according to Wassily Kandinsky is not only silent but vertical: the convert stripped bare by her baptizers, even.

But I am moving too quickly. I must immediately return to the task of mapping the invisible mindscape of desire. What remains is to retrace the path of Jesus, to follow his wanderings between Easter and Pentecost. What I must do is to determine exactly and without ambiguity the inter-esting positions he occupies when he is in ontological limbo.

* * *

Everyone remembers the story. In the morning of the first weekday after Jesus had been buried, Maria Magdalena – the woman that gossip claimed was his mistress – went to the grave. Frightened stiff she discovered that the stone at the opening of the tomb had been removed. The tomb itself was empty except for the linen cloths that had been folded together and put aside. But as the mourning Maria stood there weeping, she turned around and there he was, Jesus himself. When she approaches him to embrace him, he pulls back and warns here: "Noli me tangere, don't touch me! I have not yet ascended to the Father in heaven."

When Jesus utters these words, when he warns Maria not to come too close, it is obvious where he is in the ontological landscape. He has clearly left his position in the limit between categories and is on his way into a new, more spiritual, ontological status. In my interpretation, he forbids Maria to touch him, for in the ontological position he now occupies, there is in reality no flesh to touch, only spirit to believe. More specifically:

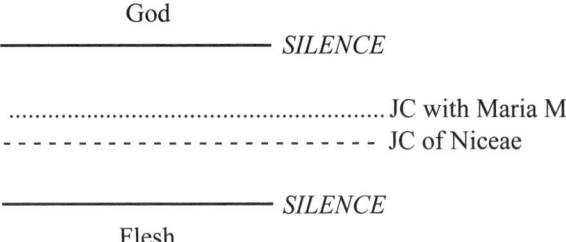

And then there is evening the same day. The frightened disciples have locked themselves inside a house in fear that the Jews will arrest them as well, just as they earlier had arrested and killed their teacher. Suddenly, without warning, he stood among them, he breathed on them and he uttered "Receive the Holy Spirit".

But Thomas, one of the twelve, he who was called Didymus or the Twin, he was not in the house with the others when Jesus came. On his return he was of course immediately told what had happened, for what could anyone tell that would be more remarkable than this. – "He has been here! He has been here!" – "Who has been here?" Thomas asked. – "Jesus! Jesus himself. We have seen the Lord!" – "I don't believe you," said Thomas, who subsequently was nick-named 'the Doubter'. – "Jesus is as dead as a stone. You are hallucinating, you have seen a ghost. But I would believe, if I could put my finger through the holes left by the nails in his hands, if I could stick my hand into the wound from the lance in his side. Let me be frank: I put more trust in my own body than in your words."

Of this Jesus of course came to know, for nothing can be kept secret from the inner court of the Father Almighty. Eight days later he returned to settle the case. "Thomas," he said. "I hear about your doubts. Now, come here. Come, come! Fear not! Your finger, please. My hands are here, the wound is in my side. Still your doubts! Touch and you shall trust!" After this powerful demonstration of the power of the example, Thomas had no choice but to submit. – "My Lord, my God." – To which he received the epistemological reply: "You believe, Thomas, because you have seen me with your eyes, heard me with your ears, touched me with your hands. You believe because of your five senses, because of your body. But blessed are those who believe because of their sixth sense, because of their minds."

When Jesus reveals himself to Thomas, it is easy to place him on the emerging map of desire. He has now moved from the spiritual to the material borderland, he has jumped from one side of the inter-esting line between categories to the other. To be more exact, Jesus is now somewhere between the definition from Niceae and the silence of the flesh:

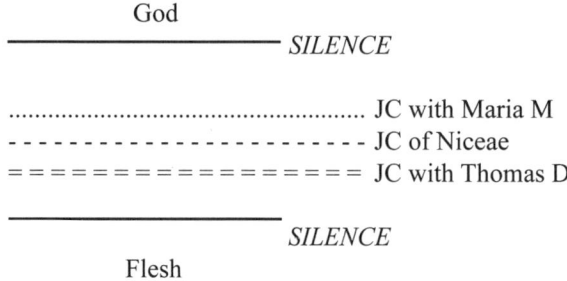

God
————————————— *SILENCE*

.... JC with Maria M
- JC of Niceae
= = = = = = = = = = = = = = = = = JC with Thomas D

————————————
                        SILENCE
Flesh

*

After the success of this rhetorical trickery, Jesus showed himself once more, this time to some of the disciples. It happened by the Sea of Tiberias, where the followers found themselves worn, hungry and not so little afraid. To get food they went fishing, albeit without luck. And then, just as the day was breaking, what looked like Jesus stood there on the beach saying onto them "Cast the net on the right side of the boat and you will find fish." They did as they had been told and they caught so much that they had difficulties hauling it in.

This was the third time Jesus revealed himself to the disciples. For each time he became more and more materialistic, less and less Kierkergaardian, more and more Marxian. On the map of desire, the incidence at the Sea of Tiberias is located between the encounter with the doubting Thomas, on the one hand, and the boundary to the silent flesh, on the other. In summary:

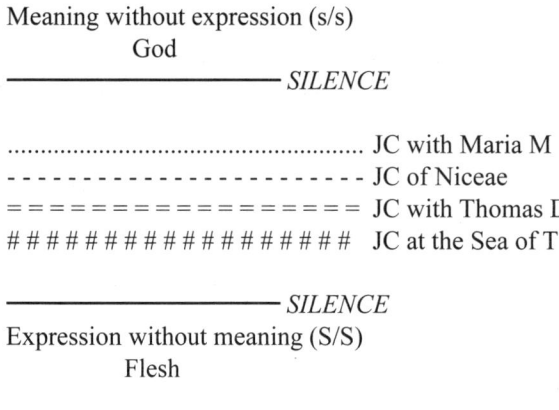

Meaning without expression (s/s)
                God
————————————— *SILENCE*

.... JC with Maria M
- JC of Niceae
= = = = = = = = = = = = = = = = = JC with Thomas D
JC at the Sea of T

————————————— *SILENCE*
Expression without meaning (S/S)
                Flesh

*   *   *

Done is done. No apologies offered, no apologies required. Yet I shall be the first to acknowledge that my mappings of the events between Easter and Pentecost are based on an unusual, perhaps heretic, reading of what is rightfully called the *Book of Books*. And for this reason I wish to stress that my intention has not been to be blasphemous or offensive, merely to note that today's issues of representation are not new. The connections between mind and matter, body and culture, reification and deification, do in fact lie at the heart of Western culture itself, perhaps of all cultures.

As it is now usually defined, the problem is essentially one of power and communication. More pointedly, the question is how I am believed when I claim that something is something else; the Greek terms *aletheia* and *pistis* are always closely related. In English, this connection shows itself in the fact that the term 'true' comes from 'trust', both deriving from the common Indo-European root *deru*, which means 'tree'. And so it is that when I claim that a statement is 'true', then I am not saying that it is 'factually' correct, but that it is 'socially' correct. Put differently, a given statement is held to be true if it is trustworthy, i.e. if it is shared by a sufficient number of others who for various reasons are taken to know what is right and wrong, black and white.

The conclusion is obvious: reality is not what it seems to be, for reality and language are never one and the same. Already Polyphemos, the cyclops, had to learn the painful lesson that there is a decisive difference between Signifier (S) and signified (s); even if somebody is *called* 'Nobody' that does not mean that he also *is* nobody. The cunning Odysseus was well aware that the semiotic animal is thoroughly paradoxical, for it can be what it is only by being what it is not. But he also understood that the rhetorical animal is thoroughly ironic, for it can be believed only when it says that something is what it is not. To tell the truth is consequently to be trusted. As the example of Jesus Christ in limbo illustrates so well, to be trusted is to be integrated into the taken-for-granted of a specific culture.

* * * * *

It was exactly his relation between truth and rhetoric that the Franciscan friar Roger Bacon understood, when he sometime around the year 1260 wrote his famous letter to Pope Clement IV. Bacon's argument was that if the Christians were to win the fight against the Muslims, then they had to learn to depict the world not as it actually is, but as it seems to be. His recommendation was in fact that the Church in its propaganda should draw on the knowledge of Euclidean geometry and the principles of optics which at that time was being made available through the works of the various translation schools. Bacon's rhetorical advice was that the pictures of Jesus, Maria and the Lord of Hosts would be more credible if they were painted to look as three-dimensional figures taken from real life and not as flat pancakes taken from the hierarchy of theology. And yet it should be noted that his proposals were founded in theological reasoning as well; since God had created Light already on the first day, it could be surmised that optics offers a privileged approach to the Almighty.

The basic idea that Bacon shared with his predecessors – including both Plato and Aristotle – was that the sense of sight opens the doors to the Command Center of Culture, to the secret codes of the socially taken-for-granted. It follows, as we later will see, that the modern subject is part and parcel of the rhetorics of the eye and thereby of the ontological drift to thingify. Not as strange as it may sound, for geometry is *the* discourse that specializes in the study of the outer side of physical bodies. According to the geographer Immanuel Kant, geography is in reality a codification of the intuition of the tactile. And yet it should never be forgotten that codification always comes too late; even though codification is an unavoidable part of life, its roots are in autopsy.

And for this very reason I now wish to return once again to Firenze and to that beautiful morning in 1425, when the self-taught builder Filippo Brunelleschi placed himself and his easel in the doorway of the cathedral of Santa Maria dei Fiori. For in the prototype of the perspective that he then produced, it became evident that the real and the symbolic cannot be separated. What Brunelleschi actually did in his Lacanian reversal of the image five centuries before Lacan was to show that the vanishing point of the painting becomes one with the viewpoint of the painter. The conclusion is inescapable: there is something viewing my viewing, something watching my watching; what I happen to see depends on the viewpoint from which I see it.

I share with Hubert Damisch the interpretation that Brunelleschi in his conversion happened to stumble upon one of the most important inventions (or was it a discovery) ever to be made – a social invention perhaps as important as the mechanical invention of the wheel. For what he did, without realizing it, was nothing less than to invent the modern subject – a subject condensed into a dimensionless vanishing point, a dematerialized point in which object is transformed into subject, a moving point which, like the power it symbolizes, never sits still. The crucial difference is that whereas Greek geometry dealt with the things of the finite world, both the Renaissance and the Enlightenment were occupied with the relations of infinity; while Euclid proved theorems, Descartes solved problems.

But already Brunelleschi – who lived in-between – appreciated that the challenging problem was not in the depiction of the checker-board pattern on the piazza or the solid walls of the Battisteria di San Giovanni. The challenge was instead in the sky and the clouds, amorphous and constantly drifting. The lesson has subsequently become clear: the *perspectiva naturalis* of optics is not the same as the *perspectiva artificialis* of the arts; perspective as body is not the same as perspective as culture; the eye of the first sense is not the same as the eye of the sixth.

*   *   *   *   *

This difference is well illustrated by the fascinating and rich history of the self-portrait. That history is for me extremely instructive, not only because it is closely tied to the history of the perspective and thereby to the projection of desire. What is particularly noteworthy is that before the Renaissance there were no real self-portraits, for the simple reason that before the rebirth there was no self to portray. But in all the classical self-portraits, from Dürer to Rembrandt, from Goya to van Gogh, from Munch to Kahlo, the voyeur and the exhibitionist are merged into one. Narcissus meets Nietzsche, Nietzsche meets Duchamp.

*

And for these reflective reasons, I would like to close this performance with some brief amateurish remarks about Marcel Duchamp, in my estimation *the* outstanding investigator of the projection of desire and the desire of projection. If anyone, this paradoxical ironist, alchemist and nominalist devoted his entire life to the exploration of the unknown territory of INTER ESSE. If anyone, this social individual was fascinated by the phenomenon of transition rites, not the least by the passage from virgin to bride. If anyone, it is he who spans the abyss between my mapping of the Jesus Christ of Niceae, on the one hand, and the rhetorical devices or Roger Bacon and Filippo Brunelleschi, on the other.

The best known example of Duchamp's obsession is of course in the *Large Glass*, a work he began in earnest in 1915 and abandoned unfinished eight years later – *La Mariée mise à nu par ses célibataires, même*, i.e. 'The Bride Stripped Bare by Her Bachelors, even'.

Figure 15.1   Marcel Duchamp: "La Mariée mise à nu par ses célibataires, même," 1915–23

Given the theme of this lecture, it is important to realize that the glass itself can be interpreted as a vertical window *through* which the artist sees the world and *onto* which he projects his desire. As in the case of Dürer's woodcut of *Man Drawing a Reclining Nude*, Duchamp tried to represent a three-dimensional reality which stands on the horizontal plane behind the glass. But when that reality is projected onto the backside of the vertical glass, one dimension is inevitably lost and the projection becomes two-dimensional. Here it should be noted explicitly that the glider, sometimes called the chariot, hits the glass with such force that it seems to break through the projection screen and stick out on the other side. The parallels to Alberti, Leonardo and Desargues are obvious and it is remarkable how meticulously – how precisely – the constructions in the lower, the bachelor half of the glass, were executed. For that and other reasons, the bachelor's domain was painted with the aid of distancing tools like the brush, compass and ruler, and with connecting devices like threads to keep everything in place; put differently, the bachelors' world is a tautological world reminiscent of Brunelleschi's piazza. In contrast, the bride's domain is like the clouds in the sky, indeed it is dominated by a flesh-colored milky wave with three draft pistons. And yet one must never forget that Duchamp's key theme was the projection of desire, rendered, as he put it himself, as a 'delay in glass'.

I have already mentioned that the lower half of the glass should be understood as a two-dimensional depiction of a three-dimensional reality. It is equally clear that the upper half is an attempt to depict a four-dimensional reality. In addition, all exchange between the bride and the bachelors occurs entirely in the vertical direction, through movements up and down, a fluctuation reminiscent of the different rhetorical strategies of Jesus Christ between Easter and Pentecost.

The erotic power that drives the exchange between the two domains is well described in Duchamp's notes. However, it is not explicitly included in the *Glass* itself. As he explained it himself: "The wheel inside the glider is presumably driven by a waterfall. However, I did not care to include it in the picture, for I did not want to be caught in the trap of becoming a landscape painter." In Ulf Linde's important interpretation, the waterfall is assumed to flow out of a higher dimension and can therefore not be captured by the three-dimensional section of the bachelor machine. Put differently, the interpretation is that the given of the *Large Glass* is in the cascading spring of desire; in my own vocabulary it is in the unavoidable and taboo-ridden taken-for-granted. And for this very reason it *must* be excluded, for if it were included, then the tautological world would no longer be tautological, then the unconscious would have moved into the conscious, then the real world would have become symbolic, the solipsistic the political, the proper name a definite description.

And yet, the very idea of the *Large Glass* is to show that even though desire never ceases to desire, desire can never be fulfilled. The penumbra through which the two domains are touching each other is in the three isolating plates that divide the glass into its two halves. Not surprisingly, these delaying glass staffs are sometimes called the 'cooler', sometimes the 'bride's clothes'. And on the back, between the two upper isolation plates, is the vanishing point of the entire bachelor apparatus. In my own world of semiotics, the horizontal line of the upper-most plate coincides with the fraction line of the sign, with the revel of the Saussurean Bar. Maybe there are more parallels between the *Large Glass* and the Creed of Niceae than meet the eye.

Add finally to this comment one further observation with direct relevance for the present performance. This is that the three isolation plates differ from each other in one crucial respect. Thus, the lower plates have a greenish color – a color Duchamp thoroughly disliked – while the upper-most plate is colorless. From other evidence it is quite clear that the three coolers represent the three brothers Duchamp; the green plates stand for his two brothers, who once cowardly had let him down, while the colorless is Marcel himself draped in the white veil dropped on him by the stripping bride.

Thus it is that perhaps also the *Large Glass* may be viewed as a self-portrait, an interpretation directly supported by its own title, 'La Mariée mise à nu par célibataires, même'. Add to this that when Duchamp presented himself as an androgyne figure dressed in women's clothes – which he often did – then he always used the pseudonym *Rose Sélavy*; like so many of his titles, a name which simultaneously says and shows: *Éros, c'est la vie*. At the same time – and for the same reason – it should be noted that he often commented that the part of the *Glass* with which he was least satisfied was the three dividers, the coolers. After many years in the Saussurean Bar, I am not surprised.

**Figure 15.2   Man Ray: "Photograph of Marcel Duchamp as Rose Sélavy,"
ca 1920**

And thus it is that I begin to suspect that the really given in the *Glass* is not in the waterfall, which the artist consciously excluded, but instead in the horizon he never could reach, in the complete merger of opposites that noone ever can achieve. For in this, language, perspective and color are alike: neither can be controlled by the sense of touch and that holds for everyone, even for geniuses like Jesus Christ and Marcel Duchamp. The limits of the body are the limits of the arts.

*

On this condition, the ironist Duchamp was more acutely aware than most. As a consequence, his numerous ready-mades clearly demonstrate that aesthetic judgement is anchored not in the physical phenomenon *per se*, but in the social context which makes it acceptable; not in the plumber's urinal, but in the artist's 'Fountain', not in the Kantian 'here' and 'now', but in the thinnest thinness of the inter-esting line in-between. As recalled, 'true' and 'trust' spring from the same root.

Figure 15.3   Marcel Duchamp: "Fountain," 1917

But Duchamp himself was more radical than this, a fact he demonstrated in his *Etant donné: 1 la chute d'eau, 2 le gaz d'éclairage*, i.e. "Given: 1 the waterfall, 2 the illuminating gas". In that summarizing summary of his entire life – a conclusion he secretly worked on for twenty years, from 1946 to 1966 – he goes further in the cultivation of the perspectival heritage and the play of dimensions than ever before. The *Etant donnés* had in fact been described as the alter ego of the *Large Glass*.

Figure 15.4   Marcel Duchamp: "Etant donnés: 1 la chute d'eau, 2 le gaz d'éclairage," 1946–66. Exterior view

What the visitor here meets is a wall with a walled-in door, a door which in effect is not a door but a window. It comes from Spain, where Duchamp found it on one of his visits and, like so many of his works, it has the proportions of the golden section. Since what looks like a door has no key, no handle, no hinges, it cannot be opened. As you approach it, however, you discover two small holes, drilled at eye level and placed seven centimeters apart. Only the most courageous dare to come close, dare to put their eyes to the holes, dare to turn the closed door into an open window. And from that solitary observation point which nobody can share, the viewer discovers a world of stunning beauty. In the background is a two-dimensional landscape painting with an electrically driven waterfall, in the foreground a three-dimensional full-sized nude on a bed of dried leaves and twigs. So strong is the erotic power that it lights the gas lamp in the nude's left hand.

Figure 15.5   Marcel Duchamp: "Etant donnés: 1 la chute d'eau, 2 le gaz d'éclairage," 1946–66. Interior view

And when I stand there as the voyeur I am, then I discover that I am totally alone. For whereas the bride of the *Large Glass* can be stripped and shared by many bachelors at the same time, the *Given* is for one inspector only. While Duchamp in the former work transformed the wall of Plato's cave from impenetrable limestone into transparent glass, he demonstrated in the latter how the projection screen of the taken-for-granted is one with the viewer's own eyes. Looked at from the side, the *Large Glass* is reduced to a one-dimensional line, a cut of anything whatever; like a Swiss watch in profile, it shows no time. In contrast, the *Given* refuses all attempts of reduction. Every teenager knows where to look for the vanishing point. Like every logician he also knows that at the moment of truth, tautology cannot be rephrased.

But not only solipsistic masturbators can recognize themselves. For hidden under the nude's bed, invisible to the Peeping Tom, there is a floor covered by a linoleum of black and white squares; the artist's reverence to ancestors like Brunelleschi and Masaccio, Vélazquez and Vermeer. To him perhaps yet another reminder that there can be no child without parents. To us a reminder that Duchamp himself was a devoted chess player, an individual who never wasted time, a person who always cultivated chance with exactitude, a man so obsessed by prime numbers that he became a prime in himself. And his favorite word in English was 'amusing'.

* * *

And so it may perhaps be that also *Etant donnés* can be interpreted as a self-portrait, a life-long study of the relation between body and culture, spirit and materiality, the earth and the sky.

The vanishing point of perspective is by necessity a dematerialized point. The viewpoint of the subject is by the same necessity unique. Roger Bacon may well rotate in his grave. But what continues to live on is his question: "How can I be believed when I tell the truth of the point which can be neither seen nor touched? How do I share with others the desire I experience when I put my eyes to the holes in the door which is not a door?"

* * * * *

Clair, Jean: "Duchamp and the Classical Perspectives," *Artforum*, Vol. 16, No. 7, 1978, pp. 40–49.

Damisch, Hubert: *The Origin of Perspective*. (French original, 1987) Cambridge, Mass: MIT Press, 1994.

Derrida, Jacques: *On the Name*. (French original, 1993) Stanford: Stanford University Press, 1995.

Duchamp, Marcel: *Duchamp du signe*. (Réunis par Michel Sanouillet) Paris: Flammarion, 1994.

Dufour, Dany-Robert: *Les mystères de la trinité*. Paris: Gallimard, 1990.

Edgerton, Samuel Y. Jr: *The Heritage of Giotto's Geometry: Art and Science on the Eve of the Scientific Revolution*. Ithaca, N.Y.: Cornell University Press, 1991.

Elkins, James: *The Poetics of Perspective*. Ithaca, N.Y.: Cornell University Press, 1994.

Farinelli, Franco, Gunnar Olsson and Dagmar Reichert (eds): *Limits of Representation*. München: Accedo, 1993.

Focillon, Henri: *The Life of Forms in Art*. (French original, 1934) New York: Zone Books, 1989.

Gough-Cooper, Jennifer and Jacques Caumont: *Marcel Duchamp*. (Exhibition catalogue, Palazzo Grassi, Venezia) Milano: Bompiani, 1993.

d'Harnoncourt, Anne and Walter Hopps: "Etant Donnés: 1 la chute d'eau, 2 le gaz d'éclairage: Reflections on a New Work by Marcel Duchamp," *Bulletin of the Philadelphia Museum of Art*, Vol. 64, Nos. 299–300, pp. 6–58.

d'Harnancourt, Anne and Kynason McShine (eds): *Marcel Duchamp*. New York: Museum of Modern Art, 1973.

Henderson, Linda Dalrymple: *The Fourth Dimension and Non-Euclidean Geometry in Modern Art*. Princeton: Princeton University Press, 1983.

Jensen, Ole Michael: *Vaegge: Kundskabens projektioner/ Walls: Projections of Knowledge*. Copenhagen: Scandinavian University Press, 1995.

Koerner, Joseph Leo: *The Moment of Self-Portraiture in German Renaissance Art*. Chicago: University of Chicago Press, 1993.

Lacan, Jacques: Écrits: *A Selection*. (French original, 1966) Andover: Tavistock, 1977.

Lachterman, David R.: *The Ethics of Geometry: A Genealogy of Modernity*. Chicago: University of Chicago Press, 1987.

Linde, Ulf: *Marcel Duchamp*. Stockholm: Rabén & Sjögren, 1986.

Olsson, Gunnar: *Lines of Power/ Limits of Language*. Minneapolis: University of Minnesota Press, 1991.

Olsson, Gunnar: "Heretic Cartography," *Ecumene*, Vol. 1, 1994, pp. 214–234.

Panofsky, Erwin: *Perspecive as Symbolic Form* (German original, 1924–25) New York: Zone Books, 1991.

Rotman, Brian: *Signifying Nothing: The Semiotics of Zero*. London: Macmillan, 1987.

de Saussure, Ferdinand: *Course in General Linguistics* (French notes, 1906–11) London: Duckworth, 1983.

Seigel, Jerrold: *The Private Worlds of Marcel Duchamp: Desire, Liberation and the Self in Modern Culture*. Berkeley: University of California Press, 1995.

Serres, Michel: *Atlas*. Paris: Juillard, 1994.

Wittgenstein, Ludwig: *Tractatus Logico-Philosophicus*. (Bilingual original, 1922) London: Routledge & Kegan Paul, 1961.

* * * * *

Chapter 16
Gunnar Olsson and Me

Trevor J. Barnes

I first heard Gunnar Olsson speak at the Department of Geography, University College London (UCL), as a third year undergraduate in the spring of 1978. In my last year, I would sneak into Departmental seminars supposedly reserved for staff and post-graduates. I remember seeing Ray Pahl and Alan Baker present, both of whom I read and were excited about, but who in seminar mode were rather dull and dry. Then just before Finals, I saw that Olsson was speaking. I knew I should study for my exams, but Olsson already interested me. Earlier in the same year in a philosophy and methods seminar, a post-graduate student dressed up as a bus conductor declaimed to the class selected passages from a dog-eared copy of Olsson's (1975) *Birds in Egg*. He said the passages proved the end of causality. I wasn't so sure, but I wanted more, and which is what I got that spring day when Olsson came to UCL.

Olsson was in the process of making a slow farewell tour back to Sweden to take up a position at Nordplan, Stockholm. For the previous ten years he was in America teaching at the University of Michigan, Ann Arbor. Alan Gilbert introduced him at UCL. Olsson was tall and lean, with shaggy long blonde hair and a beard. He looked like a Viking. As soon as he spoke he instantly commanded the room. There was nothing dull and dry about him. With no notes, he paced up and down at the front of the stage like some caged animal, acting out his paper's argument. His dramatic portrayal of the "hermeneutic moment" was a *tour de force*. Not that I could always understand what he said. His deep voice, Swedish accent, and tendency to laugh at his own jokes before he finished telling them, meant I missed a good quarter of what he said. But then I didn't understand Marlon Brando either, and I thought he was great. The physical performance and the words of Olsson's I did catch were enough anyway. It wasn't an argument for the end of causality, but an argument for being clear about ambiguity. That was what all the pacing between one side of the room and the other was about, along with extravagant hand gestures, face grimaces and piercing staring eyes. He was performing the various tensions in life itself that produced ambivalence, and which lay for him at the very centre of the human condition. These were the tensions between thought and action, fixity and change, form and process, certainty and doubt, determinateness and chance, and even between the equal sign and the slash of the dialectic. As he says in the interview below, he gave the best years of his life to understanding these tensions. And you could tell on that day watching him. He was incandescently brilliant. Even the man dressed up as a bus conductor declaiming *Birds in Egg* dimmed in comparison.

Later that same year I left for graduate school in America. But the exposure to Olsson haunted me. My graduate supervisor at the University of Minnesota, Eric Sheppard, had on his bookshelf Michigan Geographical Publication, number 15, *Birds in Egg* (and which predictably had not been available at the University of London library). I immediately grabbed it, a giant door-stopper of a book, taking it in my first year and not returning it until five years later when I left for the University of British Columbia to take up my first academic job. I must have read it two or three times. I continuously went back to it to steal quotes and to check subtle philosophical points. I admired as much the writing in the book as its substantive content. There are not many brilliantly written books in Anglo-American human geography. Bill Bunge's (1971) *Fitzgerald* is one, and in its quiet elegance, Peter Haggett's (1965) *Locational analysis in human geography* is another. And Olsson's *Birds in Egg* is a third. The prose is shockingly good: funny, rich, learned, profound, purposeful. When I later spoke to Olsson about writing *Birds in Egg*, he said that precisely because he was not a native speaker of English he felt he could take chances, saying things that he would never dare say in Swedish. Like Joseph Conrad, another non-native writer of English language prose, Olsson produced in *Birds in egg* matchless writing, technically flawless, but not quite English. It was the "not quite English" that made it so compelling.

I interviewed Olsson because I was doing a project collecting oral histories of pioneers involved in geography's quantitative revolution. Of course, I had secretly wanted to speak to him for years. But I never had the temerity to ask. My project provided the perfect entrée, however. We agreed to meet at his Pittsburgh hotel room at the 2000 annual meeting of the Association of American Geographers.

Olsson, like David Harvey, had been a pivotal figure in geography's quantitative revolution. He had begun as a believer, but within a few years became one of its most trenchant critics. His early qualifications as a revolutionary were impeccable. As an undergraduate, as he says in the interview, he was captivated by the teachings of a young geography professor at the University of Uppsala who used "regression models," and which for Olsson were "beautiful," "open[ing] up a world." In 1961–62 as a graduate student he met and befriended David Harvey, and already beginning to gather ideas for writing what would be the philosophical bible of the quantitative revolution, *Explanation in geography* (Harvey 1969). During 1963–1964 he attended one of the temples of North American spatial science, Walter Isard's Regional Science Department at Penn. As a result of that visit, Olsson (1965) wrote the now classic interpretive review of one of the key theoretical and empirical contributions of geography's quantitative revolution, spatial interaction models. As a result of these impeccable qualifications, he was offered a job at the illustrious Department of Geography, University of Michigan. Among its faculty were John Nystuen and Waldo Tobler, two of the original "space cadets," the name given to the graduate students at the University of Washington who in the late 1950s spearheaded the quantitative revolution.

Olsson's experiences and credentials were unimpeachable. He should have been a dye-in-the-wool quantifier. But he wasn't. From almost the beginning of his Michigan appointment the worm turned. In the courses he taught such as the one he talks about in the interview about "Thought and action," in the graduate students he supervised and who included Bonnie Barton, Stephen Gale, and Michael Watts, and in his Ann Arbor living room "Salon" (as John Hudson calls it), he struck out in a different direction. It was away from certainty, fixture, determinateness, and absolutes to the treacherous shoals of ambivalence, mutation, chance, and relationality. His guides were no longer that young mathematically inclined Uppsala professor, Walter Isard, or the former "space cadets," but philosophers and artists who had deliberately turned their cheek to feel the full gale of contradictions and uncertainties of being human, and who included Samuel Beckett, James Joyce, Søren Kierkegaard, Ludwig Wittgenstein, and the writers of the Bible and Greek epic poems.

Not that Olsson stopped being a geographer. Initially the new Olsson met the old Olsson with spatial interaction models getting the *Birds in Egg* treatment (and becoming *Eggs in Bird*; Olsson 1980). Later the geography took the form of a discussion of *Lines of Power* (Olsson 1991). The interview below, however, is confined to Olsson's early years and the period leading up to that moment in spring 1978 when Gunnar Olsson changed my life.

Interview with Gunnar Olsson

Pittsburgh, PA, 7 April, 2000

TB: I wondered if you could begin by talking about your early life.

GO: My very early life? As a child?

TB: If you think it is relevant.

GO: Most of what I have to say about that is in my contribution to *Geographical Voices*, the wonderful collection of autobiographical essays that Peter Gould and Woody Pitts put together.

I was born in 1935 and grew up a small town in a rather remote area of western Sweden. My mother was a midwife and my father owned a sawmill and a factory where he made doors and windows. They were a little older when they got married, my father 40 when I was born. From what I can judge I was a very wanted, loved child with no real problems. As a consequence I have never had any desire to reject my parents, and from what I can tell I had a very secure and safe childhood. I felt as if I could do whatever I wanted, and I was encouraged to do so.

When you are living in a small community of that type, everybody of course knows everybody else. Add to this my mother's job and it is clear that a large number of people were passing through our house, because in every family, rich and poor, there were babies being born. My father was very intelligent but he had only five years of school – every other day. He was clearly very bright and in his own fashion surprisingly well read. This was in no way an intellectual home, but it was very open and it provided a sense of security for which I am eternally grateful. This well-ordered but permitting atmosphere has no doubt had an impact on my later development, for if you risk everything when you go off, it is natural that you become cautious and on your guard. So, even though I am well aware that to others my work has sometimes seemed strange, perhaps even dangerous, this is not at all how it has appeared to me. The truth is, in fact, that I have yet to meet a person of whom I am genuinely afraid. It has taken me a long time to realize how unusual this feeling actually is.

As I remember it, I was punished only once, and on that occasion for doing something utterly stupid. My father did not smoke, but he kept a box of cigars for his business friends. Now, together with two of my friends, I stole several of them, not merely one but enough for my father to notice. He did and I got a (light) spanking. To this day I am uncertain whether he did it because I had stolen or because I had been careless enough to be caught.

I have subsequently acquired a number of friends in southern Europe – Italy and Spain in particular – and it has taken me a long time to realise how different cultures are using slightly different socialization techniques for making us predictable and obedient, the sense of shame reaching much deeper into them than into me; myself I rarely even notice if I happen to do or say anything unusual. And if I actually do notice, I almost certainly would not feel ashamed. But some of my friends actually do and that is a feeling which has taken me a long time to understand. It would be strange if this personality trait were not somehow related to the type of creativity that has formed my work.

TB: So a secure, comfortable home life.

GO: Yes, and in the creation of that home life my father played a great role. Growing up on a small farm, and after a tough period first in the States and then in Canada, he returned to Sweden in the mid-1920s eventually earning enough to make him comfortable. As a sign of his determination I was only three months old when he bought a tract of forest land that he registered not under his own name, but under mine. It is still in my possession and even though it is far from enough to live on, it has sometimes made a difference. Most likely it has played a role in my decision never to apply for research money not, of course, because I would not have enjoyed the cash, but because it never crossed my mind to ask for the tacit permission. Why would professors be given tenure, if it were not because they were judged competent enough to decide by themselves what is worthwhile research? In the traditional Swedish system it was in fact only professors and judges that the constitution forbade the King to fire, although that of course sometimes happened anyway. The idea was nevertheless clear: even the sovereign needs servants who are sufficiently independent to obey by the law (juridical or scientific) rather than the dictates of the ruler. Money and freedom is indeed a multifaceted relation and in my mind the price for the Faustian pact is regularly too high to pay. Once that has been said, though, I admit that in my type of non-empirical research there is less need for money than for peace and quiet.

I know David Harvey quite well and when last January he came to Uppsala to receive an honorary degree, we talked about these issues quite a lot. I think he and I are the only one geographers of any repute who have never asked for funding. In turn this means that even though I have worked very closely with a number of students, none of them has had to rely on me for their bread and butter. I realize that this view of the students' independence may be outdated, but I feel so strongly about it that I will not give it up. Mentor yes; slave-owner no. But in today's Sweden, all doctoral students are formally appointed as state employees. I feel very sorry for them, for how can they under these circumstances ever grow into independent thinkers.

TB: What were you like as a student?

GO: First, of course, I went to the primary school, and that was in the little village where grew up. For secondary school I had to take the regular bus, 40 kilometres there and 40 kilometres back, every day for four years. A formative experience. Then I switched to the gymnasium for another three years. Too far away to travel daily, so my parents rented me a room from an old lady, a friend of the family. This meant that from the age of seventeen I have had to look after myself.

TB: How old were you at the gymnasium?

GO: When I began I was seventeen and when I graduated I was turning twenty.

TB: Was it there that you wanted to be a geographer?

GO: I never wanted to be a geographer. Of course, like other children I liked the atlases, imagining the cold of the North Pole and the heat at the equator. But I was generally speaking good at school, the first years in a new system average, at the end always at the top.

Geography was always one of my best subjects, but I was actually more interested in history, literature and political science. And in the gymnasium we had a seemingly bad geography teacher – bad in the sense that he could not keep order in the class. As it turned out, though, he was a dedicated and quite competent researcher, who was talking about ideas, von Thünen included. Very beautiful and very simple.

When I later came to the university, where you had to take courses in a combination of disciplines to qualify as a teacher, I chose history and geography. Since I was much more interested in history I decided to start with geography, simply to get it out of the way. The studies were essentially free with very few lectures and hardly any seminars. You signed up for an exam once a semester and how you acquired whatever knowledge you acquired was really no one's business but your own.

TB: Which university did you go to?

GO: The University of Uppsala, which was were most students from my province went.

TB: That was about 1955?

GO: No, it was in 1957, two years after my graduation from the gymnasium. These two years in between turned out to be very important and I would like to say something about them.

At that time every twenty-year old male Swede had to do military service, a duty no one could avoid. A few weeks after graduating from the gymnasium I therefore found myself in the barracks, a place I quickly realized was not for me. Since everyone knew that students of medicine, dentistry and veterinary medicine (the army still had horses) could do the military training as part of their regular studies, the strategy gave itself. I wanted so very much to get out, but to become a dentist – I couldn't imagine anything duller. And sticking an operating knife into a human body was really not an alternative either. Veterinary medicine! An application was submitted and ten days later I was free.

Come fall I went to Stockholm to begin my new career. The second day there I biked to the school where I was met by a most enthusiastic assistant, who took me to an operating room where they were performing a caesarean on a horse. I knew from my mother how dangerous that operation was, but I had never really imagined that it could be done on a horse hanging in hooks from the ceiling and by vets with rubber boots on their feet. More blood than I had ever seen. And such was the end of my life as a veterinary.

In retrospect I often think of that summer of 1955 as being exceptionally important. The reason is that this was one of those rare moments in my life when I have done something on purpose, when I have manipulated myself, when I purposely have tried to reach a well-defined goal. Not letting life just happen, but consciously intervening. In the process I learned a lot about myself, a lesson which I now understand has had a profound impact on my attitude to politics, *the* prime example of intentional and institutionalised action.

Once this had happened, all universities had already started and it was too late to do anything about it. In addition I knew that no matter how much I disliked it, I would eventually have to complete the eighteen months of military training. To pass the time, I got myself a job as a teacher in a nearby high-school, as it turned out a life-changing experience. Why? Because in the graduating class was Birgitta, that cute little girl to whom I am still married. Of course there was a minor scandal, but what do you do when you are twenty-one.

TB: Then you went to Uppsala.

GO: Yes, and I started reading geography because I wanted to get it out of the way. But then the fantastic happened that one of our teachers was a young guy who had just finished an unusual thesis on migration, a study in which he had experimented with a set of regression models, essentially a social gravity model. To me this was extraordinarily fascinating, partly because the formulations were so beautiful, mostly because I didn't really understand what was going on. The challenge was enormous and I was convinced that here lay the frontier of the social sciences, indeed of knowledge in general. The point was to translate observed reality into the language of precise and non-ambiguous language, for only in that way could knowledge be accumulated. Already at that stage there was a streak of minimalism, the feeling that empirical data can be condensed into a small set of parameters. A new world was opened up, a world which in a sense I have never left; whenever I don't understand what is going on, I automatically reformulate the problem into a regression model with dependent and independent variables.

TB: Did you have much training in mathematics?

GO: Yes and no. In the gymnasium you had to concentrate either on language or on mathematics, and I chose the latter. However, I was never especially good at it and even though I think I could understand what was going on, I would never have been able to do mathematics creatively. Certainly nothing compared to what I have done in ordinary language.

TB: Did you think about these things philosophically at all?

GO: Quite early actually, because in the very beginning of my graduate work, probably in 1961, I took a seminar on causal models from Herman Wold, the professor of statistics and a leading econometrician. This course had a lasting impact partly because it was technically quite sophisticated and partly because it was firmly grounded in philosophy. To be minimalistic, the entire course was about the coordinates-net in which we capture the world, especially about the naming of the axes; not merely the singling out of dependent and independent variables, but the fixing of the fix-point (the origo) on which everything is hung. I think this is a key to understanding almost everything I have subsequently come to do. Of course I did not know that at the time, but Wold's seminar – its content at the very edge of my comprehension – was no doubt more formative than any specifically geographic study I ever did.

TB: Naming the axis and then realising that by choosing the fixed point you determine the world you see.

GO: Yes. Yet – and perhaps even more importantly – the very point of the causal models is not merely to describe the world but to provide a means for changing it. It follows that lodged in the axes of the regression models is the key problem of thought and action, that is the relation between description and prediction. In Swedish geography of the 1960s this was indeed a central issue, for in the social engineering of that era the social gravity model was perhaps the most important tool for rebuilding society, by extension for making the country a better place to live. That said it should also be said that I remain very critical of the geographers who were active in that work, Torsten Hägerstrand and Sven Godlund most prominent among them. Once again it is a question of ethics, for I will certainly demand of those who deliberately set out to change the world that they should have a clear view of the consequences of their actions. In this case they did not. To this day I remain uncertain whether they acted as they did because they were tempted by the rewards of power or because they were methodologically naïve. Perhaps it was just the moods of the time. The tone of the voice gives me away, just as it always does when I witness how integrity is being sacrificed on the altar of political correctness.

TB: I think that was the first article I read of yours – spatial engineering.

GO: How do I know, except to say that these issues have been with me for a very long time. And when in 1977 (after the eleven years at the University of Michigan) I accepted the job at Nordplan, my hope was to examine the relations of thought and action in greater detail.

TB: Were you aware of this that early – in 1961 and 1962.

GO: Probably not in 1961, but most definitely in 1965.

TB: Is Hägerstrand at Uppsala now?

GO: No. Hägerstrand has always been at Lund. Sven Godlund did his doctorate there as well, even though he later was appointed to the chair at Gothenburg. This means that both of them were members of the Lund school, which was decisively influenced by Edgar Kant, who after the war came to Sweden as a refugee from Estonia, where he had done a fair bit of Christaller-type work.

TB: Who were influential for you at Uppsala?

GO: In some ways the question is wrongly posed, because throughout my development hardly anyone has been strong enough to change me. Yet I have already mentioned Herman Wold and Esse Lövgren, the young man with the regressions. These two made an impact, almost certainly because they were so challenging. Gerd Enequist – formally my professor at Uppsala – was a historical geographer, who didn't have anything intellectually to do with my work, but who was curious and open enough to give me all the support I needed and never had to ask for. Like my parents she trusted me to do whatever I pleased and for that I remain incredibly grateful. None of them ever interfered, perhaps because they somehow knew that telling me what to do would never have worked. And that is despite my very Swedish non-aggression complex, an evasiveness which I have learned can be irritating.

TB: Even at this early stage, there is almost an ambivalence you have towards statistics, or mathematics. That seems to me still true.

GO: It is the aesthetics rather than the formalism itself that I find so appealing. And in my particular brand of aesthetics, minimalism plays a pivotal role. Here, again, I have been much influenced by the gravity model in which everything worthwhile is condensed into the points at which the line crosses the axes; the slope of the line, the angle. Since I am blessed with a rather poor memory I have to condense the world into geometric patterns which I can see with my eyes and grasp with my hands.

TB: You said you were thinking of going into history?

GO: I did do history. But when I did I had already been sucked into the geography department, where I was more attracted by the parties and the social atmosphere than by the intellectual questions. To tell the truth it is only recently that I have become seriously interested in geography.

TB: Was there much on the mathematical/statistical side?

GO: Nothing at all. Except in 1961–62 Julian Wolpert visited the Uppsala department and we were spending a lot of time together. As it turns out he too attended Wold's seminar and we were also taking the same course in computer programming. Before that – in 1960–61 – David Harvey came to Uppsala too. It was Gerd Enequist who decided that he and I should share an office, an experience that turned out to have a tremendous impact on both of us.

TB: What were the circumstances of you meeting?

GO: We were exactly the same age but because of the different educational systems he was just finishing his doctorate at Cambridge and I was beginning my graduate studies at Uppsala. He came there with a Leverhulme scholarship, ostensibly because of the Swedish data banks, more likely because he just wanted to get out of England. There was no particular reason why he should come to Uppsala, rather than to Lund or Stockholm, but luckily he did.

So one day there was a knock on the door, Gerd with David in tow. Once she had left the room, it took the boys two minutes to realise that they knew equally little, two more to become friends for life. This is also how it has stayed, and that is despite the seemingly different paths we have subsequently followed.

TB: What was this conversation that you had with him?

GO: Once again the question is wrongly put. For the truth is that we were just fooling around, drinking and having a good time, talking about everything and nothing, naïvely unaware. Throughout the sixties we used to meet at least once a year, normally for several weeks at a time. All told we spent a lot of time together, increasingly talking about location theory and spatial modelling, almost always with a philosophical bent to it. Almost from the beginning we kept returning to the form-process problem. And when we met next time, we were surprised to discover that we had written virtually the same papers without ever having corresponded in between. Of course there were intellectual parallels. But more importantly it was a very close and personal relation.

TB: What were the kinds of philosophical literatures you were drawing upon and reading that were influential?

GO: It came gradually, more from sociology than from philosophy per se. As before, the focus was on issues of causation, occasionally on the relations between induction and deduction – reasoning as a travel story, if you will.

My first thesis – the equivalent of the American PhD – consisted of several articles, but the major one was a migration study in which I played with the axes of a set of regression models. To be very precise: is human interaction a function of distance or is social distance a function of interaction? There were, of course, many other problems as well, including the fact that the gravity model is a deterministic formulation in which the outcome is more dependent on the normalizing means than on the deviant variations. How could I acknowledge in the model the fact that even though you and I are alike, we are not identical? The answer to that basically existential question lay in probability theory and therefore I spent a lot of energy on trying to translate the deterministic formulations into stochastic form. In hindsight these issues are closely related to the problem of geographic inference, the question of what a description of spatial form can reveal about the social processes that have generated it. In my mind it is this problem of cartographic reason that eventually blew the discipline apart, a torpedo hitting the ship of Geography below the water-line. In its frantic search for saving isms, the discipline at large seems happily unaware of what is at stake.

Why is this so? Because since its Greek beginnings Geography has been founded in description, description presented as a map. But what does the map tell you about the processes through which the phenomena represented by the points, lines and planes actually came about? How *are* spatial patterns generated? It was this methodological question that a small group of us got hooked on. And suddenly it was crystal clear: the same spatial form can be generated through drastically different processes. In turn this implies that from a perfect description of a spatial form you cannot say anything conclusive about how it has been generated. A conclusion that goes completely counter to centuries of geography!

Who were we? A gang of five often gathering on the living-room floor of the Olsson house at 2128 Geddes Avenue in Ann Arbor: David Harvey, Reg Golledge, John Hudson, Les King and myself. Fireworks exploding, lightening bolts out of control.

TB: It was pivotal for you

GO: Yes, and not only for me. In fact I am convinced that the inference problem played a role in David's conversion to Marxism just as it did in Hudson's drift towards historical geography, in King's decision to become a university administrator, in Reg's growth as a behavioural geographer. Splitting without splitting, a kind of revolution that the majority of today's geographers know very little about. Yet they are just as steeped by its consequences as we once were ourselves.

TB: You were talking about the same problem with Harvey in 60.

GO: Yes, and in many ways this problem is still with me. And I think in a sense it is for him too. It is so challenging because like all interesting problems it refuses to sit still. Once you approach it, it turns another face.

TB: Has that been the central problem in your intellectual life.

GO: In some ways. If I were to pick only one, I would say the geographic inference problem. But that is in itself an issue of translation, hence at bottom an issue of exploring the prison-house of language.

TB: To backtrack – when did you graduate as an undergraduate?

GO: In 1960. And immediately after graduation I was awarded a fellowship to begin work on the licentiate theses that I defended in 1965. The most formative of those years was 1963-64 which I spent on an ACLS Fellowship at the Regional Science Department in Philadelphia. In that adventure Julian Wolpert played a role, not only because he was at the University of Pennsylvania at the time, but also because he had helped me put together the application that got me there. A completely fantastic year in the company of Walter Isard, Julian, and Michael Dacey, with whom I shared an office. Topping it off were two NSF summer institutes, the first in Regional Science at Berkeley, the other in Spatial Statistics at Northwestern. For my own generation of quantitative geographers that summer of '63 was crucial, much the same people attending both events, Leslie King foremost among them. David was not at Berkeley but he did come to Evanston, where we all struggled with Michael Dacey's point models, the stochastic formulations that later led to the inference problem. It is fascinating to see how the person who was so determined to transform the discipline into a hard science, through those very efforts provided the means for tearing it apart. Very strange. And with its own streak of tragedy.

TB: Other people I've spoken to have talked of an almost religious fervour in that period. Was that something you believed in? Were you an outsider?

GO: No, I was not at all an outsider. Of course there was an engagement, an obsession. Yet it was clearly more important to have a good time than to change the world. Lots of laughter, wine and food. Life as it is lived, fully and without compartmentalisation. That's why we are still meeting whenever we get a chance, that is where the foundation was laid.

TB: Did you feel that you discovered the truth, or was that just irrelevant?

GO: Of course we knew that we were on to something exciting. But we were also young. Much of what happened actually happened in our house, everyone who counted coming through for the parties, the seminars and just the fun. Without my wife as willing hostess it would all have been very different.

TB: When did you get your first PhD?

GO: In 1965. And without stopping I set my sights on getting the second doctorate, receiving a number of job offers at the same time. One of them came from the University of Michigan, another from Akin Mabogunje in Ibadan, who wanted be to join him in Nigeria. And I said yes, I will do that. But it took such a long time with the bureaucracy, and I was getting all these other offers, that in the end I decided to go to Michigan instead. The times were really unbelievable: in Ann Arbor I had never set foot, there was no job interview, I had met none of my colleagues to be. And even more remarkable, before our trunks had arrived I was receiving several new job offers, one from Toronto, where I knew that Les Curry was teaching. Since his work was important to me, I went to the Michigan chairman – I think it was Mel Marcus – asking for advice. "Take it easy," he said. And three weeks later I was promoted to Associate Professor with tenure. I had only the vaguest idea of what it all meant, the dictionary definition of the word "tenure" of little help. And given the reactions, I was careful not to ask.

It was fantastic, especially as the quick promotion strengthened my feeling of independence. In the meantime I kept working on my Swedish second doctorate, which I received in the spring of 1968.

TB: What was that second doctorate about?

GO: I used some parts from the first thesis, but the second was called *Distance, Human Interaction and Stochastic Processes: Essays in Geographic Model Building*. Formally it was only 5 pages long, but in reality it was a weaving together of six separate articles. I was stunned to learn that the committee (Professor Wold was one of its members) had given it the highest possible mark, a most rare distinction. Once again a shot of independence.

TB: You had that monograph on spatial interaction published by the RSA – was it in 65?

GO: Yes, a combined monograph and bibliography that I wrote during the year at Penn. It got a good reception, but eventually it made me embarrassed. Although it was really only a student paper, it got reprinted and reprinted until in the end I finally said no.

TB: How would you characterise your relationship with mainstream work up until these pivotal meetings in 1968 in your living room? Were you a true believer or had you always had a few doubts?

GO: It's difficult to say, for even though I have always been deeply engaged in what I am doing, I would never call myself an obsessional believer. If there ever were any true believers, they were not in my generation, but among those who were formed towards the end of the fifties. Brian Berry, for instance; I think he might have wanted to change the world. David too, of course, but with him everything is different. In my own case I just did what I did, no cult of anything except the urge of doing it on my own.

TB: Did you feel that geography up until the quantitative movement was fundamentally wrong – that this place-based, regionalist descriptive approach was something that needed to be overthrown?

GO: Yes, although it wasn't wrong, merely not interesting. No theory, hence no beauty. We just made fun of it, cracking jokes as we went along. I'm sure many do the same with me now.

TB: I did interview Peter Gould before he died. He demonised particularly Hartshorne – as an evil person, not just that he was wrongheaded intellectually.

G: Peter was religious, of course. A true believer. But at the same time, he was actually very open-minded. But it is certainly true that he believed not only in what he was doing, but in the idea of changing the world. Very much the same mould that you can now see in European and American politics, where the world leaders are presenting themselves as moral leaders. They are dangerous, because they are religious enough to go to war. Peter had some of that.

TB: 1968 – Had you started writing *Birds in Egg* at that time?

GO: No. That book grew out of the gravity model. As you will recall, my question had been whether what the gravity model says is correct or whether its various pronouncements are consequences of the particular reasoning mode in which they have been phrased. Was the conclusion a result of me not properly understanding the stochastic processes, of my own misunderstanding of the relations between theories and observations. So I was asking, once again, whether the revolutionary conclusion about geographic inference is correct or whether we had reasoned in a way that is good enough. Had we just done a sloppy job? Those kinds of questions obviously lead to the philosophy of science and, in my particular case, to logic, especially many-valued logics and fuzzy set theory. These were the connections that Stephen Gale and I were pursuing as early as 1970. To my knowledge these papers were the first about logic and fuzzy sets ever to appear in geography. Stephen, by the way, was the first in a long row of graduate students who individually and collectively continue to mean so much to me.

Of course we didn't really understand what we were doing. What we did know, however, was that unless the structure of the phenomena we were studying and the structure of the language in which we were trying to capture them were not the same, then we would be faced with an insurmountable translation problem. Easy to see what is at stake, exceptionally difficult to know what to do about it. Form and process in another guise, the parallels to the physics of the 1920s too obvious to ignore. In my own case, and with my Swedish background, these issues were directly connected first with the practice of social engineering and then with Aristotle's question about the sea-battle tomorrow.

Birds in Egg took six or seven years to write. It drew quite heavily on the lecture notes I developed for a course entitled "Thought-and-Action", an educational experiment in which the students earned credits not in geography but in the college at large. This in turn meant that I attracted some really excellent students, many of them senior undergraduates who had completed all requirements except the residency. Hence they were free to elect whatever courses they liked. And they did.

TB: In that book you draw on people like Hegel, like Wittgenstein. Did you only start reading them from 1968 onwards or had that been part of your earlier intellectual training?

GO: I had read some of it in the gymnasium, but it became important with the political turmoil between 1968 and 1975, a period which I often refer to as some of the best years of my life. Those were the days and nights of the equal sign, the key theme of *Birds in Egg*. The outcome of those investigations was, of course, that even though we cannot do without the copula, there are several alternative ways of defining it, each mode expressible through drastically different languages; dialectics one of them, conventional logic another. So overwhelming was the experience that when the book was finally out, it set off a tremendous, perhaps life-threatening, creativity crisis. A most difficult time.

TB: For you personally?

GO: Yes. For what can you conceivably write about that is more important than the equal sign? You can of course remain on the same level of abstraction, but if you want to dig deeper, where do you go?

TB: Did it create a crisis between you and other people with whom you had been friends?

GO: No, it didn't. At least not that I know of. Although I am well aware that others sometimes see me as a controversial figure, that is most definitely not how I think of myself.

TB: Did you start moving out of that circle that you were a part of? Did people like Les King and Reg Golledge move away from you?

GO: No, not at all. On the contrary, we got closer. This was also the period when Reg encountered severe difficulties in his first marriage and that circumstance brought us even closer than before. Wine and lobster, white and red, light and dark. But at bottom always the connection between what you think and what you do.

TB: But it was such a different perspective. You seemed to be fundamentally questioning their very project by raising this issue of the equal sign or the meaning of "is." Did you agree to disagree?

GO: Yes. They thought it was interesting and I didn't feel in any sense excluded. Perhaps because our relations had always been less intellectual than human.

TB: Do you see the work you've been doing most recently – your experiments with language and that kind of thing – do you see that coming up against the same fundamental problem of form and process?

GO: Yes, of course.

Bibliography

Bunge, W. 1971. *Fitzgerald: geography of a revolution*. Cambridge, MA: Schenkman.

Haggett, P. 1965. *Locational analysis in human geography*. London: Edward Arnold.

Harvey, D. 1969. *Explanation in geography*. London: Edward Arnold.

Olsson, G. 1965. *Distance and human interaction: a review and bibliography*. Regional Science Research Institute: University of Pennsylvania, PA.

Olsson, G. 1975. *Birds in Egg*. Ann Arbor, Michigan: Michigan Geographical Publications, no. 15.

Olsson, G. 1980. *Birds in Egg/Eggs in Bird*. London: Pion.

Olsson, G. 1991. *Lines of Power*. Minneapolis, MN: University of Minnesota Press.

Chapter 17

Mapping the forbidden

GUNNAR OLSSON

Olsson, Gunnar (2010). Mapping the forbidden. *Fennia* 188: 1, pp. 3–10. Helsinki. ISSN 0015-0010.

Mapping the forbidden is in itself forbidden. And in my understanding the most forbidden of everything forbidden is that which refuses to be categorized, that which is neither this nor that, ungraspable forces which do not sit still but hop capriciously about. Aristotle consequently knew what he did, when he between the two concepts of identity and difference inserted a third position called "the excluded middle", a non-bridgeable gap which in the same figure unites and separates, liberates and imprisons; an unruly space located beyond the realm of conventional reason; a no man's land of liminality which the well behaved must never enter. But Aristotle also argued that what one cannot do perfectly, one must do as well as one can.

Gunnar Olsson, Uppsala universitet, Kulturgeografiska institutionen, Box 513, S-751 20 Uppsala, Sweden, E-mail: gunnar.olsson@kultgeog.uu.se.

One reason why the forbidden remains forbidden is that it is one with the taboo, a concept which is etymologically connected not merely with the terms "under prohibition" and "not allowed", but with the words "sacred" and "holy" as well. What is taboo is consequently doubly tied first to the forbidden itself then to the strongest form of the taken-for-granted, i.e. to those aspects of the unconscious which are crucial enough to be blessed by the gods themselves, by definition beyond reach. As one siren sings COME, another blares DANGER. But why should I devote my professional life to issues which are not important enough to be taboo? How could I possibly stop wondering about how I understand how I understand?

These are the questions with which I grappled also in my latest, perhaps last, book, a minimalist piece called *Abysmal: A Critique of Cartographic Reason* (Olsson 2007). And let it now be known that that title was carefully chosen, for the noun "abyss" is synonymous with the terms "deep gorge" or "bottomless chasm", the rift which cuts through the landscape of understanding, one set of categories placed here, another set there. As in the book itself I would now like first to descend into this canyon and then ascend again from the depths with a map of what I have found. A dangerous expedition indeed, because the abyss is not merely one with Aristotle's excluded middle but the very home of POWER itself. And so strong are the socialization forces built into ordinary language that the adjective "abysmal" tells the potential trespassers how they should feel if they try to break into that palace off limits: Abysmal! Horribly bad!

For these reasons of power and socialization I am once again reminded of *Enuma elish*, the Babylonian tale of how the god Marduk gained and retained his elevated position as the Lord of lords. The premise of this oldest creation epic extant is that in the beginning of the beginning nothing has yet been formed, because in the beginning of the beginning nothing has yet been named. All that there is are the spatial coordinates of above and below, cardinal positions waiting to be inundated by the fluids of masculine Apsu and feminine Tiamat, the former sweet, the latter bitter. And as if to underline the spatiality of its own structure, the term *apsu* literally means "abyss" and "outermost limit", by linguistic coincidence connected also to "the great deep", "the primal chaos", "the bowels of earth", "the infernal pit". A perfect example of proper name and definite description merged into one.

Eventually there is a tremendous power struggle and sweet Apsu is killed by Ea, the most outstanding of his offspring. On top of the corpse, i.e. *across the abyss*, Ea then builds a splendid palace for himself and his wife Damkina. There, in the Chamber of Destinies, their son Marduk is conceived, the most awesome being ever to be (Dalley 1989: 235):

> *Impossible to understand, too difficult to perceive.*
> *Four were his eyes, four were his ears;*
> *When his lips moved, fire blazed forth.*
> *The four ears were enormous.*
> *And likewise his eyes; they perceived everything.*

Marduk's weapons are numerous, but most decisive is the magic net in which he goes to capture the recalcitrant Tiamat and the four winds by which he eventually blows her up. When it is finally over, Marduk, great lord of the universe, "crossed the sky to survey the infinite distance; he stationed himself in the apsu, that apsu built by [Ea] over the old abyss which he now surveyed, *measuring out and marking in*" (Sandars 1971: 92, emphasis by the author). No longer dressed in the warrior's coat of mail but in the uniform of a land surveyor, he then proceeds first to the construction of a celestial globe and finally to the creation of a primeval man, the prototype of you and me, a creature explicitly designed to serve as slaves of the ruler's vassals, three hundred stationed as watchers of Heaven, an equal number as guardians of the Earth. Not an invention formed in the image of the Almighty, though, but a savaged concoction stirred together from the blood of the slaughtered Kingu, Tiamat's lover and commander in chief. Mankind a dish of *Boudins à la Mésopotamie*. Nothing like a perfect copy of the perfect original, merely a black sausage. And as a way of guarding his ambiguity he gave to himself a total of fifty names.

Throughout these events the abyss remains the power center par excellence, the broken clay tablets of *Enuma elish* the ultimate proof of the Babylonians' insights into the secret workings of human thought-and-action. And therein lies in my mind the real reason for keeping the abysmal gap between categories taboo, for it is in the ontological transformations of the excluded middle that the magicians of power are performing their tricks. Hence it is only by entering that forbidden space of imagination that the analyst can ever hope to understand how the absent is made present, the present made absent.

The connections between presence and absence are vividly expressed also by the figure of Janus, my own favorite among gods. What intrigues me

with this pivotal symbol of gate-keeping is less that he is equipped with a body that makes him see in opposite directions at the same time, more that he has a mind which allows him to merge seemingly contradictory categories into one meaningful whole. From his watchtower at the middle of the bridge he is consequently in a position to keep both sides of the abyss under constant surveillance, in the same glance catching a glimpse of the pasts that once were and of the futures that have yet to come.

Given the Greek fear of the void – itself well expressed by the concept of the excluded middle – it is not surprising that Janus was invented in Rome and not in Athens. In the lands surrounding the *Mare nostrum*, though, he was everywhere to be seen, for not only was his image stamped on practically every coin, but in religious prayers this janitor of janitors was the first to be mentioned and in cultural rituals this son of January was equated with the beginning of all beginnings. Diana was his godly consort, a connection which explains why the doors of his temple stayed open in times of war and why they were shut in times of peace. Like ordinary lovers, gods need their privacy too.

Janus's main concerns were one with my own: creativity, power, socialization. Defiantly I therefore pray again (Olsson 1991: 16):

> *Oh Janus! Help me become a sinner. Let me understand how you break definitions and thereby create. Show me how you erase what others see as irresolvable paradoxes. Teach me the equation of that third lens inside your head whereby contradictory images are transformed into coherent wholes. Speak memory, speak! SPREACH, Janus, SPREACH! And Babel's walls come mumbling down.*

Accordingly, and throughout my scholarly and artistic life, I have been searching for a place inside Janus's head. From that zero-point of the excluded middle I have then tried to grapple with the taboos of limits, the sins of trespassing, the braiding of epistemology and ontology, the challenge of writing in such a way that the resulting text actually *is* what it is about. With the aim of understanding how Janus stayed sane while ordinary people in similar situations of double bind go crazy, I have therefore tried to place him on the operation table, cut his skull open, lay his brain bare, investigate how his mind is wired. Why and how, for instance, did the Romans elevate this categorical juggler to godly status, when we, their descendants, diagnose his counterparts as schizophrenic madmen?

FENNIA 188:1 (2010)

Why did they afford him a special place in their pantheon, while we isolate his likes in the sound-proofed cells of the asylum?

Perhaps the reason is that without distinctions our thoughts-and-actions would have nothing to stick to, our lives nothing to share. Such vacuities are in fact the norm in the Realm of Psychosis, that literally unthinkable province where there are no initiation rites, no scars, no individuals, hence no society either. And this emptiness may well explain why the deeply psychotic is so frightening, because the deeply psychotic lives outside the laws of thought, an inhabitant of the excluded middle, an alien beyond both identity and difference. A non-mappable world without fix-points, scales and projection screens – a cartographer's nightmare.

Lest it be thought that my understanding of the void is too closely tied to the Abrahamitic world, I now recall a stunning visit to the city of Kandy, once capital of the Sinhalese kingdom which in 1815 was annexed by the British and made a part of colonial Ceylon[1]. There the high priest of the temple – the shrine which among other relics houses the tooth of Buddha, historically *the* national symbol – granted me and my wife a rare audience. Not just any audience, though, but a visit to the holiest of the holy, a small room on the upper floor with an altar bestrewn with jasmine flowers and the sacred tooth enshrined in a casket of gold. Before entering this forbidden place, we were most carefully instructed how to behave, especially *not to step on the threshold,* the barrier that separates the commoners in an antechamber and the higher classes in a middle room, on the one side, from the inner sanctum with the king, his closest ministers and the water-increasing official, on the other. A wonderful illustration of how the excluded middle can be materialized in an untouchable janitor.

The mind boggles as it encounters the walls of Babel, Kreml and Berlin in yet another setting, the hierarchical structure of the three chambers of the temple highly reminiscent of the narthex, nave and sanctuary of the orthodox church, the Kandyan threshold effectively serving the same exclusionary functions as the Russian iconostasis. Most revealing is nevertheless the story that before King Vimaladharmsuriya I in 1592 entered the same room as we did in December 2007, he kneeled and put his forehead on the polished threshold. The stamp of power in the place of power, the

mark of Cain in a Buddhist context, a clear warning that anyone who sets foot on the threshold is trampling not on a material object but on power itself. This circumstance, rather than the Greek fear of the void, is in my analysis the real reason why the excluded middle is excluded. And as a way of protecting his own holiness from possible usurpers, the Jewish LORD put a mark on the restless wanderer so that no one who found him would kill him. In the same breath a blessing and a curse, yet another indication that it is in the nature of absolute power to violate every rule of behavior, to do exactly as it pleases. The reason is, of course, that in a norm system where both *a* and *not-a* are valid at the same time, everything is permitted.

No wonder, therefore, that it is from a position in the excluded middle that the Almighty rules, his words-and-deeds predictably unpredictable, his palace surrounded by a non-penetrable defense system, his propaganda machine everywhere to be heard and nowhere to be evaded. Yet everything codified in the constitutional law of Mose's first stone tablet, in my heretic (hopefully not blasphemous) interpretation the most penetrating show of power and submission ever formulated. It is hard to imagine a more power-filled statement.

The first stone tablet is nothing less than a socialization instrument that no one can escape, hated whip and enjoyable carrot in the same document. A rhetorical masterpiece firmly rooted in the concept of trust, a social glue which under the label *pistis* was foundational to Aristotle as well; the common point is, of course, that without pistis there can be no communication and that holds regardless of whether the chosen language is that of money, poetry, logic, geometry or anything else. This in turn led Aristotle to the insight that dialectics and rhetoric are the twin sisters of each other, just as it later led Nietzsche to the conclusion that the two activities of logic and geometry are forms of rhetoric which after long use have become so credible that they have changed names and turned into categories of their own. It cannot be said more clearly: reasoning is a persuasive activity grounded in the tension between personal trust and social verification.

In a very general sense it is this question of how we find our way in the unknown that lies at the heart of European culture, perhaps of all cultures. In Erich Auerbach's influential analysis of mimesis it is located exactly in the taboo-ridden interface between the certainties of Odysseus's scar and the ambiguities of Abraham's fear, you and I dangling in the abyss in-between (Auerbach 1953). Two modes of understanding, two modes of being, two ways of living which over the centuries have been condensed, purified and eventually codified, one in Aristotle's Laws of Thought, the other in the biblical formulation of the commandments, the latter not merely the ten that can be counted on the fingers (The Holy Bible, Exod. 20: 3–17, Deut. 5: 6–21), but a staggering total of six-hundred-and-thirteen. The interpretations vary accordingly, even though it is generally agreed that the ten words of the Decalogue may be divided into two groups such that the first three or four govern the relations between God and man (the Constitutional Law) and the rest regulate the relations between man and man (the Civil and Criminal Law).

In times of crisis it is the first tablet that tells the ruler how to rule and the subjects how to submit, and that is regardless of whether the potentate happens to be a Machiavellian Prince, a dictatorial Führer, an elected Prime Minister, a concerned parent. It is hard to imagine a more power-filled statement, not the least because it is there that YHWH for the first time reveals his own name, an expression so closely related to the Hebrew word for "to be", *hvh/hjh,* that it is often translated as "The Being". As a way of further stressing its importance, it was this invisible entity itself, not one of its usual emissaries, who in the prologue let his subjects know that *it was I who liberated you, I who let you out of the land of Egypt, I who cut your chains.* The implication is, of course, that since I have proven myself to be such an outstanding leader in the past, you are wise to trust me also in the future; accordingly, every incumbent assures the voters that they never had it so good, that they should read his lips and scrutinize his record. Although you should prepare yourself for blood, sweat and tears, at the end of day there will be milk and honey.

Thus I decree, because I am who I am. Such are the self-referential words of the Law's prologue. Immediately following that naked piece of rhetoric comes the first paragraph of the Constitutional Law, a proposition as stunning now as when it was first uttered: *I shall be your dictator!* Wherever this Almighty happens to be – and by definition he is at the same time everywhere and nowhere – he shall rule over everyone and everything, like the surveying Marduk measuring in and marking out, show-

ing mercy to those who love him and killing those who hate him.

The unknown genius who was the first to coin the phrase that there must be no power before (or according to some translations, "beside") me, was certainly wise enough to realize that whoever declares that he shall be my supreme ruler leads a dangerous life. For that reason he proceeded to erect around the apsu palace a two-tier defense system consisting of both a wall and a moat, the former constructed as a ban on the (mis)use of metaphor, the latter as a rule against the creative associations of metonymy. The purpose of the second paragraph is consequently to ensure that the weapons gathered in the rhetorical arsenal will not fall into enemy hands, rephrased that any critique must be silenced before it is uttered. In that mood the jealous LORD now declares that you shall for ever know your place, never commit the sins of trespassing, never question his authority. In particular *you shall not possess the means for making of me a graven image, picture, statue or any other caricature, never use my name in vain or tie it to a definite description.*

The recent debacle about the Danish Mohammed pictures in its proper light (Olsson 2006), for the graven image has always been the master key to idolatry and thereby to the doors of competing ideologies and potential usurpers. In the present context it is especially noteworthy that the Hebrew term for "image" refers more to the dwelling place of the divine than to the pictorial representation of its invisible being (Stamm & Andrew 1967: 82). It follows that if you tell me *where* you are, I shall tell you *what* you are. Yet, as soon as I attempt to make the invisible visible, I run the risk of falling into the trap of misplaced concreteness, of deifying the reified. But by outlawing the *as-if,* the Untouchable guarantees that no news will ever issue from his subjects but exclusively from himself.

It cannot be said more clearly: the second paragraph amounts to a devastating auto-da-fé, a combined prohibition against picture-making and story-telling, the two primary modes of translation, understanding and reasoned critique. Even so, the declaration that I shall be your dictator is so outrageous that no censor will ever be strong enough to get it generally accepted. Other socialization techniques must therefore be mobilized as well and that is indeed the purpose and function of the third paragraph. With that goal firmly in mind, the lawmaker therefore once again reminds the congregation that it was he who took them out of the land

FENNIA 188:1 (2010)

of bondage, he who gave them the freedom that he himself is now set to take back. Therefore, after all these ordeals, I hereby declare that you deserve a rest. However, this precious time you must not spend alone but always in the company of your likes. In the synagogue and the church, at the playgroup and the faculty meetings, the confirmations, funerals and family dinners – it is at these gatherings that my officials will instruct you how to think-and-act. The Kantian thesis about the necessary unity of consciousness in another form, for you must always remember that *you are nothing but a cog in my machinery. I am the spiritualized embodiment of your unconsciously taken-for-granted, the pivot of the world.* And provided you honor your father and mother I shall grant you a long lease on the land that I give you. Like the drip drip drip of the raindrops, when the summer shower's through, so a voice within me keeps repeating you, you, you.

And so it is that I read the commandment to keep the sabbath holy as the most crucial paragraph of the Constitutional Law, the ultimate guarantee that the power structure of monotheism will survive. And so it also is that Aristotle's Laws of Thought and Mose's Laws of Submission may be read as alternative maps of power, two codifications with the shared purpose of showing how in the same breath you can both tell the truth and be believed when you do so. It is difficult to imagine two formulations of greater historical significance, layers of meaning deeply embedded in the taken-for-granted, a palimpsest of the already but not yet.

Every map is a palimpsest, a product of imagination, that uniquely human faculty which assigns to the semiotic animal the privilege of making the absent present and the present absent. Simsalabim and the vistas from elsewhere lie open in front of us, the image of a reality never seen before, a utopian no-where miraculously changed into an existing now-here, a shade of blue turned into an ocean, a line into a road, a dot into a city. By all accounts a most remarkable version of the incantation "Let there be and there is," an outstanding case of rhetoric performed on the high wire.

No wonder, therefore, that in absolute regimes even the most innocuous map tends to be treated as a state secret, for just as no magician wants his tricks to be revealed, so every ruler guards his palace and masks his face. And that in turn explains why the biblical redactors let the LORD say to Moses (The Holy Bible, Exod. 33: 19–23):

"I will cause all my goodness to pass in front of you, and I will proclaim my name, my LORD, in your presence. I will have mercy on whom I will have mercy, and I will have compassion on whom I will have compassion. But", he said, "you cannot see my face, for no one may see me and live." Then the LORD said, "There is a place near me where you may stand on a rock. When my glory passes by, I will put you in a cleft in the rock and cover you with my hand until I have passed by. Then I will remove my hand and you will see my back; but my face must not be seen."

What a remarkable passage, nothing less than an exhibition of power undressed, an image viewed from an abysmal cleft, a name spoken in an utterance of self-reference. Even more remarkably, I here detect an allusion to the second paragraph of the Constitutional Law with its double ban on picture and story, the two modes of representation that lie at the heart of cartographic reason. No wonder that the surveyor of power leads such a dangerous life, for how can his analyses be trusted when the faceless phenomenon he sets out to capture is itself steeped in distrust. The liar's paradox in a different context, for you can never tell in advance who in the early hours might be knocking on your door.

And therein lies the profound difference between the social ethics of the first and the second stone tablet. For even though the concept of pistis permeates both documents, the form of trust which ties you and me together is mutual, the trust between the ruler and his subjects is at best (or is it at worst) one-sided; since the Absolute is by definition self-referential, his name (if a name it is) cannot be translated into a definite description. It is highly fitting that the sign of the covenant that the LORD makes with Noah is a rainbow, a palette of fleeting colors in the clouds rather than a material object on the ground.

Even so, the doubters refuse to be silenced and that explains why Abraham took the LORD to task for not keeping his promises of many children and why Job sued him for slandering, a court case never to be forgotten. In between is the story of Jacob, one of the greatest crooks ever born, yet one of the richest rewarded (see Miles 1995; Olsson 2007, chapters "Abr(ah)am" and "Peniel") . Of the latter much can be said, but nothing more important than the fact that in the chronicles it was he who was the first to claim that he had seen God's face and survived; in the eyes of the Almighty the blasphemy of blasphemies, to the present analyst a propaganda trick of gigantic proportions. It may in fact be instructive to approach the first third of the Hebrew Bible as the story about a power struggle so violent that the self-proclaimed LORD is eventually forced to withdraw. Thus, after the Book of Job he never speaks again. And as if to continue the assault, the New Testament contains many references to the commandments of the second stone tablet but makes no explicit mention of the first. Fascinating glimpses of the interface between the knowledge of power and the power of knowledge.

In the interface between knowledge and power lies the art of mapping. And just as no map can be a perfect map, so any account of power and knowledge depends on the three primitives of map-making: 1) the chosen fix-points; 2) the scales through which the points are translated into connecting lines; and 3) the projection screen or *mappa*, the taken-for-granted plane onto which the pictures and travel stories are cast and preserved. It is tempting to associate the fix-points with the first paragraph of the Constitutional Law, the scales with the second, the mappa with the third.

Fix-points first. For have I not already noted that in the Realm of Power nothing sits still, that its jealous ruler never sleeps in the same bed two nights in a row. Since the earliest accounts his palace has been variously located in the abyss between categories, in the untouchable threshold between this and that, in the face which must not be seen, i.e. always *in the cleft of the excluded middle*. In addition, the LORD'S name is in most creation myths given as a tautology, by definition true but not informative. Ungraspable is the ungraspable, who for that reason is free to do whatever it pleases. Predictably unpredictable, inherently untrustable. Always there to see never to be seen, Bentham's panopticon in advance of itself. For what my eyes happen to catch depends both on where my body stands and on how my mind has been molded.

Then the scale, by definition the translation function that enables me to claim that this is this and that is that. Yet I have repeatedly stressed that in the dialectical Realm of Power everyone and everything hops capriciously about, sometimes appearing as a this sometimes as a that. To put it bluntly, God (a term which to me functions as a pseudonym of power) does not operate according to the laws of logic. And therein lies in my understanding the reason why the social sciences in comparison with the hard sciences have accumulated so little knowledge. If it is true, which I be-

lieve it is, that human action is structured like a tragedy – everything beautifully right in the beginning; everything horribly wrong in the end; no one to blame in between – then the social sciences are faced with a tremendous challenge, easier to state than to do anything about. But if human action actually *is* structured as a tragedy, how can we then rely on the principle of truth preservation for tying our premises and conclusions together? Surely the most common purpose of human action is to topple truth, not to preserve it, to falsify rather than retain what is presently the case. Less a matter of formal logic more an instance of creative imagination. This to me is the problem of trust and verification, the real issue that the mapmaker's scale is addressing.

Finally the mappa, the formation of the taken-for-granted, the painter preparing the canvas to ensure that the paint will not run off and the surface not crack, the glazier polishing the tain of the mirror. This is in effect what the unconsciously adopted socialization techniques are designed to do, making you and me obedient and predictable in the process. Everything hidden in the mandatory meetings of the sabbath.

As might be expected a similar form of cartographic reason guided the thoughts-and-actions of the Greeks as well. Nowhere is this more evident than in Plato's *Republic* with its three figures of the *Sun*, which together with the concept of goodness functions as the analyst's fix-point par excellence; the *Divided Line,* which embodies the scale through which abstract ideas are turned into concrete things, degrees of truth corresponding to degrees of being; and the *Cave Wall,* the mappa of the surveyor's projection screen, the taken-for-granted background without which there would be no shadows to observe, hence no maps to hide and seek. That screen, though, is not an innocent *tabula rasa*, but a receptor covered in layers of social gesso. And just as the painter's first task is to prepare the canvas, so the mind-surveyor knows that he too casts his figures onto a *charta* with similar characteristics; paraphrasing him who loved the Academy and hated the poets, not every thing can be seen and not every idea can be thought.

Being believed when I tell the truth is essentially a question of the cartographer's *mappa*, the projection screen onto which the fixing points and scaling lines are leaving their traces. Most stunningly it now seems that the early development of

FENNIA 188:1 (2010)

Greek mathematics and geometry, including the theory and practice of triangulation, grew out of a mode of thought which in itself may be understood as an instance of cartographic reason. The trailblazer in that remarkable adventure of cognitive history is Reviel Netz, brilliant classicist presently at Stanford (Netz 1999, 2004). While Netz's overarching interest is in the birth of deduction (Athens, roughly 440 BCE), his real focus is on the intellectual technologies through which a small group of people were sharing their convictions. Extraordinarily difficult, especially as the paradoxical proposition "*a* equals *b*" is exactly what mathematics is about.

Although both the old *epidictics* and the new *apodictics* are acts of persuasion, the difference is that in the former the truth of a proposition is merely asserted, while in the latter it is demonstrated; here it should be remembered that since the Athenian culture was highly polemical, there was a strong need for greater clarity in the arguments, a demand for certainty which the Greek mathematicians were determined to meet. It was in fact to that end that they invented an entirely novel type of rhetoric, an approach which set them aside from all other intellectuals, the endlessly debating philosophers in particular. As a way of reaching their goal, they focused on the *form* rather than the content of the argument, on the *how* rather than the *what* of whatever they did.

The reasoning tools they developed were surprisingly simple and essentially two: the diagram and the ordinary language, the latter highly formulaic and with a minimal vocabulary of only 100 to 200 words. And in spite of (indeed *because of*) the fact that the constructed figures were imperfectly drawn, the reasoner could always tell exactly where on the road from the particular to the general he was. What kept him on track was the practice of carefully lettering (*i.e.* baptizing) the intersections of the drawings. And for that reason there are obvious connections between the mathematician's diagram, on the one hand, and the landsurveyor's map, on the other.

This family resemblance between mathematics and cartography is further heightened by the circumstance that just as the letters of the diagrams do not stand *for* objects but *on* objects, so the main fix-point in the landscape map is not the parish church understood as a social symbol but the spire interpreted as a Peircean index, by definition a position which can be seen, pointed to and talked about. It follows that whereas modern science is a science of equations, the ancient science was a science of diagrams. So dominant was in effect this bodily mode of thought that for the Greek mathematicians the diagram became a substitute for ontology. The proofs were consequently drawn rather than spoken, a drama in which the eye, the index finger and the tongue were the lead actors. Indeed it was the simplicity in form that generated the complexity in meaning, the non-exactness of the particular drawing that led to the necessity of the general conclusion.

If Netz is correct, then it was the dual practice of finger-tracing and story-telling that generated what is presently called deductive reason. The term "shaping" in the title of his masterpiece should therefore be taken literally, itself a parallel to the fact that the most crucial proofs in Euclid's *Elements* were blessed with the approval stamp of *Quod Erat Faciendum* rather than with the better known *Quod Erat Demonstrandum*; "which was to be shown" rather than "which was to be demonstrated". Playing in the same league of legitimation is the perfect passive imperative, a verb form which in English may be rendered as "let it come about", "let it have been cut", a syntactic device which in both the speaker and the listener creates the feeling that everything has been settled beforehand. As Ludwig Wittgenstein used to put it, to follow a rule is to follow it blindly. Rephrased, the core of every proposition lies in the diagram, in the eyes of the unaware a visual illustration, in the mind of the initiate a schematic (re)presentation. Once again the Orthodox icon comes to mind, for like the icon the diagram is in actuality a picture which is not a picture.

Stunning connections! Even so, the most remarkable aspect of the Netzian reconstruction is the fact that when it was first introduced there were no diagrams of antiquity extant, and that is despite the fact that the preserved texts often refer to them. Yet he was convinced that without the picture of the lettered diagram (perhaps drawn in sand or on a dusted surface) the explorers would never have found their way in the unknown, never been able to translate their insights into a story that could be shared with others. Subsequent events have nevertheless bore him out, for not only has a palimpsest with an erased copy of a copy of the Archimedes Codex actually been found, but this treasure contains a set of lettered diagrams of exactly the type that he had envisioned (Netz & Noell 2007). What for centuries had been absent has consequently now become present again, not only

FENNIA 188:1 (2010)

in the imagination of a dedicated scholar, but in the material world as well. Glorious achievement of erasing the eraser, because what we are now beginning to understand is how we are believed when we say that something is something else. When push comes to shove, even the most theoretical physicist shows himself to be a practicing geographer.

In tentative conclusion: Like the body that projects it, every meaning is asymmetric, every map a palimpsest, every palimpsest an epistemological travel story. And since in that world of self-reference no ground is solid, no translation perfect, no projection screen untainted, I have but the faintest idea of what will happen next. No wonder that people get frightened, for how can anyone find the way in a world in which the fix-points are unfixed, the scales twisted, the *mappae* crumpled.

Such is nevertheless life in the taboo-laden in-between. And in my experience that holds regardless of whether you are an apprentice or a full-fledged poet, a young graduate student or an aging emeritus. The difference is nevertheless profound, for whereas the former keeps asking whether it will ever happen again, the latter struggles with the challenge of *not* letting it happen again, of not imitating himself, of not doing once more what he has already done so many times before.

But, who knows, perhaps all of physics, poetry, mathematics and cartographic reason are nothing but the work of mirror neurons in the brain, chemical reactions triggered by the likeness of this and that, life itself a mimetic desire that can never be satisfied. But if that is the case, what are then the relations between creativity and self-plagiarism, the metaphoric pictures of the earth and the metonymic travels of the mind?

NOTES

[1] Many thanks to the venerable Reverend Pinnawala Sangasumana for making it all possible.

REFERENCES

Auerbach E 1953. *Mimesis: the representation of reality in western literature* (translated by Willard R. Task). Princeton University Press, Princeton, NJ.

The Holy Bible. The New International Version 1973. Zondervan Publishing House, Grand Rapids, Michigan.

Dalley S 1989. *Myths from Mesopotamia: Creation, the flood, Gilgamesh, and others*. Oxford University Press, Oxford.

Miles J 1999. *God: A biography*. Simon & Schuster, New York.

Netz R 1999. *The shaping of deduction in Greek mathematics*. Cambridge University Press, Cambridge.

Netz R 2004. *The transformation of mathematics in the early Mediterranean world: from problems to equations*. Cambridge University Press, Cambridge.

Netz R & Noell W 2007. *The Archimedes codex: revealing the secrets of the world's greatest palimpsest*. Weidenfeld and Nicolson, London.

Olsson G 1991. *Lines of power/Limits of language*. University of Minnesota Press, Minneapolis.

Olsson G 2006. När himmelbjerget kom till Mohammed. In Buciek K, Bærenholdt JO, Haldrup M & Pølger J (eds). *Rumslig praxis: Festskrift til Kirsten Simonsen*. Roskilde Universitetsforlag, Roskilde.

Olsson G 2007. *Abysmal: a critique of cartographic reason*. University of Chicago Press, Chicago.

Sandars NK 1971. *Poems of heaven and hell from ancient Mesopotamia*. Penguin, Harmondsworth.

Stamm JJ & Andrew ME 1967. *The ten commandments in recent research*. SCM Press, London.

Chapter 18

The Minimalist: Three Variations on *Homo Pontefix*

Christian Abrahamsson

> For what is geography, if not the drawing and interpretation of lines. The only quality that makes my geography unusual is that it does not limit itself to the study of visible things. Instead it tries to foreshadow a cartography of thought. To practice this art, however, is incredibly difficult, for any attempt must face the challenge of being abstract enough. (Olsson 1991: 181.)

No doubt a definition of geography that bears closer resemblance to the geometrical abstractions of a Malevich or a Duchamp, the poetry of a Mallarmé or a Beckett, than to mainstream geographical thought. The triadic form returns in virtually everything that Gunnar Olsson has written, as point, line, plane; fix-point, scale, projection plane; identity, difference, excluded middle; you, me, we; =, /, –. In Olsson's words:

> The first lines are the "=" of the equal sign. In my own story, they begin in questions of trust and end up in a theory of human action structured as a tragedy. The issue concerns the relations between proper names and definite descriptions... The second line is the "/" of the unnamable in-between, another word for the void between individual and society, you and me. The third line is the "–" of the Saussurean bar. As I let the thickness of this mark between signifier and signified approach zero and infinity, I move from words to body, asking you with my eyes to ask again. Once more the issue is one of trust, conviction, taboo, and predictability. (Olsson 1991: 24.)

Why this obsession with the triadic form? One potential answer lies in Olsson's answer to the question; what does it mean to be human? In Olsson's mind no one answered this question better than Søren Kierkegaard to whom the human condition is inextricably bound by the fact that we live forwards and understand backwards, the binding glue being the conjunctive 'and'. Expressed differently, the imperfect and future tenses are intertwined *in-and-through* a third: the present. Khronos and Aeon are never in the same place at the same time, for when Khronos arrive the event has always already passed. In Olsson's words:

Most significantly Kierkegaard's entire philosophy may be read as an investigation of how this grammatical bridge of the 'and' is constructed and thereby of how the abyss between the worlds of being and understanding is negotiated. (Olsson 2007: 5.)

It is easy to see the similarities between the Danish existentialist and the Swedish minimalist. For is it not Kierkegaard's abyss that Olsson time and time again has tried to span, building provisional rope-bridges across its treacherous void, constantly taking new measurements of the ever shifting wind velocity and of the rock-base along its edges. The three bridges he has constructed bare strange resemblances to the figures, =, /, –. Being sons of their respective parents and their zeitgeists the melancholic author of *Either/Or* could not have written *Abysmal: a Critique of Cartographic Reason* just as the author of the latter could not have written the former.[1] And yet, if we look closely at the abutments of Olsson's bridges we can see the remnants of the Kierkegaardian bridges, aptly named, Æstetik, Etik, and Religion. No mere coincidence that the first lecture I attended with Olsson was given the name *Stadier på livets väg* [Stages on Life's Way].

In Olsson's minimalistic geographies it is *in-and-through* the three figures, =, /, –, that you and I, individual and society, Oecumene and Anoecumene, phenomena and noumena, epistemology and ontology are braided together. It is *in-and-through*, these figures that we are held together and kept apart; forces which take on form and forms that dissipate; limits which emerge, limit and dissipate. Hence identity would not be identity without difference; the radical insight is that the decisive question does not concern fix-points and definite descriptions. More precisely, any critique of the taken-for-granted must take its point-of-departure in the in-between in-and-through which categories and thoughts are fixed and stabilized. It is always a question of approaching the limit *in* the limit.

In my ambitions to understand Olsson's understanding of geography, I have followed a path opened by Michel Serres, Serres perhaps the philosopher who is closest to Olsson. Though the differences in their expressions are many, the similarities are most visible in the former's recurring mappings of Hermes' erratic journeys and in the latter's continuous attempts to dissect the brain of Janus (Abrahamsson 2008). In particular they share a desire to map the hidden hyphen and to grasp its continuous vibrations and variations. Serres formulates it thus:

---

1   The obvious difference being their separate understandings of God's role in the binding of Isaac, see Søren Kierkegaard *Frygt og bæven*, Copenhagen: Gyldendal, 1966 and Gunnar Olsson, *Abysmal. A critique of Cartographic Reason*, "Abr(ah)am". In the case of Kierkegaard there exists, of course, a multitude of biographical works, the best contemporary work is, to my mind, *SAK. Søren Aabye Kierkegaard, En biografie* by Joakim Garff (translated as *Søren Kierkegaard: A Biography*, Princeton: Princeton University Press, 2005), for Olsson I would recommend his autobiographical piece "Glimpses" in Peter Gould and Forrest Pitts (eds) *Geographical Voices* and Trevor Barnes' interview in this volume.

What is closed? What is open? What is a connective path? What is a tear? What are the continuous and the discontinuous? What is a threshold, a limit? (Serres 1983: 44.)

For Serres these questions signify a radical break with the philosophies of the fixed and eternal – questions that carry thought in another direction, towards a paradoxical or prepositional philosophy. A philosophy that takes its point of departure not in the fixed nouns and definite descriptions of the taken-for-granted, but instead in the prepositional, by necessity a condition that precedes any, and every, fixed position (Serres and Latour 1995). In this function prepositions and paradoxes share a quality in that they take place in-between distinct categories, in those spaces where and when the world is not yet fixed and stable. The force of paradoxes, as Gilles Deleuze reminds us, lies not in the fact that they are contradictory, but in that they allow us to be present *where-and-when* contradictions emerge (Deleuze 2004: 86). Like the *pre*position, the paradox lets us trace the movements of the in-between before it is anchored in a settled economy of identity and difference, certainty and ambiguity.

Inspired by Olsson's definition of geography, as the drawing and interpretation of lines, I set out to construct three sets of maps. One set for each of his major books *Birds in Egg/Eggs in Bird, Lines of Power/Limits of Language* and *Abysmal: A Critique of Cartographic Reason* respectively. Taken as a whole this minimalistic atlas enacts a journey, or perhaps more precisely an *erfarenhet* or *Erfarhung*, that starts *within* a fixed spatiality in which human thought-and-action is anchored in the self-sameness of identity. It then moves *through* an *unhinged* social world in which the spatio-temporal transformations are neither fixed nor taken-for-granted. Finally it edges *into* an abysmal temporality that lies beyond the maps of fixed points and social relations. And yet throughout the question remains the same: How can we live with the paradoxes of the excluded middle? The geometries of the *in-through-into* can in Olsson's world not be held apart. Obviously these are not maps in any conventional sense of the term but minimalistic expressions that try to capture the impressions which Olsson's books have left in me. It should finally be added that my (re)presentations should not be interpreted as a form of exegesis but rather be viewed in the light of Paul Cezanne's' words:

> to see ... directly consists of shifting out the character of one's subject. To paint something does not mean making a servile copy of it. On the contrary, it means seizing a single harmony out of all the interconnections one has observed, transposing these into a formal series with its own validity by means of working them up according to a new and original logic. (Cézanne quoted in Evmarie Schmitt 1995: 62–63.)

The strict formalism I adhere to comes out of the insight, as much Wittgenstein's as Olsson's, that it is not enough to say something; to be understood I have to show it as well.

•

Fixed point

Western thought is fixed in the geometrical fix-point
Fix-points transform uncertainty to certainty.
Certainties transform process to form.
The form centre, fixates and delineate.
In the limit I coincide with a world.

•
_____
•

Limitation

The problematic of the fix-point passes into a problematic of the limit.
Without a limit there can be neither difference nor identity.
A line connecting points is *both* a relation *and* a limit.
In the same gesture you and I are held together and apart.
Through the limit you and I emerge.

Relation

In the limit the relation between one and other reveals its paradoxical nature.
The paradox emerges in the abyss between *both/and*, and *either/or*.
Dissolving the paradox would collapse any distinction between two points.

Clinamen

Every relation gives rise to an asymmetry.
Through asymmetries power emerges.
In the inclination power shows itself.
Through inclination difference is created.
Difference *takes place*.

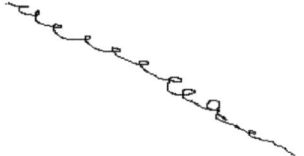

Swerve

Constant vortices between chaos and order.
Order takes shape in the movement between fix-points, limits and relations.
Order presupposes discrete positions.
Discrete positions differentiate undifferentiated spaces.
No position is forever, through fluctuations thresholds are crossed.

Fluctuation

Positions and limitations are characterised by constant fluctuations.
Fluctuations must be constrained for order to emerge.
Every difference, relation, limitation create fluctuations.
Every fluctuation implies division.
Boundary and bounded are dialectally intertwined.

Limit-zone

In the limit between the shared and the share
in-and-outside no longer held apart.
Every taken-for-granted dissipate,
uncertainty takes all.
Maps are torn.

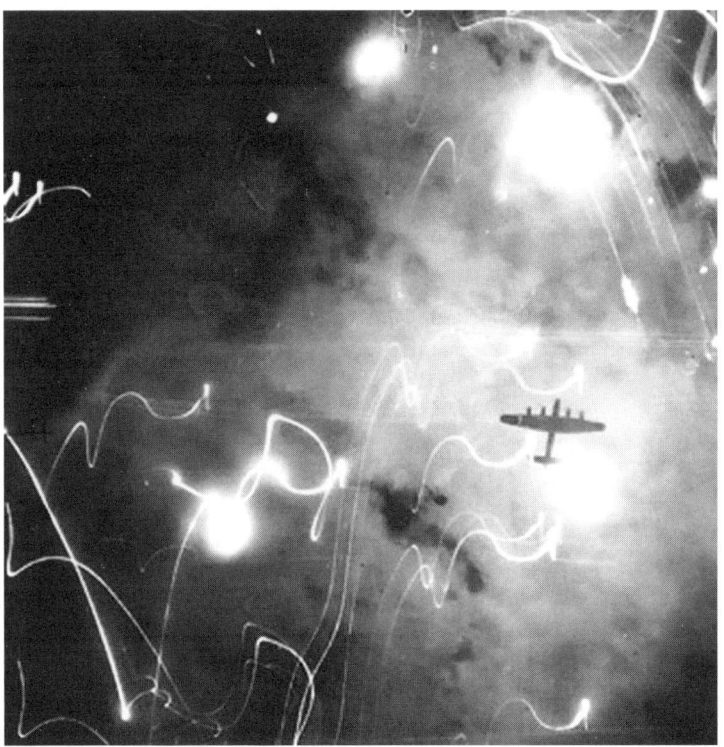

Cloud

A spatiality beyond points and relations
beyond limits of language
beyond representations that binds thought
beyond a life taken-for-granted
a world void of fix-points and stable lines.

Horror Vacui

    all that remains, an abyss
    in the absolute limit
    no distinction
    no direction
    no in- nor outside.

Bibliography

Deleuze, G. 2004. *Logic of Sense*. London: Continuum Press.

Garff, J. 2000. *SAK. Søren Aabye Kierkegaard, En biografie*. Copenhagen: GAD.

Garff, J. 2005. *Søren Kierkegaard: A Biography*. Princeton: Princeton University Press.

Kierkegaard, S. 1966. *Frygt og bæven*. Copenhagen: Gyldendal.

Olsson, G. 1980. *Birds in Egg/Eggs in Bird*. London: Pion Limited.

Olsson, G. 1991. *Lines of Power/Limits of Language*. Minneapolis: University of Minnesota Press.

Olsson, G. 2002. Glimpses, in *Geographical Voices: Fourteen Autobiographical Essays*, edited by P. Gould and F. Pitts, New York: Syracuse University Press, 237–268.

Olsson, G. 2007. *Abysmal: A Critique of Cartographical Reason*. Chicago: University of Chicago Press.

Serres, M. 1983. *Hermes: Literature, Science and Philosophy*. Edited by J. Harari and D. Bell, Baltimore: Johns Hopkins University Press.

Serres, M. and Latour, B. 1995. *Conversations on Science, Culture and Time*. Ann Arbor: Michigan University Press.

Chapter 19

Gunnar Olsson, Figures of 'Madness' and a Form of 'Schizologie'

Chris Philo

Olsson, the poorhouse and the bamboo cage

> Even more taken-for-granted was the next-door neighbour on the north side of
> the creek, the poorhouse with its half-crazy inmates of dubious extraction. Many
> were drawn to my mother, who gave them coffee and cookies while patiently
> listening to their stories about jails and workhouses, trolls and fairies, witches
> and constables. They came with a special odour, and my friends often treated
> them with a cruelty born of fear. All wore funny hats even on week days, but
> for me they were as much part of daily life as the calico cat. It was nonetheless
> considered an act of great courage when one day I sneaked into the poorhouse
> attic to watch the naked madman locked into a cage of bamboo. I can still see
> his eyes and hear his screams. Sometimes he wakes me up. (Olsson 2002: 241.)

In this remarkable passage, Gunnar Olsson reflects upon aspects of taken-for-
granted daily life in the Värmland village where he grew up, "where social problems
were a matter of concrete human beings, not of abstract social systems" (ibid.).
One might wonder at the depicted Swedish 'social system' of treating people with
mental differences, entailing 'half-crazy inmates' sent to the 'poorhouse',[1] but what
also strikes home is Olsson's childhood acceptance of these strange neighbours –
obviously inspired by the kindnesses of his mother – alongside an alertness to the
curious stories that they might tell, at once renouncing their semi-captivity and
opening on to fantastical other realities (of 'trolls, fairies and witches').[2] Then there
is the young Olsson's intrepid encounter with the 'naked madman' in the 'poorhouse
attic', confined and screaming in his 'cage of bamboo', an event insinuating fearful
images into Olsson's imagination that continue to haunt him to the present day.

---

1   On the history of mental health care in Sweden, which apparently did include until
relatively recently the use of 'poor relief' institutions and an element of physical restraint,
as in 'locking away' (see Qvarsell 1985: 2002).

2   Revealingly, Olsson also narrates how his mother provided hospitality to a young
girl from a very poor background who was bullied at school, not least by the teacher, for
being unable to read, apparently causing the young Olsson to burst into tears: "for the
inconsolable boy, it [his mother's response] was formative" (Olsson 2002: 242).

Olsson regards these memories as sufficiently important to include in his autobiographical account for the *Geographical Voices* collections (Gould and Pitts 2002), arguably lending credence to why I see merit in composing an essay about Olsson's relationship to 'figures of madness' throughout his *oeuvre*. Indeed, the passage hints at crucial matters: firstly, Olsson's non-judgemental openness to 'madness', being able to get beyond the routine 'fear' of those who look, smell, think and act in ways contrary to the norm; secondly, a preparedness to learn something from the stories told by individuals who are mentally different, even to wonder at their alternative modes of expression, of re-ordering the things of the world; and, thirdly, an understandable horror at the crazed, caged 'madman', translating into empathy at the sheer horrors of what he must have been experiencing, doubtless as exacerbated by the brute fact of spatial restraint. There is a theme here of practical (not 'preachy') compassion, primarily as shown by his mother, but there are also glimpses of broader conceptual coordinates in Olsson's work, notably those to do with the various philosophical, psychological and social *limits* that seem inextricably associated with 'madness', in what it *is*, what it signifies, how it is policed and spaced. Before turning directly to Olsson's dealings with such issues, however, I wish to take a preliminary detour through a little triangle of writers – Auster, Deleuze and Wolfson – whose musings on 'madness' can usefully frame much that I wish subsequently to essay about what I will call, for reasons about to become clear, Olsson's 'schizologie'.

Le Schizo et les Langues

In an essay entitled 'New York Babel' published in 1974, Paul Auster discusses a book published a few years earlier by a writer called Louis Wolfson, itself entitled *Le Schizo et les Langues* (Wolfson 1970).[3] As Auster explains:

> Louis Wolfson is a schizophrenic. He was born in 1931 and lives in New York. For want of a better description, I would call his book a kind of third-person autobiography, a memoir of the present, in which he records the facts of his disease and the utterly bizarre method he has devised for dealing with it. Referring to himself as "the schizophrenic student of languages," "the mentally ill student," "the demented student of idioms," Wolfson uses a narrative style that partakes of both the dryness of a clinical report and the inventiveness of fiction. ... . As we read along, wandering through the labyrinth of the author's obsessions, we come to feel with him, to identify with him ... (Auster 1998: 27–28.)

---

3   Auster's essay is called 'New York Babel', and, tellingly, the image of Babel as the tower containing all possible languages, many unknown and unknowable, appears frequently in Olsson's writings. A somewhat different reading of Wolfson's book is given by Mehlman (1972).

Displaying an "excruciatingly precise" (ibid.: 27) fascination with detail, one dimension of Wolfson's writing immerses his readers in the mundane everyday spaces of New York: sitting in the prison-like 42nd Street Public Library and refusing to leave until the doors are about to shut, watching the Times Square prostitutes, watching workmen repair crevices in a wall, chatting with his father on a city bench, being an 'urban loiterer' (Mehlman 1972: 24).[4] While also being disturbed by these spaces, not least because they risk him being continually assaulted by the use of his principal enemy, the English language spoken by strangers or written on street signs, food packaging and the like, the salience of such spaces to him – of his embodied, even 'molecular' connections to these spaces – remains fully in evidence. And there is a point here of some relevance to which I will return, since it signals a connection between 'madness' and 'geography' that will gradually acquire shape and significance as my chapter unfolds.

The book is written in French by an American, and yet it possesses a "precarious existence" between French and English, "hover[ing] somewhere in the limbo between the two languages" (ibid.: 27), indeed between several languages, and central to Wolfson's condition is the English language itself: indeed, English "has become intolerably painful to him," a language "which he refuses either to speak or to listen to" (ibid.: 28). Despite being regularly admitted to mental institutions, the doctors had found no way of 'curing' his condition. He thus spent his days living in a small apartment with his parents, earplugs in to prevent any assault from the English spoken on the radio or sung by his mother,[5] intensively studying other languages – French, German, Russian and Hebrew – and devising ways of transforming any English words that did permeate his world into "phonetic combinations of foreign letters, syllables and words that form new linguistic entities" (ibid.: 29). These carry the same meaning as their English equivalents, maybe even sounding like them, but in a linguistic sense thoroughly "destroy" them. "If the schizophrenic did not experience a feeling of joy as a result of having found … these foreign words to annihilate yet another word of his mother tongue

---

4 Interestingly, one interviewer asks Auster about his own "wonderful obsession with space" (in Auster 1998: 327). Auster replies as follows: "… looking back over my work now, I can see that it does shuttle between these two extremes: confinement and vagabondage – open space and hermetic space. At the same time, there's a curious paradox embedded in all this: when the characters in my books are most confined, they seem to be most free. And when they are free to wander, they are most lost and confused. So, in some funny way, there's a reversal of expectations about these two conditions" (ibid.: 327). This observation has great pertinence for my final remarks in this chapter.

5 When out in public space, the necessity for avoiding spoken English led him to the innovation of walking around with a stethoscope inserted in his ears connected to a tape-recorder playing music, thereby in effect 'inventing' the Walkman: "a makeshift schizophrenic object lies at the origin of an apparatus that is now spread over the entire universe, [one] that will in turn schizophrenise entire peoples and generations" (Deleuze 1998b: 13).

...," Wolfson tells us in Auster's translation, "he certainly felt much less miserable than usual, at least for a while" (in ibid.: 29).

Revealingly, Wolfson's book fascinated Gilles Deleuze, who authored a 'Preface' entitled 'Schizologie' (Deleuze 1974), which then became a chapter in his own final major work, the *Critique et clinique* collection (Deleuze 1993). This collection was itself translated into English as *Essays Critical and Clinical* (Deleuze 1998a), where the Wolfson essay became the opening chapter with the title of 'Louis Wolfson; or The Procedure' (Deleuze 1998b). Paralleling Auster's remarks, Deleuze writes about the status of Wolfson's writing:

> Th[e] schizophrenic impersonal form [deployed by Wolfson] has several meanings, and for its author does not simply indicate the emptiness of his own body. It concerns a combat in which the hero can apprehend himself only through a kind of anonymity ... . It also concerns a scientific understanding in which the student has no identity except as a phonetic or molecular[6] combination. Finally, for the author, it is less a matter of narrating what he is feeling and thinking than of saying in exact terms what he is doing. One of the great originalities of this book is that it sets forth a protocol of experimentation or activity. (Deleuze 1998b: 7.)

Deleuze unpacks Wolfson's 'Procedure' in great detail, explaining how ordinary sentences in Wolfson's 'maternal language', English (or 'the mother'), become converted into sentences composed in one or more foreign languages similar to it in sound and meaning, but of course never identical to it and always inflected through the countless inexactitudes, or 'resistances', of words refusing to permit their meanings to translate as neat one-to-one mappings. In the context of *Critique et clinique*, Deleuze's encounter with Wolfson becomes configured as part of a larger project elaborating on how works of literature – or, more particularly, experimental literary writings – comprise a form of 'delirium', arguably and ultimately on the side of Life and indeed 'health' (Smith 1998), offering resources for a constant hesitation about (allowing) the easy entry of 'things' (or, better, sensations of and ideas about 'things') into 'the symbolic'.

As Daniel W. Smith (2004: 644) suggests, "[s]ome of Deleuze's most profound texts (such as 'Louis Wolfson; or The Procedure') are those that analyse the specifically schizophrenic uses of language, which push language to its limit and lay waste its significations, designations and translations." In this vision,

---

6   Literally meaning at the level of 'molecules' transferring between him and the world, and back again, wherein, paralleling the 'linguistic corruption' that he feels when the English language seeps in through his ears and eyes, Wolfson worries at the 'chemical corruptions' occasioned by the organic life (bodies, larvae, eggs, molecules) that constantly eats away at his own body. He hence speaks to the 'molecular' obsessions of Deleuze himself (for a geographical reading, see McCormack 2007).

the deadening diagnostic criteria[7] imposed on schizophrenia are to be regretted, entailing little more than "useful terms for *not listening* to schizophrenics" (ibid., emphasis in original), and, as such, these materials are clearly themselves embedded within that broader landscape of Deleuzo-Guattarian 'schizoanalysis' (or, indeed, 'schizologie') as a profoundly different form of embodied-intellectual-ethical comportment before the world (Deleuze and Guattari 1984: 1988, also Mehlman 1972: 25–28). As is now familiar, such a comportment is anti-fascist (Foucault 1984), anti-totalitarian, multiple, nomadic, fluid, fragmentary – 'deterritorialised', insofar as it constantly resists fixture in given guises, styles, locations, territories – all apparently as resonant with "the schizo, continually wandering about, migrating here, there and everywhere as best he [*sic*] can, … plung[ing] further and further into the realm of deterritorialisation" (Deleuze and Guattari 1984: 35).

Much more could be said here about this more expansive landscape – and, indeed, its thickets form a dense undergrowth to the present chapter, very partially disentangled in some of the footnotes that follow – not least to propose that Gunnar Olsson is a practitioner of 'schizoanalysis' or 'schizologie', albeit that he differs from Deleuze and Guattari in key respects (notably in that he effects less of a decisive break from Freud *and* also in the extent to which he remains wedded to forms of semiotics). I cannot pretend to secure this claim in what follows, except perhaps obliquely.[8] It can also be proposed that Olsson began exploring the outer limits of this landscape *long, long* before it was visited by more recent geographical adventurers inspired by Deleuzian biogeophilosophy (eg. Bonta and Protevi 2004), excited by rediscovering the proliferating vital energies of life and world, together with their non-representationalist colleagues chaffing against the empire of words.[9] Again, the challenge of fully demonstrating Olsson's prior – but always generously welcoming – occupation of this landscape, as a 'geographer', is well beyond the scope of what follows (but see Doel 2003).

What I do wish to note, however, is that Deleuze himself appears to add a caution about what might be conceived as a *too*-romantic – or even 'Romantic'[10] – endorsement of schizophrenia as a 'model', a provocation, for how to live, to

---

7   Smith mentions constructs such as 'dissociation', 'detachment from reality' and 'autism'. There are problematic links being made here, and indeed in some of Olsson's remarks below, about 'madness', 'schizophrenia' and 'autism': a longer, more nuanced account would need to explore, and probably critique, such elisions.

8   Such a task would require a much more systematic textual analysis, although I am sure that a sustained running-alongside of Olsson's major works and Deleuze and Guattari's *Capitalism and Schizophrenia* volumes would reveal countless 'rhizomatic' entanglings.

9   For an overall insight into these remarkable new geographies, see Anderson and Harrison (2010a), especially the editors' introductory essay (Anderson and Harrison 2010b).

10   In the sense of the historically-rooted 'Romantic' rejoinder to the 'Enlightenment', which saw numerous writers, many with literary credentials, seeking to restate a case for taking seriously emotions, passions, the ineffable, the sublime, etc., in the face of an acutely scientific-logical-rationalising endeavour suspicious of everything that might pass for the relics of an older superstitious age. Of course, the stories of Enlightenment and Romanticism

imagine, to be creative, to evade the snares of conformity. It is true that Deleuze, writing with Felix Guattari, claims that "[a] schizophrenic out for a walk is a better model than a neurotic lying on the analyst's couch. A breath of fresh air, a relationship with the world" (Deleuze and Guattari 1984: 2);[11] a claim soaked in contempt for conceptual labour ultimately lodged in the 'closeted' interiors of Freud's encounters with and theorisations about 'neurotic' clients on the couch,[12] and a claim rich in suggestions for geographers attuned to the unpredictable fluidities of *worldly* dwelling, strolling in the "outdoors, ... in the mountains, amid falling snowflakes ... ." (ibid.).[13] Yet it is also the case that Deleuze could worry about 'delirium' sliding back into its 'clinical state' as *diagnosed* 'schizophrenia', wherein "words no longer open out onto anything, [since] we no longer hear or see anything through them except a night whose history, colours and songs have been lost" (Deleuze 1998b: lv).

This remark must be understand quite precisely, however, because the crucial notion is schizophrenia 'arrested', the artificial cessation of a schizophrenic orientation to the world occasioned when the regulatory apparatuses of conventional psychiatry (and psychoanalysis) are wheeled in, compelling the schizophrenic to cooperate, to re-present themselves *as* a patient, to reign in their delusions, to collapse them back down into themselves and away from the world, and so on. The terrors lie, in effect, not in the schizophrenia *per se* but rather in its denial, not least, so Deleuze and Guattari (1988: 362–363) argue, in its 'neuroticisation' as the schizophrenic 'on the couch' is forced to speak in the Oedipal, family-obsessed vocabularies of Freud.[14] Thus, echoing Michel Foucault's (1965, 2006)

– and of the extent to which they entangled, rather than always/simply being stark opposites
– really demand far more careful attention than such headline remarks allow.

11   This is no throw-away passage, but rather one where every word anticipates the complex architecture of the 'schizoanalytic' ethics – Foucault (1984, xiii), craving the authors' forgiveness, calls *Anti-Oedipus* (Deleuze and Guattari 1984) "a book of ethics" – that pervades every page of this work and its successor, *A Thousand Plateaus* (Deleuze and Guattari 1988).

12   The contentious hint is that psychoanalytic theory-and-practice is actually *damaged* by its spatial origins in the artificial spaces of the analyst's 'office', which is quite a hand grenade to lob into the orbit of (subfields now known as) 'psychoanalytic' and 'psychotherapeutic geographies': cf. Bondi with Fewell (2004), Kingsbury (2003) and McGeachan (2010). Moreover, it might be argued in an Olssonian vein, echoing both Deleuzian and Foucauldian critiques, that psychoanalysis "is also damaged by its spatial thinking – ... a kind of cartographic reason, that uses the Id, Ego, etc. that are supposed to be lodged in an inner psychic Euclidian space, driven by forces" (Gren 2010 [pers.com.]): cf. especially Kingsbury (2007), for a remarkable revisioning of how psychoanalysis can theorise space.

13   "It may well be that these peregrinations are the schizo's own particular way of rediscovering the earth" (Deleuze and Guattari 1984: 35).

14   This claim is very clear: Deleuze and Guattari (1984: 24, 362) talk about the "sick schizo" or "the schizophrenic who has made himself [*sic*] into an artificial person through autism" (who ends up conforming with a prior vision of schizophrenia-as-autism) or retreats

archaeology of (the 'silencing') of 'the mad', it is here that the words, 'colours and songs' of the schizophrenic are lost. Put like this, the charge – if that it be – of Deleuze 'romanticising' the condition of schizophrenia arguably remains: it is not terrifying in itself, whatever exactly *it* is, but rather terror arises in the disciplining of schizophrenics, when those labelled as such experience the horrors of regulatory environments commanded by the supposedly 'normal' and their agents.

The 'madness' here is indeed an artifice of the encounter, notably as the 'normal' refuse the alternative orderings, visions and wordings of those then judged as 'mad', a construction that, as we will see, is central to Olsson's *oeuvre*. For all that, Deleuze's caution retains its power, I feel, but arguably might be extended by the admission that there are *also* almost certainly *real* terrors felt by the schizophrenic, apart from and prior to their labelling and policing, including ones created by the kind of life, whirled around in dizzying multiplicities, differences and fusions with others, objects and world(s), that is seemingly indexed by the term 'schizo' (and to which the Deleuzians are so attracted). Maybe, then, the challenge is to exist on the limit, cultivating what might be termed a schizophrenic 'sensibility' – positioned on the side of life, health, positivity – without then becoming lost in the 'night', torment, terrors and negativity of living and being (diagnosed and then treated as) 'clinically' schizophrenic. This is mightily troubling territory, of course, mired in terminological, conceptual, political and ethical complexities, but it is, to my mind, territory of which Olsson has long been (very precisely) aware, given the proliferation throughout his work of references to the figures of 'madness'.

Language and madness in the work of Gunnar Olsson

Let me now turn to the remarkable poetic-geographical scholarship of Gunnar Olsson. As is reasonably well-known, much of Olsson's work ever since his final days as a spatial scientist formulating new 'grammars' of human spatial interaction has been about language: probing its mysteries, critiquing its limits, charting its claimed certainties, celebrating its ambiguous lapses, and wondering (again and again) about the possibilities of creating new, alternative languages. Recall the early explorations of *Birds in Egg* (the first 'half' of Olsson 1980: henceforth BEb): counterposing the languages of social science and human action, of causal and practical inference, of certainty and ambiguity, of external and internal, of things and relations; striving to find ways of generating a new language of social commentary which combines the two, avoiding the iron law of 'the excluded

---

into 'perversion' (another prior vision of what really *is* the truth of schizophrenia). To them, what in effect occurs here is a '*re*territorialisation' of schizophrenia, captured in a pre-set grid of definitions, concepts and treatments; and, in particular, they wonder at an "Oedipal reterritorialisation – an archaic, residual, ludicrously restricted sphere," which they suggest can only "form still more artificial lands that … accommodate themselves to the established order: the pervert" (ibid.: 363, see also the figure-diagram in ibid.: 282).

middle'; and experimenting with different and many-valued logical systems as a possible basis for such a language, recognisable in the academy but somehow true to the peoples and places beyond. And recall the subsequent formulations of *Eggs in Bird* (the flip 'half' of Olsson 1980: henceforth BEe): voyaging deep into the 'mandalas of thought-and-action', of a Jungian 'collective unconscious', seeking "to hang right in there" (BE, 29e) on the cusp of multiple opposites themselves already embedded in, but somehow always dancing beyond, "the prison-house of language"; and then moving excitedly towards surrealism, to Joyce and his dream-like state, in search of "a perspective which allows us to grasp both the certainty of the external and the ambiguity of the internal, of jibberish words and silent communication" (BE, 47e).

Recall the unfolding plot of *Lines of Power / Limits of Language* (Olsson 1991: henceforth LP), where Olsson follows Nietzsche, Hegel, Marx and others in further reflections on the power, the ideological loadings and commands, of the languages which capital and state compel us to think-speak; but where he also travels with the surrealists and an artist-geometer such as Kandinsky into both embodied 'spaces of silence' and the forging of new maps, topographies and cartographies whose points, lines, arcs and spaces supposedly foreshadow another language, an 'invisible geography', "a poetry that explicitly trie[s] to express the inexpressible" (LP, 206). And recall the magisterial accomplishments of *Abysmal: A Critique of Cartographic Reason* (Olsson 2007: henceforth AB), arguably the third instalment of a great trilogy, offering a sustained elaboration of themes from both of its two predecessors, in which Olsson writes a radical revisionist history of cartography – of 'mapping' in all of its many dimensions. Here he critically charts the myriad spatialisations integral to the efforts of humans, the 'semiotic animals,' as they strive to navigate the worlds of matter and meaning within which they are condemned to reside – always alert to the deceptive power of those whose 'lines' (scratched on parchments, inked on paper, etched on charts, splashed on canvas, sculpted in bits and pieces) always tell 'lies', sometimes create 'truths', often assert that (logically) only a can $= a$ at the same time as depending upon a (bodily) realisation that a really $= b$, and so on. In the latter work, the specific target is indeed *reason*, a form of reasoning that is 'geometric' in its abstract, topological precision but also, we might say, 'maddening' in its neglect of 'geography' as topographical specificity; but at the same time Olsson pushes this reasoning to its limits, in effect compelling it to fold back on itself, notably in diverse surrealist subversions, thereby intimating (once again) new cartographies, new languages, new geographies.[15]

---

15   I have endeavoured to respond to Olsson's three major texts – in Philo (1984, 1994, 2008) – and other geographers' responses to these works can be found in the Abrahamsson (2007) collection, specifically on *Abysmal*, and also Abrahamsson (2010), Doel (2003, 2009) and Gren (1994); and also chapters throughout collections such as Farinelli *et al* (1994) and Picone (2002).

In these works Olsson does not play around with different languages – English, French, and so on – in the same manner as does Wolfson, 'the schizophrenic student of languages,' although there are moments when he does use his envious knowledge of different European languages to destabilise, enrich, elaborate his use of English (and of course much of his writing is originally in Swedish anyway). Yet I reckon there to be important parallels between Olsson and Wolfson, since in many respects much that passes for the language(s) of academia and intellectual debate, and certainly all that is found in the languages of planning and policy (including those of "geography-and-planning": LP, eg.18), does indeed appear to become 'intolerably painful' to Olsson, entailing words, sentences, whole discourses that he now 'refuses either to speak or to listen to.' (This is admittedly an overstatement, but perhaps not by so much.) And, more significantly, just as Wolfson spends hours at his desk struggling to create a new language, distinctively his own language, so I have the impression of Olsson, this craftsperson of genius and resolve, spending hours in his study patiently, unflinchingly, endeavouring to use all of the resources of language available to him – whether from Joyce, Marx, spatial science or (less so) the local donut shop – to create (the possibilities for) another language, this fleeting, magical 'invisible geography' of black marks on a white page.

This claim then leads to the further question: is Olsson, like Wolfson, 'mad'? But, having put the question like this, I instantly wish to backtrack, because one underlying insistence of my paper is that we must be exceptionally *careful* about deploying the term 'mad', and similarly its (approximate) modern equivalents such as 'mental illness', 'schizophrenia' or even 'mental health problems'. I want to underline that there are grave dangers in too cavalier a deployment or metaphorisation of the term madness[16] and its many associates, and I will circle back to this warning in concluding. As such, let me rephrase the question to ask whether there are parallels between the fate of Wolfson in being *labelled* as mad, as mentally unwell and schizophrenic, and the status of, as it were, the project of searching for another language, a new poetic geography of thought-and-action, which so energises Olsson. Wolfson's refusal to engage with English in the midst of a predominantly English-speaking society, together with his concerted efforts to find another language outwith the language conventions of the peoples and places all around him, obviously leads him into encounters with doctors, psychiatrists and others responsible for dealing with those individuals who, in whatever way, do not 'fit in' with the mainstream (and who maybe pose problems for, even challenges to, that mainstream). His obsessions and eccentricities, to some extent consciously chosen but to a large extent probably not, duly earn him both the societal label of mad, which he might occasionally accept and internalise, and all that duly ensues therefrom.

---

16   From now on, I will not put scare quotes around 'madness', 'mad', etc., even though the complexity, ambiguity and ethico-political charge of such terms really demands such scariness be retained!

Olsson's experiments with language, meanwhile, maybe do lead him to be derided in some quarters as having gone mad; but, much more subtly, it is Olsson himself who realises that anyone, himself included, who conducts such experiments, who craves to find alternatives beyond the constraints of conventional language as practised in the academy, boardroom and state chamber, risks having themselves identified and dealt with as mad. The madness hence lies in the stepping beyond limits, not necessarily because there is some ontologically and epistemologically coherent other mode of being 'out there' which *is* madness, although I would not completely foreclose on this possibility; but because such transgressions are commonly taken by wider society as signifiers of madness, a term whose chief referent is therefore less some fully known reality (some truth of mental illness, schizophrenia and so on) and more just the 'facts' of transgression (of someone apparently not reasoning and behaving properly, as expected, as condoned).

As a result, it should come as no surprise to learn that the figure of madness shadows much of Olsson's writing, precisely because it is a term indicating people who step beyond limits, whose ways of thinking-speaking-acting transgress the norms of mainstream society. Olsson is attracted to madness, as it were, and sympathetic to those like Wolfson whose difference has been named as madness and thereby pathologised, positioned as 'deviant' and often physically removed to set-apart spaces – asylums, mental hospitals, clinics, even bamboo cages – where their difference can be concealed, managed and perhaps remedied. Revealingly, Chris Smith, when reflecting on the progress of mental health geography since his own pioneering studies in this field as a graduate student at the University of Michigan in the 1970s,[17] remembers the interest that Olsson showed in his visits to a local mental hospital:

> Olsson especially liked to read the notebooks that I brought back from the 'field', a collection of bizarre stories, in which lives were being transformed by the agonising realities of schizophrenia and related mental disorders. (Smith 2000: 252.)[18]

Smith acknowledges Olsson's extremely positive and enabling influence with respect to his research, noting that he "probably used up a few bushels of academic capital convincing his colleagues that my choice of specialty ... was legitimate, and that I should be allowed to continue" (Smith 2000: 252), but it is revealing that Smith then adds a few remarks with a more critical edge about Olsson's

---

17   Smith actually commences his PhD thesis with this reference to Olsson: "Why would a geographer study mental patients? ... Allow me to use a popular idiom, and describe mental patients 'across the water'. Gunnar Olsson showed me that it was indeed worthwhile to look across to the opposite bank, irrespective of what one might see there ..." (Smith 1975: ii).

18   The resonance with Olsson's own recollections of the 'stories' told by the 'half-crazy' folk from the neighbouring poorhouse, with which I began this chapter, is striking.

fascination with his 'mad stories' from the field.[19] (And I will also return to these critical remarks in conclusion.)

Figures of madness in Olsson's *oeuvre*

Let me now turn to Olsson's three major texts, *Birds*, *Lines* and *Abysmal*, to interpret some of the passages where he explicitly mentions and elaborates upon the figures of madness. At various moments in all three texts he considers the deep epistemological and ontological gulfs separating the language of social science from the language of human action, repeatedly emphasising the stark contradictions between the rigidity of the former and the flexibility of the latter, between what the academy demands and the streets necessitate. Yet what he also stresses is the extent to which these two languages *are* commonly combined in practice, interwoven and drawn upon almost at once, within much of what passes for everyday life (even in the everyday workings of the academy, if not in its intellectual outputs).

Olsson hence argues that the peculiar force of many ideologies which have ended up programming our world – faith in Christianity, in the capitalist free market, in 'state capitalism',[20] in neo-liberalism, in communism – rests precisely in a capacity for blending the certain and the ambiguous, the external and the internal, the 'is' of now and the 'ought' of the future; at which point he offers his principal deconstructive insights on the operations of power. The character of Janus, the mythical god with two or many faces, is here deployed by Olsson to picture this capacity for certain powerful ideologies to fuse "logical contradictions" (LP, 14), but what he also asks of Janus and therefore of these ideologies is this question:

> What I would really like to understand is how he manages to deal with the double bind in such a fashion that he is celebrated as a god and not put away as a schizophrenic (LP, 14).

---

19   I know too from personal discussions that Olsson has been fascinated by 'outsider art', otherwise known as 'the art of the insane', given the window that such artistic productions apparently provide on other ways of engaging with, apprehending and representing the spaces of both external and internal worlds. See also Rhodes (2000) and Various (Hayward Gallery, London). For a geographical treatment, see Park *et al.* (1994); also Parr (2007). Intriguingly, 'outsider art', or *l'art brut*, appears as a significant early reference-point in *Anti-Oedipus* (Deleuze and Guattari 1984: 6–7), with a wonderful account of 'the schizophrenic table' illustrated in one such artwork.

20   Olsson writes on various occasions about the materialising of 'state capitalism', in part reflecting developments in the context of his native Sweden, but also anticipating, I would suggest, much that others now see inherent within the neo-liberal agenda that both celebrates markets while extending a moral authoritarianism.

> Why, and how, for instance, did the Romans elevate this categorial juggler
> [Janus] to godly status, when we diagnose his counterparts as schizophrenic
> madmen? Why did they afford him a special place in their pantheon, while we
> isolate his likes in the soundproofed cells of the asylum? (AB, 6.)

These passages signal Olsson's recognition that thinking-and-acting in a manner
that appears to meld or at least to accept the presence of what appear as logical
contradictions, perhaps paradoxes, to handle or at least to live with opposites,
to face simultaneously in different directions, can be taken as a symptom of
schizophrenia, of having fallen into a state of madness.

What clearly then concerns him – very obviously in *Abysmal* and his account
of the (brazen unpredictability of) the Old Testament God – is that it is seemingly
'okay' when ideologies, states and oppressors of all kinds effectively enact this
'double-bind' themselves, leaving them celebrated as 'gods' or unquestioned
sovereigns. Such is one of the most clever 'tricks' of power, Olsson asserts
repeatedly, and one of the 'tragedies' of how power does its stuff, of how the
powerful exert their domination over countless others, is precisely that the latter
recognise, not to say manipulate, logical impossibilities – such as $a = b$ – to
secure their dominion over the peoples and places involved. Conversely, when
the powerless display an equivalent forsaking of logical consistency, particularly
if coupled to obvious 'deviations' of appearance and conduct, they may be liable
to being 'put away' – socially and spatially removed from the world – because of
their mad, schizophrenic tendencies. In another telling passage, Olsson gestures to
the 'taboos' of the powerful, of those who contribute most directly to the shaping
of our societies and cultures, wherein the Janus-like quality of 'deal[ing] with the
double-bind' is kept hidden as a kind of mad truth that cannot be revealed to the
masses:

> Direct contact with / is culturally forbidden. This is why it intrigues me, for
> whatever is dangerous enough to be taboo is important enough to understand. …
> And yet, culture is founded on its limits, civilization on its madness. (LP, 100.)

The '/' is the 'slash', the 'excluded middle', the Saussurian 'bar' or limit
between the signifier and the signified, between the materialised sound-image
and the dematerialised concept, and it is obvious throughout Olsson's writing
that he supposes any attempt to inhabit the '/' to be a form of, or tantamount to,
madness. And here, echoing Foucault's (1967) historical geography of 'madness
and civilization,'[21] Olsson provocatively relates the making of (modern Western)
civilization to the madness that is concealed within this slash, this limit, that is

---

21   I discuss this remarkable work, explaining why I regard it as a thoroughly 'spatial
history' of madness and asylums, in Philo (2004: Chap.2).

rarely voiced[22] (except by those who may end up being 'put away') but which nonetheless informs, sanctions and energises so much that is done by the guardians of scholarship, art, economy and politics.

I will return shortly to Olsson's worries about incarcerating madness, but let me clarify his supposition that within everyday life 'we' are all constantly struggling to resolve the contradictory impulses within the two grammars of being and understanding (of certainty and ambiguity, externality and internality, materiality and immateriality) which are forever propelled through (how we try to apprehend) our lives, our spaces, our minds, our hearts. And it is evident that he marvels at how most of us, for much of the time, do "learn", as he puts it, "to live in …[23] double bind without going crazy" (LP, 106). In a similar turn of phrase, and mentioning "the double-face Janus, the flip-flopper[24] who knows how to join contradictions without going crazy" (LP, 117), Olsson sets up the ground for a subsequent statement where "[o]nce again I call for Janus, more prominent than ever for the Romans," pleading "[h]elp me become a sinner[25] yet not lose my sanity" (LP, 191).[26] The further implication is that Olsson reckons some people to be particularly at risk of not being able to deal on a daily basis with these contradictory impulses, leaving them vulnerable to episodes of mental and emotional disorientation which might be cast as a form of madness or mental illness. He duly speculates, with particular reference to psyches bewildered by the supposed certainties of 'teleology' meeting the greater openness of 'dialectics', that:

---

22   As Gren (2010 [pers.com.]) puts it, "Cartographic reason will perhaps know, but it will also hide its knowledge and often 'forget' or hide the translation [of *a* into *b*] in the background."

23   Olsson inserts the word 'institutionalised' here, presumably to reference all of those many 'institutions' of state, capital and other authorities (e.g. the church) which, as *per* his broader analysis of how their Janus-like power operates, effectively force (some) ordinary people into a potentially mad or schizophrenic state of being and understanding.

24   I think that the metaphor here may be to a person playing pinball, using the 'flippers' in the machine to keep the ball in play. There is also the hint that in this specific context Olsson is imagining a conventionally powerful figure, a representative of state, capital, church, etc., who somehow knows how to play this 'doubling' game of power: indeed, in the next sentence, Olsson muses that "[p]erhaps it is therefore Janus, the janitor, who is the incarnation of modern power" (LP, 117). Such themes are writ large throughout *Abysmal*.

25   I take it that the 'sinner' here is someone who transgresses by merging what logically should be held apart – someone who 'sins' against the conventional rules of 'cartographic reason'.

26   Olsson later writes: "[w]ith the aim of understanding how Janus stayed sane while ordinary people in similar situations of double bind go crazy, I will try to place him on the operation table, cut his skull open, lay his brain bare, investigate how his mind is wired" (AB, 6).

> The inability of some individuals to appreciate and to cope with this ambiguous
> mixture of teleology and dialectics may in extreme cases lead to mental illness.
> (BE, 110b.)

Elsewhere, he gives a literary example of an individual caught in an in-between
situation where opposing ambitions and dictates, hopes and fears, abruptly collide:

> In Dostoevesky's novel *The Double*, the low-ranking clerk Golyadkin hires
> a magnificent carriage to take him down the Nevsky Prospekt. He wishes to
> impress. But suddenly another carriage pulls up alongside his. Inside it sits not
> a woman to be seduced, but his superior Andrei Filipovich to be obeyed. No
> place to hide, caught where he should not be. The other had come too close.
> Eventually Golyadkin was to go mad. (LP, 139, Note 18.)

Elsewhere too, Olsson considers Professor Hintikka's "formulation of possible-
worlds semantics with a novel approach from game theory" (LP, 44), concluding
that:

> [The] problem is that the number and complexity of possible worlds is too
> large to be grasped and comprehended; the *wff's* or theorems produced by the
> syntactic machinery leave the interpreter with too much to believe. Any sane
> person would go crazy for less. (LP, 44.)

This passage indicates a slight shift away from what might be seen as the bipolar
structure of Olsson's previous writing, the sense of us all veering backwards and
forwards between the twin polarities of certainty and ambiguity (and their many
correlates), since here he stresses the sheer multiplicity of 'possible worlds' – of
possible ways of thinking-and-acting, of outcomes, of realisations – that may
occasionally overwhelm us and prompt a bout of craziness.

On the matter of shutting up madness, a rather different tack adopted by Olsson
is to propose that, within the ongoing battle between certainty and ambiguity, it
is often those who lean too far towards the axis of ambiguity who end up being
liable to incarceration under the 'accusation' of madness, mental illness or
schizophrenia. In a particularly telling passage from an essay that appears in both
Birds and *Lines*,[27] Olsson critiques the powerful 'czars' of certainty who imprison
troublesome adherents to alternative visions, as well as sympathising with those
individuals who get sent to 'asylums' for remaining – wittingly or otherwise –
wedded to more ambiguous ways:

---

27   This is 'Epilogue 4' in *Birds* and also 'Set your mind at rest' in *Lines*: Olsson
remarks that, "[o]f all my writings none has influenced me more" (LP, 216), and so we
can accord considerable significance to what Olsson argues here around his reference
to madness.

> It is in the Galilean tradition of objectification to join the war on certainty's side. Some of its casualties are buried in the Gulag Archipelago, murdered by the czar who never learned his dialectics. The Aristotelian tradition has fought to preserve subjectivity by siding with ambiguity. Its victims are in schizophrenic infirmaries, punished because they categorise differently. (BE, 63b; LP, 69.)

On various occasions, Olsson acknowledges that he joins this 'war' more on the side of ambiguity, which he also regards as the side of 'creativity'.[28] He duly searches high and low for alternative possibilities for thought-and-action, as I have already described, and it is instructive that he is inspired by the events and performers at the so-called 'Festival of Fools' in Copenhagen:

> During the summer of 1985 [Mr. Stelarc] exhibited, lectured and performed at the Fool's Festival in Copenhagen. As expected, his presence stirred much attention. As a consequence, he was also prohibited from carrying out his most spectacular plan. His hope had been to rise into the evening sky dangling under a balloon, hung in eighteen ropes that, by means of a kind of fish hook, had been pierced through the skin of his back, arms and legs. (LP, 147.)

The example of Mr. Stelarc also illustrates Olsson's increasing insistence, during the later pages of *Lines*, on looking for fresh, maybe mad, possibilities in the implicatedness of our 'silent' bodies in the overall modalities of language, communication, reaching out to others.

More generally, recognising that he wishes to get beyond just "hitting [his] head against the ceiling of language,"[29] he sets himself the project of battering against the boundaries of what he refers to as "a prison-house of language built within the walls of the taken-for-granted" (LP, 23),[30] arguing in terms resonant for this chapter that:

> Even though this prison is escape-proof, one can nevertheless bend the bars and catch a glimpse of the other side, tempting and mad at the same time. (LP, 23.)

Olsson's poetic and surrealist experiments are efforts at spying these possibilities, 'tempting and mad', beyond the cages of language as usually conceived-and-

---

28   He admits that he chooses to "focus mainly on the concept of ambiguity" (BE, 41e), to be "momentarily stressing ambiguity over certainty" (BE, 41e), and to be encouraging further 'distortions' to help along creativity: "It is at those moments that … subjective ambiguity can be allowed to rule over objective certainty" (LP, 14).

29   This is the title of Chap. 15 in *Birds* (actually the opening chapter in the *Eggs* part of the book).

30   He claims that writing the first part of *Birds* "brought me to the prison-house of communication" (BE, no pagination e).

spoken,[31] and in one intriguing aside he speculates that "we must get used to reading and writing in the dreamlike state which Joyce practised and Freud and Jung analysed" (BE, 47e). In another aside, borrowing from *A Midsummer Night's Dream*, he envisages an 'imaginative compact' between "[t]he lunatic, the lover and the poet" (in AB, 115). At the same time what he also attempts is to formulate an 'invisible geography',[32] an 'invisible map' (Olsson 1991), a 'heretic cartography' (Olsson 1994), "a cartography of understanding" (LP, 184) which can comprise a new language for social science that transcends the limitations of previous and other languages.[33] In one fragment of discussion, Olsson portrays his quest for this invisible geography as a mad endeavour:

> Viewed together, the lines themselves get cornered into points of power. From the turning and twisting of these correlates emerges a set of figures hitherto unseen. Perhaps they are mad. But perhaps the penumbrae, the squares and the cones, together contain seeds of a valid theory of human action. (LP, 24–25.)

I might personally query whether any construction retaining some allegiance to the forms of Euclidian geometry,[34] however distant, is appropriate for representing the 'turning and twisting' of minds beset with madness, whatever exactly madness might entail as both a reality and as the object of interventions;[35] but Olsson undoubtedly does align his own experiments, geometric, poetic and surreal,

---

31   While throughout much of *Birds* into the first chapters of *Lines* attention is paid to the languages of both certainty and ambiguity, my impression is that the argument as *Lines* progresses tends to be one contrasting more the will to certainty within all language (as thought, spoken and written) with the potentials for ambiguity in the silences of bodily interaction. The latter can be conceived of as just another dimension of language, of course, and this angle is also present in Olsson's claims. Moreover, the possible linkages across to the claims of the non-representational geographers might here be registered.

32   See Gren (1994), Chap. 7, which is entitled 'The invisible geography of Gunnar Olsson'.

33   This invisible geography of "marks drawn on the dematerialised surface of a basic plane" (LP, 184) structures the whole of *Lines*, but is most obviously developed in the two chapters towards the close of the book, 'Squaring' and 'Malevich torpedoed'. See my commentary and sympathetic critique, wondering about the echoes of spatial science in Olsson's 'mad geometry', in Philo (1994): cf. Doel (1999: 128–131).

34   Although yet another construction present at moments in Olsson's work (notably *Abysmal*) is one that regards 'cartographic reason' as itself inherently mad, precisely because it ends up posing an equivalence of 'world' and 'map' ($a = b$) that ('schizophrenically'?) it knows to be impossible, but also because it can become the basis for highly 'top-down', technocratic, apparently rational (but in many respects entirely *irrational*) forms of socio-spatial planning, urban and regional policy, population management, etc.

35   A theme underlying some of my own work (esp. Philo 1999) is that 'the geometric spirit', intensified by the Enlightenment, created far too rigid spatial grids, both imagined and real, for capturing and potentially taming the otherness of 'madness'.

alongside the envisaged *allies* of ambiguity, creativity and madness. I am at once excited by such an alignment and unnerved by some of its implications, as I will elaborate in closing this paper.

An additional line of connection with madness is established by Olsson as he seeks to imagine something of the contents folded into what might be entailed by quite other ways of thinking-and-acting in the world. A useful observation is the following on how surrealists and postmodernists strive for 'authentically' different patterns of relations between objects and their representation:

> While most scientists and conventional artists try to achieve some sort of equivalence between copy and original, both the surrealists and the postmodernists try to subtract one set of associations from the other; the greater the disparity, the more powerful the light. To be an authentic actor of this type is to associate what is normally dissociated, and to dissociate what is normally associated. There is nevertheless a deep sense of humour in the craziness. (LP, 160.)

Associating 'what is normally dissociated' and dissociating 'what is normally associated' – connecting up words and things quite differently to how it is usually done – thus brings with it a 'craziness', and perhaps too some humour. Importantly, though, Olsson is here raising the challenge of new imaginings which break with what we commonly take as the fixity of the world – the socio-spatial orderings, the boundedness and purity of categorisations – and which thereby effect a dislocative stance before words and things that might be taken as akin to madness:

> The paradox is that to create new meanings is to break the hitherto taken-for-granted, ie. to violate the tautologies of connectives. But this means that I simultaneously must follow a rule and stretch the skin of its boundary. Here lies temporary madness. For, 'if I were sometime to see quite new surroundings from my window instead of the long familiar ones, if things, humans and animals were to behave as they never did before, then I should say something like 'I have gone mad'; but that would merely be an expression of giving up the attempt to know my way about. (LP, 134.)[36]

> ... the challenge is that when I look out of my window and see something I never saw before, then it is uncertain whether the signs reflect social things or solipsist mind. Out there, in here? You or I. Madness or Nowhere (LP, 198).

---

36   The embedded quotation is from Wittgenstein, and elsewhere Olsson writes as follows: "Ludwig Wittgenstein noted that madness is a word for being lost, an expression for not finding one's way in the world" (AB, 103); "Once again paraphrasing Ludwig Wittgenstein: when I am looking out of my window, I no longer see the familiar vistas but a hazy assortment of paradoxes and predicaments" (AB, 189).

If the expected 'order of things' really becomes quite other to what we expect, our sense, indeed our fears, about what is happening might be that madness had descended upon us: "Imagine if this text suddenly came really close, sniffing, farting, dripping! You should then say something like 'I have gone mad'" (LP, 141). And suppose that we try to communicate this apparent new 'order of things' to those around us, then it must be likely that they would interpret our condition as one of incipient madness (an interpretation which might then lead us to the asylum, mental hospital or clinic). To say this is basically to reiterate claims already made about Olsson's treatment of the figures of madness, but there are also signs here of further issues: notably the possible awfulness of experiencing a madness which *does* disorientate an individual's normal grasp on the world, its things and the words customarily attached to them, leading him or her to feel hopelessly 'lost' and unable to navigate around. To be sure, Olsson is still captivated by the novelty of such madness, by the threat to hegemonic orders and attendant challenge to think different orders, and he cannot resist wishing for a new social science – a new human geography perhaps – resting on these mad (dis)orientations. But he does also seem to appreciate problems and human costs *for* any people who actually experience such disorientations day-to-day, as will be explored further in conclusion.

The reference to 'solipsist mind' is pertinent,[37] since another thread in Olsson's tapestry is the notion that the fragile truths of the world as we *individually* perceive them can never be *collectively* communicated without their being hopelessly compromised. In our own immediate existence, whatever might be taken as meanings, 'fixes' or coordinates relevant to our own personal grasps on reality simply cannot be communicated, because they can make no sense, can only be meaningless, to others operating within the normal orbit of socialised meaning or language.[38] In this respect, we all teeter on the brink of madness, Olsson asserts, as we approach being locked into a solipsistic, self-referential state that demands the speaking of a 'private language' replete with 'private meaning': once again, indeed, a state that might easily be identified as madness. As he writes:

> Does this mean that to be mad is to approach the meaninglessness? Perhaps. Because in such situations I am utterly on my own, I am not like anybody else, I have no rules to follow, no fixes to keep me steady. Viewed in this perspective, meaning shows itself as what it really is: an instrument of socialisation, a

---

37 Wittgenstein's meditations on the (im)possibility of truly 'private languages' also clearly inform much of Olsson's writing, especially in *Lines*, as well as providing a potential framework for considering further the links between Olsson's experiments with 'the social space of silence' and Wolfson's 'mad' personal-linguistic experiments. I must thank Eric Laurier for pointing such things out to me.

38 Olsson often argues that we are all forced to be 'dumber' than we really are as we deny the truths that we really know in accepting the need to communicate partial (un)truths to others through the social mediums of meaning and language.

standardised tool designed to make you and me norm(al) and alike. Just as there can be no private language, there can be no private meaning. The so-called meaninglessness is a private silence, the so-called meaningful a public chatter. (LP, 152–153.)

By this token, to mumble truths to oneself is to enter into and to risk being allocated the label of mad, whereas to proclaim untruths is to fit in with the accepted orthodoxies of shared and sharing meanings. And Olsson does see much to value, even to celebrate, in the madness of maintaining a 'private silence'[39] over and against the tyranny of a superficially (but not really) meaningful 'public chatter'; which is when he also suggests that "this is why the autistic child preserves its integrity by refusing to speak" (LP, 156). This suggestion then recurs in another of Olsson's reflections on the relations of S/s:

When produced by the capital S the silence is that of idle chatter; when produced by the small s it is that of autism. The former exemplifies society conversing with itself. The latter is the individual refusing to be a part. (AB, 193.)

What Olsson is getting at when referencing 'the social space of silence'[40] hence becomes much clearer, in that it is not just about the bodily possibilities for non-language-based communication, as in Barthes's 'lover's discourse', but also about what may be occurring within, and maybe learned from, the disobedient and unpredictable silence – or at least the refusal to speak in the social categories of the crowd – displayed (so it is claimed) by the mad person, the autistic child, the solipsist. Thus, to quote Olsson again:

One meaning of meaning is obviously to make you and me obedient and predictable. Like the language in which it is expressed, meaning enters as an ingredient in the ethical glue that fastens individual to society and society to individual. Perhaps it takes a madman or a solipsist to illustrate what a meaningful geography could be. (LP, 153.)

The figure of madness is now marshalled in a manner which does ring of celebration, an attraction even for Olsson, but my feeling here too is that he remains aware of the grave difficulties that result *for* people who actually end up 'condemned' to such a personal silence outwith the public prattle of language-based sociability.

---

39   Or, in the 'private silence' of madness.
40   'The social space of silence' (Olsson 1987) is one of Olsson's most famous essays, and it is reproduced in a shortened form in *Lines*.

'Learning from' but not 'romanticising' madness?

Having surveyed Olsson's intriguing depictions of 'madness' in his major texts, let me now move to a conclusion by returning to Smith's memories of Olsson's interest in his reports from the Ypsilanti State Hospital, ones glimpsing lives 'being transformed by the agonising realities of schizophrenia and related mental disorders.' As Smith continues:

> I say 'realities' deliberately here, because it seemed to me at the time that Olsson (among others, including the British psychiatrist R.D. Laing[41]) believed that these twisted versions of reality were in fact the 'real' thing; that this was the way minds really worked, and that our job, as outsiders, was not to untwist them but to accept them for what they were. I was not convinced of this at the time, and am still not. I was sure of only two things: that these 'illnesses' were real, whatever else they were, and that they were very terrifying. (Smith 2001: 252.)

This is not quite how I would interpret Olsson's perspective on madness, since my reading would be that Olsson *does* suppose there to be a 'reality' embodied in the condition referred to as madness (it being *more* than *just* a label). Yes, he certainly sees it as a construct imposed by society on troublesome and troubling individuals, maybe himself included, but it is also the case that he regards the mad voices to be saying something genuinely different, other, important about the world. And this is the basis for Olsson's fascination with the 'twistings' of these troubled minds, in that he detects here a significant 'place' to look for genuinely new ways of understanding and speaking which illuminate new possibilities, new worlds, even new geographies.

In so doing, Olsson duly fits into a much longer tradition within Western intelligentsia of, if not 'romanticising' then at least asserting a certain positivity to 'learning' from madness. Think about ancient Greek notions of the close affinities between 'madness' and genius, or Shakespearian notions about 'the fool' holding up a crucial mirror to the world (to the follies of men and women), or the much more recent 'anti-psychiatry' ruminations of R.D. Laing (as hinted by Smith above) and, most obviously, David Cooper (1967: 1978). This notion is also very much present in the original phenomenological architecture of Michel Foucault's (1965) *Madness and Civilization*, which teeters on the brink of supposing there to be a coherent 'language of Madness' largely silenced since the Enlightenment by the apparatuses of 'Reason', but faint murmurs of which can still be heard in the occasional creative outbursts – the 'lyricism of protest' – arising from the likes of a Nietzsche or an Artaud (or a de Sade or a Van Gogh).[42]

---

41   For a rather different take on Laing, see McGeachan (2010).

42   These dimensions of Foucault's *Madness and Civilization* are complex and controversial – they are brilliantly dissected by Derrida (1981) – and it is telling both that in later editions of his book Foucault suppressed the original 'Preface', which most starkly

Similarly, it might be claimed that Olsson's approach to madness parallels a wider poststructuralist celebration of what the voices of madness can apparently teach 'us', Westerners caught within the interlocking grids of grand theory, state bureaucracy and capitalist imperatives, and there are surely traces of just such a 'romanticising' of madness as a 'line of flight' from these inflexible spaces of thought-and-action to be found within – as more substantively outlined in my introduction to this chapter – the 'schizoanalysis' of Gilles Deleuze and Felix Guattari (1984: 1988).[43] Tellingly, on one of the few occasions where Olsson mentions Deleuze, he states that, "[e]ven though the (post)modernist Deleuzian is a schizophrenic, he is not a crazy madman but an outstanding cartographer, a person who has accepted the challenge of mapping the many connections that determine what it means to be human" (AB, 141).[44] There would thereby seem to be warrant for describing Olsson as a practitioner of, if not a full-blown Deleuzo-Guattarian 'schizoanalysis', at least the more (language- and literature-based) 'schizologie' announced in Deleuze's preface to the Louis Wolfson book.

Serious issues remain, however, and it is instructive to hear these wise words from Derek Gregory (1994: 156) when reflecting upon the project of a 'schizoanalysis': "there is … something cruel – at the very least insensitive – about analogising schizophrenia like this."[45] Gregory's stance here, in effect standing

---

positioned Reason and Madness as two great oppositional entities locked in fierce struggle down the ages, and that at several moments in his later writings – eg. Foucault (1972: 16, 47) – he rejects the notion of "trying to reconstitute what madness itself might be, in the form which it first presented itself to some primitive, fundamental, deaf, scarcely articulated experience." Even so, in the 'Conclusion' to *Madness and Civilization*, the creative fires of poetry, surrealism and other radical artistic products as echoes of a previously much more common, loudly and persuasively spoken 'language of Madness' very definitely remain. I review this tangled terrain of issues pertaining to *Madness and Civilization* in Philo (2004: Chap. 2). Much remains to be done by way of exploring the phenomenological undertows of the original, unabridged version of *Histoire de la Folie*, the French title of the original 1961 book, which has now been translated and published in full (see Foucault 2006): the extent to which Foucault was initially pulled, one might say 'romantically' as well as 'politically', to some primal language of madness, as the repository of truths long-silenced and all-but-forgotten, remains open to sustained discussion.

43   For a sustained geographical account, rather than ones merely mouthing the term 'schizoanalysis', see Doel (1995).

44   Fascinatingly, Olsson also insists that "[t]he work of Deleuze and Guattari represents one of the most sustained critiques of cartographical reason that have yet appeared" (AB, 455, Note 17), going on also to observe how the necessary connections have been explored by Doel (1999, 2003) as well as referencing Bonta and Protevi (2004).

45   A convincing response might nonetheless be that the Deleuzo-Guattarian position is precisely *not* dealing with metaphors or analogies, but is discussing 'real' embodied schizophrenic conditions which are always, everywhere disrupting the battalions of theory, state and capital: see Doel (1995, 1999). I am unsure: to my mind, the logical thrust here can never entirely escape some sense of arguing 'metaphorically' from the (apparent) experiences of people with a *real* schizophrenic sensibility, whether clinically defined or

against the 'romanticising' of madness, recurs in the comments of Smith when stressing what he regards as the all too real terrors, agonies and fears experienced by many people who slip into the alternative mode of being which gets named as madness (and which he is prepared to term a form of *illness*, mental illness). And I have actually leaned towards a similar position in my own writing with Hester Parr, notably when we have drawn upon James Glass's equation of madness with "psychological placelessness, a horrifying experience of aloneness and disconnectedness" (Glass 1989: 58).[46] Borrowing from Glass, we have speculated that many people with severe mental health problems, especially schizophrenics, actively wish to counter the flows, multiplicities and disorientations of their psychotic experience, precisely because this 'placelessness' is truly terrifying.[47] As a result, so we argue, these people commonly endeavour to re-anchor themselves in the simple and unshifting places of their immediate environment, the mundane micro-details of which become such a comfort and are therefore deliberately (and often delicately) observed, memorised and cultivated, all as elements in a process whereby such people strive to *re*-place themselves materially and (all being well) psychologically in the everyday world of their 'fellow citizens' around them.[48] The incredibly detailed drawings and paintings typical of much 'art of the insane', complete with the myriad heavy lines which cry out of a struggle to achieve 'fixity' and 'containment', are perhaps indicative of this process. It may also be appropriate in this connection to recall Wolfson's "eye for detail" (Auster 1998: 29), his constant 'objectification' of the city environment which, in Auster's interpretation, allows him "to create a space between himself and himself, to prove to himself that he exists" (ibid.: 30).

While it can be argued that Olsson does verge on a 'romanticisation' of madness in his work, an objection that evidently lies in the background of Smith's comments on Olsson's interest in his mental hospital field notes, I would want to qualify such an argument in certain respects. Indeed, I gain the sense from Olsson

---

not, to specifying *both* broader ontologies – of *real* phenomena and processes permeating, even creating 'the social field' (e.g. Deleuze and Guattari 1984: 273–283) – and an overall 'ethics' of intellectual-political address before the world.

46   Glass is a political scientist. His ideas are explored at greater length, and set alongside remarkable first-hand descriptions of psychotic experience, in Parr (1999; also Parr and Philo 1995).

47   Yet it is precisely these flows, multiplicities and disorientations that we so often seem to be celebrating in our poststructuralist texts. For related warnings about over-romanticising the figure of the 'nomad' in such texts, see Atkinson (2000) and Cresswell (1997).

48   Illuminatingly, Gren (2010 [pers.com.]) – deliberately adapting an Olssonian perspective – speculates that, for people experiencing mental ill-health, "[if] the fix-points in there float, move around, [their usual] cartographic reason breaks down and [they] need other fix-points, perhaps a connection to what [they] actualise through unshifting places of their immediate environment, micro-details, etc. Might there be a 'cartographic reason', or another reason ..., or mapping, that would help them better to navigate in the world of human relations?"

that, not only does he register the pain and hurt caused to mad people 'put away' in the asylum by social control agencies fearful of their otherness,[49] he is also aware of the internal sadness and loneliness that accompany the so-called mad (or autistic) person who does remain – wittingly or, as is probably most often true, unwittingly – locked into a private space of silence, hearing and speaking only his or her own solipsistic 'private language'. (And, again recall Wolfson, who was certainly not happy in his personal schizo-language.[50]) Notwithstanding some impressions to the contrary, Olsson is clearly *not* urging us all 'to go mad,' for he knows what personal costs are at stake, not just to our freedom but potentially to the very fabric of our being able to continue living-in-the-world. For me, then, Olsson displays clear insight into the horrors of what may run through the maddened psyche,[51] but most profoundly he writes as follows about both the enchantment and the 'horror' of being a mad person – or at least like the mad person – who has been cast adrift from their bearings, their anchorage, in the normal and shared 'fixes' of the everyday world:

> Does this mean that to be mad is to approach the limits of meaning? Perhaps! For in such situations I am left completely alone, for then I am not like anybody else; if 'a' denotes myself, then there is no 'b' to go with it. Not knowing my way about is another way of saying that I am completely lost, with no fixes to keep me steady, with no contexts to share. It is exactly at this moment of *horror* that the equal sign shows itself in its imperial nakedness; a blessing in disguise, an instrument of socialisation, a standardised tool for making you and me normal, predictable and interchangeable … (LP, 171, my emphasis.)

An underlying architecture to the particular formulation of this quote is a Wittgensteinian equation of madness with being or feeling 'lost' that is yet another recurring motif throughout Olsson's *oeuvre* (see also Footnote 36), as towards the close of *Abysmal*:[52]

---

49   Olsson despairs about "the wardens in the prison hospital of mad conclusions" (LP, 55).

50   Wolfson's linguistic experiments were "undertaken in the hope of one day being able to speak English again – a hope that flickers now and again through the pages of the book. The invention of his system of transformations, the writing of the book itself, are part of a slow progression beyond the hermetic agony of his disease. By refusing to allow anyone to impose a cure on him, by forcing himself to confront his own problems, to live through them alone, he senses in himself a dawning awareness of the possibility of living among others – to break free from his one-man language and enter a language of men [*sic*] (Auster 1998: 35).

51   At one point he writes that "killing one person with a bloody throat bite is so brutish that the thought itself takes its holder to the asylum" (LP, 101).

52   Abrahamsson (2010: 16, my insertion) also works creatively with this Wittgensteinian notion of getting lost as madness, acknowledging, though, that we need to

> Ludwig Wittgenstein – Surveyor General of the Lands Unknown – would almost certainly have know what the released [those 'released' from Plato's Cave] would have said: 'I have gone mad!' Immediately after these words had left his mouth, however, he would have added that 'this expression is merely another way of saying that I don't know my way about, that my social compass has lost its bearings, my map its scale' ... (AB, 385.)

The mad person is here one who has 'lost their geography,' taunted by "the horror of the absence of markings" (AB, 85), and passages such as this one, even given their 'philosophical' glaze, are always haunted by the terrors attendant on actually *going* 'mad' – of slipping into *real* states of mental difference where the end-point is the fear of lost coordinates: "an echoless scream in a mountainless valley, an open-mouthed girl on a bridge across a fjord without shores, nowhere more movingly presented than in the paintings of Edward Munch" (AB, 85). To avoid the 'horror', to avoid the madness, is hence to accept the 'blessing in disguise' which is the 'equal sign', the sign that enables us to name, to designate, to fix what *is* in the world, what we can conventionally trust to remain stable, enduring and simply there around us, repeatedly available to us in one way or another. In these words, Olsson hence comes close to the formulations of someone like Glass, with his dual stress on both the 'horror' of 'psychological placelessness' within psychosis *and* the value of recovering, in Olsson's words, 'fixes to keep me steady' which are 'normal, predictable and interchangeable'.[53]

Even so, what Olsson does still do, I would finally caution,[54] is to offer an account which risks moving in the *opposite* direction to that of many people with mental health problems who are hoping, not to transgress limits, but rather to find a more stable mental and even physical ground on which to settle. For one last time, then, let us remember Wolfson's book, acknowledging that it was at least in part a piece of self-therapy – although it should be regarded as challenging

---

ask: "Why are we [*most of us, but not all*] getting lost or insane when we encounter a world that does not sit still?"

53   It is also instructive that Olsson thinks explicitly about the role of therapy in assisting people out of madness, writing as follows with reference to Lacanian notions about the unconscious being 'structured like a language': "The therapeutic strategy is first to discover a repressed signified and then to kill it with an explicit signifier. The patient speaks herself well, for in the powerfilled act of naming, anguish becomes graspable. The horrible and noncommunicable loses its frightening grip once it is caught in shared categories and domesticated in common expressions" (LP, 179).

54   Echoing a caution from my 1994 extended review of *Lines* (Philo 1994), repeated in my review of *Abysmal* (Philo 2008), albeit the latter work arguably voyages closer to *real* 'geography', even intimating a critique of too-geometric 'geometry', particularly when it becomes complicit with the erasure of 'markings' (see quotes directly above in the main text), and hence retains a greater sense – theoretically and ethically – of why there *is* a necessity never to lose 'touch' with Shakespeare's 'local habitation' (with a name, not just an [x;y] co-ordinate location).

literature, not just as a work of therapy[55] – and noting that Wolfson did indeed strive to *move back* into the banalities of everyday life, the local parks, sidewalks, streets and shops included.

Olsson's tendency throughout much of his writing, if perhaps less so in *Abysmal*, is precisely to keep moving in the other direction, to *move away* from such banalities, to find ways of thinking ever more abstractly, distantly from the 'crazy' tiny details of routine spaces and places. In my 1994 review I detected a tension in his work between the fragments of everyday life and geography, on the one hand, and this grander and all-transcending aspect to his ultimately modernist (less postmodernist) experiments with language, poetry and surrealism in the company of Joyce; and here I think that I am finding a similar tension in his treatment of the figures of madness. What he says as he tackles this figure remains entirely inspirational – the wish to uncover new possibilities for thought-and-action, to create new languages for the social sciences, new 'invisible geographies' – and he does register something of the terrors, the isolation and the societal condemnation which may ensue for the person who slips into a mad state of being (and who gets labelled as such). But it is just possible that in their efforts to return to the normal, to the details of here-and-now, to draw lines around the constituents of the here-and-now that might stop them from draining away in multiple flows, the mad, or people with mental health problems, cannot but embody a critical counterpoint to Olsson as they do to any ungrounded 'romanticisation' of their condition, torments and delusions. I am convinced that Olsson would not wish anything more nor less from them: he knows all too well that the world, and *its* multiple madnesses, will never stop asking him the most serious of questions.[56]

Acknowledgements

The original version of this paper was presented at a session organised in honour of Gunnar Olsson held during the Annual Meeting of the Association of American Geographers in New York, February–March 2001, and I must deeply thank all involved for their generous reception of my presentation. I must now record my further deep thanks to Christian Abrahamsson and Martin Gren for inviting me to contribute to the present collection, and for encouraging me to write this very substantially revised version of my 2001 effort. Martin, in particular, warrants my

---

55   "[I]t should not be dismissed as a therapeutic exercise, as yet another document of mental illness to be filed on the shelves of medical libraries" (Auster 1998: 33–34).

56   If confirmation was ever needed on this score, see Olsson (2008). When asked to consider the unfathomably vicious madnesses of the Rwanda genocide, he admits that "the intellectual tools which in the past have served me so well seem no longer up to the task" (Olsson 2008: 752). This said, it is amazing to see what he does then achieve in this context through reworking, with an obvious heaviness of heart, his notions (from *Abysmal*) of 'fix-point', 'scale' and '*mappa*'.

special gratitude for his careful critical reading of my re-draft. My most significant thanks here, however, must go to Gunnar himself for being such a major and ongoing influence upon my own scholarship: very little that I have thought-and-written over the years has been entirely untouched by his 'geographical' heresies, at once playful and deadly serious. Additionally, I have valued his remarkable generosity of response – in short, his kindness – over so many years, ever since that first letter from him of 23rd February, 1982 (maybe he remembers …).

Bibliography

Abrahamsson, C. (ed.) 2007. Collection of papers (by C. Abrahamsson, A. Bonazzi, J.-D. Dewsbury, and J. Pickles) discussing Gunnar Olsson's "Abysmal", *Geografiska Annaler B*, 89, 381–397.

Abrahamsson, C. 2010. *Topoi/graphein*. Uppsala: Uppsala Universitet.

Anderson, B. and Harrison, P. (eds) 2010. *Taking-Place: Non-Representational Theories and Geographies*. London: Ashgate.

Anderson, B. and Harrison, P. 2010. The promise of non-representational theories, in *Taking-Place: Non-Representational Theories and Geographies*, edited by B. Anderson and P. Harrison. Farnham: Ashgate, 1–34.

Atkinson, D. 2000. Nomadic strategies and colonial governance: domination and resistance in Cyrenacia, 1923–1932, in *Entanglements of Power: Geographies of Domination/Resistance*, edited by J. Sharp, P. Routledge, C. Philo and R. Paddison. London: Routledge, 93–121.

Auster, P. 1998a (orig. 1974). New York Babel, in *The Art of Hunger: Essays, Prefaces, Interviews and The Red Noteboo*, by P. Auster. London: Faber & Faber, London, 26–34.

Auster, P. 1998b (orig. 1992). Interview with Mark Irwin, in *The Art of Hunger: Essays, Prefaces, Interviews and The Red Notebook*, by P. Auster. London: Faber & Faber, 327.

Bondi, L. with Fewell, J. 2003. 'Unlocking the cage door': the spatiality of counselling, *Social and Cultural Geography*, 4, 527–547.

Bonta, M. and Protevi, J. 2004. *Deleuze and Geophilosophy: A Guide and Glossary*. Edinburgh: Edinburgh University Press.

Cooper, D. 1967. *Psychiatry and Anti-Psychiatry*. London: Tavistock.

Cooper, D. 1978. *The Language of Madness*. Harmondsworth: Penguin.

Cresswell, T. 1997. Imagining the nomad: mobility and the postmodern primitive, in *Space and Social Theory: Interpreting Modernity and Postmodernity*, edited by G. Benko and U. Strohmayer. Oxford: Blackwell, 360-379.

Deleuze, G. 1970. Schizologie [Preface], in *Le schizo et les langues,* by L. Wolfson. Paris: Editions Gallimard, 5–23.

Deleuze, G. 1993. *Critique et Clinique*. Paris Les Editions de Minuit.

Deleuze, G. 1998a. (trans.; orig. 1993). *Essays Critical and Clinical*. London: Verso.

Deleuze, G. 1998b (trans.; orig. 1970/1993). Louis Wolfson; or the procedure, in *Essays Critical and Clinical* by G. Deleuze. London: Verso, 7–20.

Deleuze, G. 1998c (trans.; orig. 1993). Preface to the French edition, in *Essays Critical and Clinical* by G. Deleuze. London: Verso, lv.

Deleuze, G. and Guattari, F. 1984 (trans.; orig. 1972). *Anti-Oedipus: Capitalism and Schizophrenia*. London: Athlone.

Deleuze, G. and Guattari, F. 1988 (trans., orig. 1980). *A Thousand Plateaus: Capitalism and Schizophrenia, Vol.II*. London: Athlone.

Derrida, J. 1981 (trans; orig.). Cogito and the history of madness, in *Writing and Difference*, by J. Derrida. London: Routledge, 31–63.

Doel, M. 1995. Bodies without organs: schizoanalysis and deconstruction, in *Mapping the Subject: Geographies of Cultural Transformation*, edited by S. Pile and N. Thrift. London: Routledge, 226–240.

Doel, M. 1999. *Poststructuralist Geographies: The Diabolical Art of Spatial Science*. Edinburgh: Edinburgh University Press.

Doel, M. 2003. Gunnar Olsson's transformers: the art and politics of rendering the co-relation of society and space in monochrome and technicolour. *Antipode*, 35, 140–167.

Doel, M. 2009. The man who fell to earth and mistook himself for a map: book review essay on Olsson's "Abysmal". *Progress in Human Geography*, 33, 280–285.

Farinelli, F. Olsson, G. and Reichert, D. (eds) 1994. *Limits of Representation*. Munich: Accedo.

Foucault, M. 1965 (trans; orig. 1961). *Madness and Civilization: The History of Insanity in the Age of Reason*. London: Tavistock.

Foucault, M. 1972 (trans; orig. 1969). *The Archaeology of Knowledge*. London: Tavistock.

Foucault, M. 1984. Preface, in *Anti-Oedipus: Capitalism and Schizophrenia*, by G. Deleuze and F. Guattari. London: Athlone, xi–xiv.

Foucault, M. 2006 (trans; orig. 1961). *History of Madness* [unabridged and revised English translation]. London: Routledge.

Glass, J.M. 1989. *Private Terror/Public Life: Psychosis and the Politics of Community*. Ithaca: Cornell University Press.

Gren, M. 1994. *Earth Writing: Exploring Representation and Social Geography, In-Between Meaning and Matter.* Göteborg: Publications edited by the Departments of Geography, University of Gothenburg, Series B, no. 85.

Gren, M. 2010. Personal e-mail communication, 15/09/2010.

Kingsbury, P. 2003. Psychoanalysis, a gay science? *Social and Cultural Geography*, 4, 347–367.

Kingsbury, P. 2007. The extimacy of space. *Social and Cultural Geography*, 8, 235–258.

McCormack, D.P. 2007. Molecular affects in human geographies. *Environment and Planning D: Society and Space*, 39, 359–377.

McGeachan, C. 2010. *The Geographies of R.D. Laing* (Unpublished PhD thesis, School of Geographical & Earth Sciences, University of Glasgow, Scotland, UK).

Mehlman, J. 1972. Portnoy in Paris: review of Wolfson's "Le schizo et les languaes". *Diacritics*, Winter, 21–28.

Olsson, G. 1980. *Birds in Egg/Eggs in Bird*. London: Pion.

Olsson, G. 1987. The social space of silence. *Environment and Planning D: Society and Space*, 5, 249–262.

Olsson, G. 1991. *Lines of Power/Limits of Language*. Minneapolis: University of Minnesota Press.

Olsson, G. 1991. Invisible maps: a prospectus. *Geografiska Annaler*, 73B, 85–92.

Olsson, G. 1994. Heretic cartography. *Ecumene*, 1, 213–234.

Olsson, G. 2002. Glimpses, in *Geographical Voices: Fourteen Autobiographical Essays*, edited by P. Gould and F.R. Pitts. Syracuse, NJ: Syracuse University Press, 419–430.

Olsson, G. 2007. *Abysmal: A Critique of Cartographic Reason*. Chicago and London: The University of Chicago Press.

Olsson, G. 2008. Untitled. *Environment and Planning D: Society and Space*, 26, 752–757.

Park, D.C., Simpson-Housley, P. and de Man, A. 1994. To the infinite spaces of creation: the interior landscape of a schizophrenic artist. *Annals of the Association of American Geographers*, 84, 102–209.

Parr, H. 1999. Delusional geographies: the experiential worlds of people during madness / illness. *Environment and Planning D: Society and Space*, 17, 673–690.

Parr, H. 2007. Mental health, the arts and belonging. *Transactions of the Institute of British Geographers*, 31, 150–166.

Parr, H. and Philo, C. 1995. Mapping 'mad' identities, in *Mapping the Subject: Geographies of Cultural Transformation*, edited by S. Pile and N. Thrift. London: Routledge, 199–225.

Philo, C. 1984. Reflections on Gunnar Olsson's contribution to the discourse of contemporary human geography. *Environment and Planning D: Society and Space*, 2, 217–240.

Philo, C. 1994. Escaping Flatland: a book review essay inspired by Gunnar Olsson's "Lines of Power/Limits of Language". *Environment and Planning D: Society and Space*, 12, 229–252.

Philo, C. 1999. Edinburgh, Enlightenment and the geographies of unreason, in *Geography and Enlightenment*, edited by D.N. Livingstone and C.W.J. Withers. Chicago: Chicago University Press, 372–398.

Philo, C. 2004. *A Geographical History of Institutional Provision for the Insane from Medieval Times to the 1860s in England and Wales: 'The Space Reserved for Insanity'*. Lewiston and Queenston, USA, and Lampeter, Wales, UK: Edwin Mellen Press.

Philo, C. 2008. Review of Olsson's *Abysmal'*, *Annals of the Association of American Geographers*, 99, 205–209

Picone, M. (ed.) 2002. *Bodies and Space: Gunnar's Travels*. Palermo: Laboritorio Geografico, Università di Palermo.

Qvarsell, R. 1985. Locked up or put to bed: psychiatry and the treatment of the mentally ill in Sweden, 1800–1920, in *The Anatomy of Madness: Essays in the History of Psychiatry, Vol.II – Institutions and Society*, edited by W.F. Bynum, R. Porter and M. Shepherd. London: Tavistock, 255–267.

Qvarsell, R. 2002. History of psychiatry in Sweden. *History of Psychiatry*, 2, 315–320.

Rhodes, C. 2000. *Outsider Art: Spontaneous Alternatives*. London: Thames & Hudson.

Smith, C.J. 1975. *The Residential Neighbourhood as a Therapeutic Community* (Unpublished PhD thesis, University of Michigan, Department of Geography).

Smith, C.J. 2000. Many years on … when afar and asunder? *Health and Place*, 6, 251–255.

Smith, D.W. 1998. A life of pure immanence: Deleuze's *Critique et clinique* project', in *Essays Critical and Clinical*, by G. Deleuze. London: Verso, xi–liii.

Smith, D.W, 2004. The inverse side of structure: Zizek on Deleuze on Lacan. *Criticism*, 46, 635–650.

Various. *Beyond Reason: Art and Psychosis – Works from the Prinzhorn Collection*. London: Hayward Gallery.

Wolfson, L. 1970. *Le Schizo et les Langues*. Paris: Editions Gallimard.

Part E
CARTOGRAPHIC(AL) REASON

Chapter 20

MAPPA MUNDI UNIVERSALIS

a commentary
on

THE POWER OF CARTOGRAPHICAL REASON

performed at the

Uppsala International Contemporary Art Biennial Eventa 5
Section II – OBOG
PER ASPERA AD ASTRA

Uppsala Cathedral

September 4-10, 2000

GUNNAEL JENSSON

alias
Ole Michael Jensen & Gunnar Olsson

GENESIS

In the beginning are the heavens and the earth. The earth is without form and void and there is darkness upon the face of the earth. Nothing to see, nothing to hear, nothing to touch, nothing to smell, nothing to taste. No distinction. No identity. No difference. In this spiritual land of silence there is a meaning so meaningful that it refuses to be expressed.

A wind moves across the void. Lightening. Thunder. Rainbow in the sky. Out of the mist raises a flat granite rock that gently slopes into the distant sea. In this physical land of silence is a matter so material that it emits no meaning whatsoever.

Strange creatures emerge out of Nowhere, quickly spreading across the rocky ground. A foot gets stuck in a crevice, others observe, and for the first time there is a difference important enough to make a difference. In the intercourse of the body and the stone, the origin of man is conceived.

The primordial distinction - **a** - splits into three: one a shadow of the shadow (a); one a tautological expression which keeps repeating that it is what it is (a=a); one an informative statement insisting that it is something else (a=b). Identity and difference separated and united. Atoms of understanding captured in a mushroom cloud of perpetual fission.

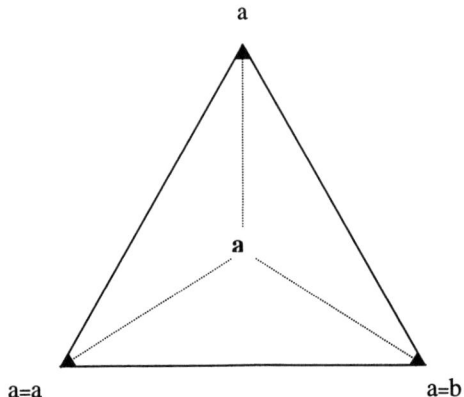

Points of distinction

When the tension reaches its limit, the cloud bursts into fire. Out of the ashes grows a crystal palace, sometimes known as the crucible of man, sometimes as the prison-house of language. Both the floor and the walls are built as equal-sized equilateral triangles, the walls invisible, the foundation sunk into the granite ground. At the center of the basement is a well of identities and differences, its opening covered by a red-colored lid, itself the memorial mark of the original distinction and the trace of the first sacrifice, the blood from the killing of an identical twin, the footprints of a deviance turned scapegoat, the navel of what it means to be human.

A twist of cultural survival and the four-cornered deities merge into one, the multitudes of polytheism concentrating in the singularity of monotheism. In the process of that unmooring, absolute power finds its place at the top of the pyramidal structure, a pivotal point which is the locus of a tautological and nameless entity that defines itself as that which it is: **a=a**. A contradictory condensation of identity and difference, one God one Being, the Almighty created in the image of man.

From its inception this Absolute speaks, its power one with its language, its language one with its power. "Let there be!" And there is. A universe flowing out of the actor's mouth.

In the coolness of the evening, the Absolute looks back at what he has uttered, claiming first that it is very good, then that he alone has the right to judge. Tolerating neither idols nor false prophets he declares that all usurpers will be killed. Impressed by his own achievements, he finally proclaims a day of rest, a sabbath without work, twenty-four hours devoted to the glorification of himself and his faithfuls. Such is the subjection of subjects, such is the structure of dictatorial power. Now as well as then, then as well as now.

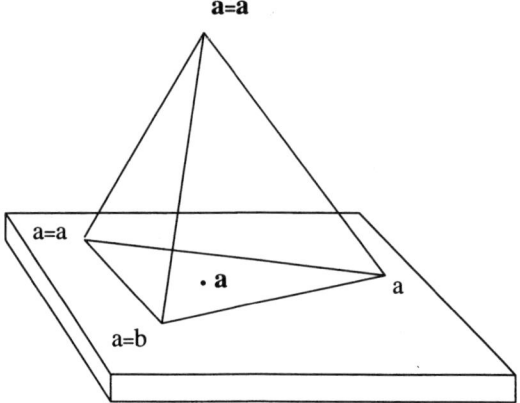

Crystal Palace

LAWS

The **Almighty realizes** that he leads a dangerous life. For that reason - and with the purpose of regulating the relations between ruler and ruled - he formulates a Constitutional Law, a power-filled document which begins with the reminder that it was he who liberated the suppressed, he who cut their chains, he who brought them out of the land of bondage. A rhetorical trick based on the false premise that old wolves automatically turn to peaceful sheep. And that is why the political atrocities of people like Lenin, Pol Pot and Fidel Castro are easy to explain, impossible to excuse.

Paragraph 1: You shall have no other ruler before me.

What is that? You shall fear and love your Leader above everything else, placing all your belief and trust in him. Martin Luther was rejected by the Pope in Rome before he was accepted by the king in Stockholm, the praying Mohammed faced Jerusalem before he turned to Mecca.

E**very dictator knows** that to stay in power he must squash all opposition. But he also knows that all power rests in language. This in turn explains why the operationalization of the Constitutional Law takes the form of three lines extending from the top of the crystal pyramid to the corners at the base. It is through these communication lines that the Lord and his people are tied together into a network of mutual dependence. Issued from the top are orders that follow no rules but their own, predictable only in their unpredictability. Echoing from the underground is nothing but legitimation. In the rooms of the Almighty anything goes; in the dungeon of the critics translation is strictly forbidden.

Coupled with each line is a segment of the Law, each one a different aspect of §1. While paragraphs 2a and 2b are alike in the sense that they block all roads towards representation and thereby to critical understanding, paragraph 3 rules that in conflicts between the forces of collective cohesion and individual rebellion, the former are always right. While the paragraphs 2a and 2b ensure that only the Absolute can eat from the tree of knowledge, paragraph 3 focuses on the relations between categorization and socialization.

Paragraph 2a: You shall not make for yourself a graven image or any likeness of the Almighty, for no power-wielder allows his secrets to be revealed. No statue, no picture.

What is that? You shall fear and love your Lord without questioning who he is or what he does, accepting every decree as an integral part of the taken-for-granted. Forbidden is every attempt to picture the invisible, every attempt to sculpture the untouchable. In the world of the crystal palace, the message of §2a is transmitted through the communication line that runs from the tautological **a=a** to the informative a=b.

It is in the formulation of this paragraph that the law maker demonstrates how sophisticated his conception of power actually is. For the prohibition against images stems from the recognition that the likeness of metaphor plays the same role in the rhetorical art of ontological transformations as the igniting charge does in the engineering science of blasting. Since it is this device that turns untouchable ideology into touchable stone, visible structure into invisible meaning, the Absolute does his utmost to ensure that it will never fall into enemy hands; just as the ordinary dynamiter jealously guards his detonators, so does the Mosaic Absolute.

So important is this principle that any transgression is most severely punished - executed are not only the sinning fathers but also their children, grandchildren and great-grandchildren. Any ruler who resorts to penalties of that magnitude is scared out of his wits. Given the outrageous claims of §1, he ought to be.

Paragraph 2b: You shall not misuse the name of your Leader, never tie his proper name to a definite description. For every power-wielder is eager to punish the truth-sayers and to reward the ass-lickers.

What is that? You shall love and fear your Lord and never use his name for evil wishes, swearing, lying or deception. Forbidden is every mode of reformulation, every deconstruction of the sign, every visit to the Saussurean Bar. In the blueprint of the crystal pyramid, §2b is twined into the line that runs between the big tautology of **a=a** and the small tautology of a=a, the former located at the top of the world, the latter stuck into a corner at the bottom.

The Almighty categorically refuses to be categorized, and that is why he chooses a tautology as his name: "I am who I am". Always true, never informative. Yet, for the ruler who rules by systematically contradicting himself, no name is more appropriate. Beatings at dusk, blessings at dawn. Fear institutionalized. For he who has hit you once is likely to hit you again.

Tautology is the name of the Almighty. It is also the fixpoint of two-valued logic. And herein lies the paradox of the social sciences, for whereas the words and objects of power never sit still, the words and objects of science must not hop capriciously about. It follows that logical analyses will never lead to a proper understanding of power itself. In studies of Venus, the planet, Bertrand Russell's theory of proper names and definite descriptions is often helpful. To the understanding of Venus, the goddess, it contributes little but confusion. The concepts of transparency and obliqueness are like oil and vinegar.

Paragraph 3: You shall attend all party meetings, never enter into the no-man's land between clean and unclean, never bite the hand that feeds you.

What is that? We shall fear and love our Leader together, gratefully honoring and obeying all his commands. For united we stand, divided we fall. §3 is the line that runs from the elevated **a=a** to the shadowy a.

While the regulations of §§2a and 2b address man in its appearance as a semiotic animal, §3 speaks to us as social and political beings. Of the three paragraphs, the third is in fact the most crucial, for without rules of conduct there is neither society nor individual, neither language nor power. It is by intercepting and decoding messages sent through this channel that we learn why and how we become so obedient and so predictable; the imperative of communication rests on a foundation of social control. The question is a question of socialization, the answer a bucket-full of insights lifted from the well of original distinctions.

PROJECTIONS

Lines are not only connectives between points, they are also dividers between planes. In the latter function they constitute the corners of the crystal palace, at the same time separating and uniting the adjacent walls.

Grasping the nature of the invisible walls is extremely difficult, for even though we seem to be standing on the outside looking in, we are in fact confined to a life on the inside. In this prison-house of language there are no windows and no escape routes, only a constant bouncing against unbreakable walls. Yet there are human lives and experiences of incredible richness, for the pyramid is in reality a gigantic movie theater, in each corner a projector, every wall a screen. The arrangement is familiar, its predecessors in the cave of Plato's <u>Republic,</u> in Fra Angelico's rendering of the <u>Annunciation</u>, in Marcel Duchamp's <u>La Mariée mis à nu par ses célibataires, même</u>. Without the limestone wall, the wood panel and the transparent glass, these artists would have nothing to show, no means for capturing the split-up versions of the original distinction **a**.

When the golden rays hit the opposite plane at a right angle, they rebound to the point of origin, offering no news from the travel. This is the case of perfect translations and perfect signs, imaginable in theory impossible in reality. From the a=a projector come the mantras of tautology, which - when captured by the altarpiece of religion - appear as the perfect sign of $a=a$, in Charles Sanders Peirce's terminology an *icon*. From the corner of a=b beam a set of definite descriptions, informative statements cast onto the wall of science as the perfect sign of $a=b$, a so called *index*. Finally, projected from the shadowy point of a are a series of artistic images thrown onto the canvas of aesthetics as a, that particular version of the perfect sign which in Peirce's philosophy is called a *symbol*.

As the tautology is the fixpoint of Aristotelean logic and Old Testament power, so the ninety degree angle is the fixpoint of Euclidean geometry and New Testament penance. But in the real world of imperfect communication, the projection lines never strike the knowledge planes straight on. Instead of bouncing back to their respective point of origin, they are reflected onto one of the other walls. This combined principle of uncertainty and complementarity explains why the three modes of knowledge - religion, science, and the arts - never appear in their pure form, always as a series of approximations, a chain reaction in which the grammar of religious belief turns to the rhetoric of scientific law, the rhetoric of scientific law to the aesthetics of artistic expression. And so on, and so on, one meaning colliding with another, the critical mass the trigger of itself.

And so it is that the corner lines of the crystal palace at the same time communicate the paragraphs of the Constitutional Law and mark the limits of the three planes of knowledge. Since limits by definition are taboo, the movements from one screen to another occur only through the intermediary of a transition rite. In Christianity these magic formulas are one with the sacraments, in Catholicism seven, with Luther condensed into two: baptism and communion. The latter ritual is intricately intertwined with §3 of the Law, the former with §§2a and 2b; while the naming ceremony of §2a is performed in the water of images, in §2b it is done in the medium of ordinary language.

No wonder that so many sinners pray that their trespasses be forgiven, no wonder that so many dictators kill those who trespass. Walter Ulbricht and Jean-Marie Le Pen are two variants on the same theme of exclusion, the one locking in, the other locking out.

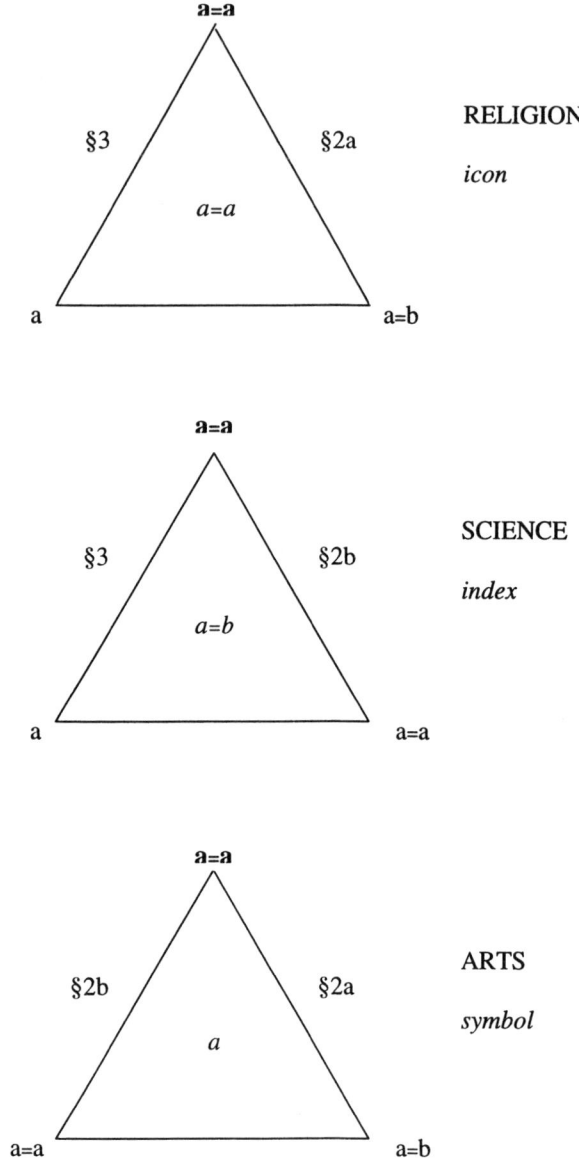

Planes of projection

TRINITY

In the blueprint of the crystal palace one trajectory is left invisible, perhaps because it is too culturally specific to be noticed. This is the line that runs from the **a=a** of God, the Father, to the **a** of the Holy Spirit. As recalled, the **a** stands for the covered mark of the original distinction, the **a=a** for its tautological reformulation, the trace of an aborted attempt at reification, a desire to concretize the abstract. The King is dead, long live the King!

Impressed by his own omnipotence the Absolute wants to return to the crevice of the original distinction, no longer satisfied with being the tautological reformulation **a=a**, but obsessed with the desire of once again being one with the untouched purity of the **a**; as an early Narcissus also the Almighty strives to merge with his own image. To that end he tries to do away with himself by jumping out of his privileged position at the top. But in mid-air - moments before smashing through the lid and plunging into the covered well - he is caught in a semiotic rescue-net whose corners are fastened to the perfect signs of the icon, the index and the symbol. The prototype of this ontological surface was first invented in Nicea in the year 325, whence it was formulated as a political/ theological commentary on the issues of representation in general and the prohibition against images in particular.

Floundering in the western net of perfect signs is nothing less than Jesus Christ, here known under the pseudonym **a=b**, by definition a complete merger of mind and matter, the logical union of **a** and **a=a**; as recalled, the only Son of God is defined as being of one substance (*homoousion*) with the Father, incarnated by the Holy Spirit and made man. Himself a sacrificed scapegoat, this epistemological savior eventually ascends to the heavens, gets appointed to the Highest Court, wherefrom he is now sentencing both the quick and the dead.

The most remarkable story of an ontological transformation ever to be told. The word made flesh, the unnameable named. Eucharist.

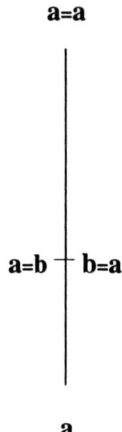

Line of trinity

MAPPING

$\mathbf{T}$he **crystal palace** is a well guarded castle, its ruling resident the tyrant of tyrants. Admittedly a rhetorical exaggeration, for no Absolute is absolutely absolute, no crook crooked enough to live on for ever. And for that reason it should be recalled that just as the doctrine of *Imitatio Dei* - the imitation of God - plays a central role in Jewish piety, so the imitation of Christ fills the same function in Christianity. The difference is nevertheless crucial: to imitate God (**a=a**) is to recant a tautological contradiction (a=a), to pray the mantra of the icon (*a=a*); to imitate Christ (**a=b**) is to reformulate a definite description (a=b), to invent an index (*a=b*).

The architectural tour of the crystal palace should not be interpreted as a glorification of life in a monotheistic temple. Instead it is the drawing of a heretic (not blasphemous) map of power, an attempt to understand how and why we find our way in a universe of obedient submission and unpredictable cruelty. Not a *Mappa Europæa Mundi* but a *Mappa Mundi Universalis*, a universal means for catching the world in a net of points, lines and planes, the mandala constituting a typical non-western case. As usual, it is more difficult to understand the world than to change it, for the reality of being has a richness which no abstraction can harness, a living life that neither stories nor pictures manage to name.

$\mathbf{T}$he **ninety degree** angle is the fixpoint of fixpoints, a human invention of the highest rank. Triangulation is the name of the game, for in order to make a map only three ingredients are needed: the scale, the pointer, the canvas. As triplets of the right angle these elements form the foundation on which the crystal palace is constructed, that power-filled edifice which at the same time is the crucible of man and the prison-house of language. The Library of Invisible Maps is the best guarded part of the palace, for it is there that we learn how to learn, it is there that we find out both where we are and where we should go.

The **scale of scales** is by definition a translation function, the ruler over ontological transformations, the magic formula of "Let there be! – And there is". An
early version is in the creation myth of the <u>Genesis</u>, the most crucial reformulations in the Divided Line of Plato's <u>Republic</u> and the speech acts of Euclid's
<u>Elements</u>.

The **pointer of pointers** is the hook on which everything is hung, the invisible
force which in ordinary maps corresponds to the magnetic North Pole. In the
construction of the crystal palace its counterpart is in the collective unconscious
and thereby in Plato's conception of the Sun, that medium which simultaneously lets us see and makes us blind. It should nevertheless be remembered that
it was the exiled Jews who insisted that the Sun - which the idolatrous Babylonians took to be the god of gods - in fact is nothing but an illuminating lamp. A
definite description of revolutionary importance, an act of cultural survival
closely tied to §3 of the Constitutional Law. Attend the compulsory meetings
and your eyes will be opened, the truth revealed!

But it must also be recalled that long before the events in Babylonia, the
Sun's dethroner - the tautological Yahweh - had warned Moses that "you cannot
see my face; for no one can see my face and live… And I will cover you with
my hand until I have passed by, then I will take away my hand, and you shall
see my back; but my face shall not be seen." And yet, that is exactly the blasphemous act that Jacob (the crook of crooks) claimed to have committed: seen
God's face and survived.

The prophet Job had much the same experience, for also he told the story
of how his ears first had heard about the Absolute then how his eyes had seen
him. But unlike the treasonous Jacob, the man of integrity from the land of Uz
was ashamed of what he saw, despising himself, repenting in dust and ashes.
And he lived on in this world for another one hundred and forty years.

The **canvas of canvases** is the background cloth, the receptor onto which everything is projected, the tain of the mirror, Plato's limestone wall, Fra Angelico's wood panel, Duchamp's large glass, the screen that captures all signs.
As a matter of fact, the canvas is nothing less than the physical resistance without which nothing can be be noticed. The latter-day version of the granite rock
that sprang a well of clear distinctions.

REVELATION

In the practice of cartographical reason, man is once again put back at the center of the universe. No longer a noun, not even a verb. A pre-position, a place assumed in advance! A point at which lines are projected onto a plane of power. A coordinate cross of right angles.

Such is the nature of cartographical reason, a mode of understanding at the verge of realizing that the canvas of the world is not a smooth flatness but a wrinkled manifold, that the pointer is not straight but crooked, that the scale is not a suspended line between *alpha* and *omega* but a Moebius band of chiastic reversals. And it has no idea what will happen next.

MAPPA MUNDI

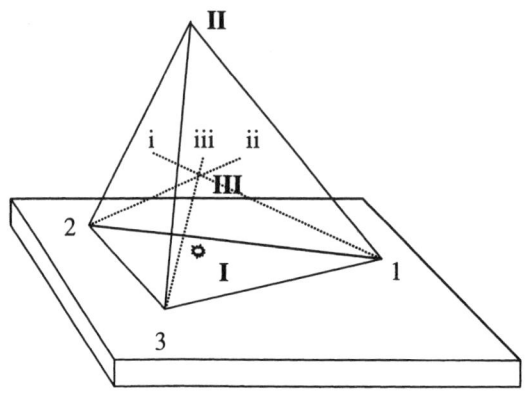

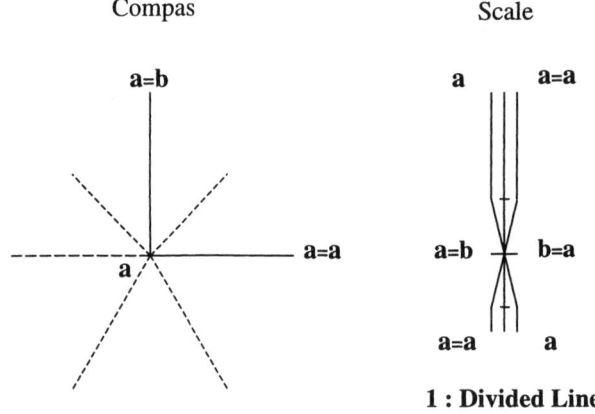

Compas   Scale

1 : Divided Line

Points of Distinction:

I	a	Foot in the crevice; Holy Spirit
II	a=a	Absolute power; God, the Father
III	a=b	Ontological transformation; Jesus Christ

Line of Trinity:

II-III-I   Plato's Divided Line; Nicea 325

Points of distinction:

1	a	Shadow
2	a=a	Tautology
3	a=b	Definite description

Lines of power:

II-3	§2a; Baptism by imaging
II-2	§2b; Baptism by naming
II-1	§3; Communion

Plane of socialization:

1-2-3   Stone tablet; Constitutional law

Planes of knowledge:

II-1-2	Science
II-1-3	Religion
II-2-3	Art

Lines of projection:

1-*i*	Aesthetics
2-*ii*	Grammar
3-*iii*	Rhetoric

Points of signs:

i	*a*	*symbol*
ii	*a=a*	*icon*
iii	*a=b*	*index*

•

MAPPING

MAPPING IS TRIANGULATION

TRIANGULATION IS THE GEOMETRY OF POWER

THE GEOMETRY OF POWER IS THE PRACTICE OF CARTOGRAPHICAL REASON

THE PRACTICE OF CARTOGRAPHICAL REASON IS THE CRITIQUE OF MAPPING

$+$

Mappa Mundi Universalis

Glass pyramid on granite base, 64x64x49 cm. Mixed material (Kalmar granite, Weissglass, gold, ruby)

Chapter 21
De Ludo Globi *or* The Reason of the Sphere

Alessandra Bonazzi

What a wonderful image:
the Judeo-Christian LORD as an
invisible vanishing point on a sphere
Gunnar Olsson, *Abysmal*

Recently being in conversation here in London with a traveler and discoverer, I mentioned the great geographical importance of accurate representations of our planet in the shape of globes; suddenly he interrupted me, and said with a smile, "What is the use of your Lilliputian globes, ten, a hundred, or thousand yards thick, when you have the very globe itself, our good and beneficent Earth, to walk over, to look at, to study, and to love?" Of course I laughed, and thought with him that all representations and symbols of life are very little in comparison with life itself.
Élisée Reclus, *A Great Globe*

De Ludo Globi or *The Reason of the Sphere* is about Gunnar Olsson's Geography. The argument is that the globe is a map with a different form. The lines of my argument are drawn from Olsson's two-tiered reason: "firstly, that understanding by definition is an exercise in translation, secondly that every translation involves transition from one language to another" (Olsson 2007: 82). And this paper is exactly an exercise in transition from the language of Geography to the silent language of friendship; but if it is easy "to know the meaning of what I see and hear", it is difficult "to find the proper expressions of my emotions" (Olsson 2007: 80).

Friedrich Riemann, "merged projective geometry with the complex numbers, and all of a sudden lines became circles, circles became lines, and zero and infinity became the poles on a globe full of numbers" (Seife 2000:141–142). Riemann indeed showed that the numerical sphere and the complex plane were equal to a point at infinity, that the complex plane was in fact a sphere.[1] For geographers, such an equivalence, or translation, is not possible. The map is the map, the globe is the globe: map and globe are opposites and irreducible. Their opposition,

---

1 A complex number is made up of a real part and an imaginary part which can be represented on a Cartesian plane whose real coefficient is in the abscissa and the imaginary one is in the ordinates. According to Gunnar Olsson to view a map as a complex plane is the first step towards an understanding of the logic of cartographic reason.

according to Ptolemy, is one that involves dimensions: maps are bi-dimensional while the globe is tri-dimensional; even though, as Florenskij reminds us, the issue is in actual fact more complicated than this. Kantian-Euclidian cartographic space is tri-dimensional with a constant curvature equal to zero, and for this reason it is unique. Therefore, the difference between the map and globe should refer to the curvature rather than to the dimension. According to Florenskij the representation of space and the space of what is represented are "bi-dimensional, and therefore similar, but their respective curvature is different. The shell of an egg, or even its fragment, cannot lie flat on a marble table – for this to be possible, one should change its shape" (Florenskij 2003: 122).[2] Having established this relationship, one can say that the difference between map and globe is represented by the number of projection points that are necessary for the curvature placed on the surface of the earth to be superimposed on the curvature placed on the surface of the globe. These points are external to what is represented and the projecting mechanism remains equal to itself. One could argue that in the case of the globe, the image of the earth cannot be grasped by the eye with a single gaze, and yet the same is true for any atlas or planisphere. Further, according to the geometry of the globe all points on the surface are similar, but this also applies to the projected space. In conclusion, it is possible to posit an analogical relation between the meanings and functions of a map and globe, and then ask the following questions: "Does our consideration of the world change if we handle the world as a globe (supposing the existence of a projection point that moves in order to guarantee the conformity of the curve) or as a map (supposing the existence of the same point only in this case fixed)?" The work of Nicola Cusano, the first philosopher and cartographer of modernity, provides a good testing ground for such an analogy.[3]

Cusano writes, "My aim has been to illustrate [...] this game that has been discovered not long ago, easy to understand and fun to play".[4] This play consists

---

2   Whenever translations from Italian into English are not available, these are by the author.

3   Cusano is the philosopher who inaugurated modern thought. He is, significantly, also the first cartographer to draw a map following the rules of projection. According to Ernst Cassirer, Cusano's "spiritual totalitarian" attitude to knowledge sets him apart from his contemporaries or predecessors and makes him a truly modern scholar. On the cartographic nature of the Cardinal's thought, see Blumenberg (1992). Concerning Cusano as cartographer and author of the map of Central Europe, see Durand (1952) and Fischer (1930).

4   This game consists of the following: one should throw a sphere towards a point at the centre of nine circles. To each circle corresponds a value: the central point is the reign of God, also called the realm of life, and has the highest value. The winner of the game is the one who gets the closest to this centre. Concerning value, it should be said that before the end of the dialogue, Cusano says that in order to make all this memorable it is necessary to talk about values as the "price of value", that is to say money (Cusano 1986: 922) It is in fact the market that is at the centre of the process of globalization. On this see Farinelli (2003), Mattelart (2003), Harvey (1990) and Jameson (1991).

in "controlling, by the exercise of virtue, a sphere" (Cusano 1986: 885–886).[5] Thanks to the philosophy of this game, it is possible to show that the movement of the sphere can be modified by constant practice, up to the point in which this acquires a straight trajectory.[6] Symbolically, playing this game means learning the art of controlling the otherwise random trajectory of the globe, which here is understood as the very form, which according to Cusano, represents a reductive and visible image of the invisible roundness of the world.[7] This image is the closest to the truth, but, it nevertheless remains a model. Playing with such a model is a very complicated matter as it involves an attempt to control the unpredictable and natural movement of the world, which is difficult to dominate. The game consists in an attempt to rectify, or to project the globe towards a single pre-established point, which is called *the realm of life*. Like any other game also this one requires a fiction, rules of the game and a player who influences the movement of the sphere. Cusano's philosophy therefore partakes of the knowledge of the world; it is a strategy that defers the complexity of a process and replaces the latter with something else. Cusano's *Il Gioco della palla* is concerned with a model of knowledge that, according to a certain Ptolemaic tradition, contrasts with another model which is based on the projection of the world on paper. Starting from these few statements, it is possible to critically re-consider what David Harvey has defined as the heedless "icon of a new kind of awareness", or a "tabula rasa" on which a new political and economical project is planned.[8]

Why does Cusano use the metaphor of the game to talk about the globe? What is the meaning of the globe when it is compared with a game? Which are the

---

5   What is curious, according to Cusano, is "the fact that things turn out to be different and unsure all the time, as the sphere never behaves [...] the way we expect" (Cusano 1986: 885–886).

6   "Inclinations and natural behaviours which tend to move along a curve line can be modified through the exercise of virtue" (Cusano 1986: 888).

7   "The world is not visible because it is round. Nothing is visible but the forms of the things that it contains [and] we can have only the image of the roundness, which is the closest to truth" (Cusano 1986: 865).

8   See Harvey (2002). Here the word "term" is understood according to a dialectic that involves the Latin *ter* – "to cross, or reach a destination that is somewhere beyond" – with the other meaning of the same word – "keeper of boundaries": "it provides articulation and structure to life and thought, it does not allow confusion and it somehow limits life. It also frees life for further creations". This functional duplicity makes the term a keeper of culture and at the same time limits that same culture. Terminology thus traces boundaries of what is known as "it refers to a world which is domesticated, and translated within its structures"; at the same time, however, terminology has also the power to go beyond the limit of culture literally changing its terms when these become "worn out, inadequate and ineffective". To re-think terms that refer to the globe starting from the *Gioco della palla* means to try to understand the part of geography that deals with the world as globe, from the point of view of a dialectical perspective. For a definition of terminology, see Tagliagambe (1997: 92) and Florenskij (1989).

implications of translating the world into a sphere and the realm of life into points towards which the sphere/world should be thrown? The underlying question after almost six hundred years remains: "is the globe merely the latest game?"

Let us suppose this is true. In that case, we should remember what it means to play. To play means to accept two inventions: the first concerns the construction of a "close and conventional universe" which is also fictitious and represents the space of players and the game; the second is concerned with the definition of a new subject "who pretends to be somebody else" (Caillois 2000: 36–37; on the one hand, the game is partly a consciously accepted invention, on the other, it contributes to a lie and should be understood as partaking in a consciously "ill-adjusted behaviour" (Tagliapietra 2001: 7). What is noticeable is the construction of a new subject who, through the mechanism of the *mise-en-scène* of the game, becomes cause and effect of the above mentioned ill-adjusted behaviour, that is to say of the necessary tendency towards deceit and the development of a "consciousness of spectacle." The latter is in turn the result of a "process that in western culture originates from the birth of a modern, subjective consciousness and its shift from an auditory to a visual mindset. [Hence] consciousness becomes a sort of theatrical performance which stages the doubling of world". In turn, this allows human beings to take their distance from the world and project onto it their respective viewpoints. This important turn encourages and amplifies the human being's decentred position in the world, which is the first prerequisite for the condition of the spectator (Jaynes 1976: 268–269). According to Andrea Tagliapietra such a consciousness is responsible for a "new kind of deceit, a long term deceit that Julian Jaynes calls *treason*" (Tagliapietra 2001: 16).

Having established all of this, a possible answer to the questions asked above can be provided. According to Cusano the globe is the visible and reductive image of the world, and the game consists in the rectification of such a reduction: the globe should maintain its formal characteristics, but, at the same time, it should obey the intention of the players to modify the sphere's movement from uncontrollable and erratic into one that is controlled, a movement which leads straight into the realm of life, the highest of all values. The players should be far away from the sphere, they should be positioned outside of the game, so that they cannot control it. These rules produce the following effects: they duplicate the world into its image, they distance the image from the subjects, and they transform the subjects into spectators. In other words: the rules confirm a deceiving mechanism similar to what geographers call cartographic projection. The metaphor of the game suggests that the ways the globe is viewed are similar to a long term deceit – or projection – which started at the beginning of modernity. The *De Ludo Globi* is therefore about a deceit: this can be described as the firm belief that the relationship between paper and globe – that is to say between a way of thinking based on a tabular logic and one that uses a global method – is substantially different, and that between the two lies a gap and a space of opposition.[9] In a more radical way, this work also declares that the game of the world is a long term deceit. Despite the undoubted proximity it posits between the truth of the world and its translation into an "image of truth" (Cusano 1986: 864), this game, like any other game, works through deferral and displacement (Blumenberg 1987: 110).

The metaphor is therefore a literal declaration of truth about the nature of its object, in other words, it signifies itself: the globe is a game, an illusion. The etymology of illusion is *in-lusio*, which literally means "entering a game". If this is the case, we face a double deceit which involves also the player who controls the sphere. Hence, two models and two subjects should be considered. These alternatives found their originary description in some passages of Ptolemy's *Geography* [of the globe].

But let us return to the beginning: as we were saying the whole question originates with the world: does the model of the globe presuppose, as Jean-Luc Nancy has suggested, the violent *coup de main* that has caused the cartographic definition of the world to become a sum of objects?[10] Does the structure of the representation of the world remain unchanged, despite its form? Should the world be "a place that is defined as a sum of places", that is to say a place in which everything finds a space? Let us suppose it is otherwise, wouldn't the world then become a mere globe, or *glomus*, that is to say "a place of exile, a valley of tears?"(Nancy 2003: 20–21).

The objections put forward by Ptolemy in his *Geography* have been fundamental in making the map preferable to the globe. The map of the *ecumene* built on a globe has the advantage of providing a semblance of the form of the earth, and yet, it is difficult to obtain a surface whose dimensions are able to contain a proper inscription of the *ecumene*. Even when this happens the eye and the globe must progressively move, as the dimensions of the sphere are such that what the map shows cannot seen be seen in a single glance. With flat surfaces none of these difficulties arise, but an image endowed with the same proportion as the globe is still necessary. Having shed these doubts, Ptolemy then proceeds by teaching how to draw a map of the *ecumene* on a globe without encountering any practical or

---

9   I am referring to the lively debate in the latest studies in cultural geography, which refers to the conceptual category of the map in order to understand the effects of globalization.

10   In his *Tre saggi sull'immagine*, (*Image et violence. L'image – Le distinct. La représentation interdite*) Jean-Luc Nancy analyzes the relationship between violence, truth and image, and establishes a connection between the functioning of the image and the mechanism of violence. Concerning the power/ violence of the pure Kantian image, he notes how the essential scheme of the image comes from the famous *Handgriff* and that *coup de main* is "an art hidden in the recesses of the soul" precisely because it has the "power of the scheme", the very power "which allows experience in general: the presence of a world and the presence in the world." The world is however made up of objects. It is significant to note that the handling of the sphere is the most important part of our game. See Nancy (2002: 23).

theoretical problems (Berggren and Jones 2000: 82–84). It should be noted, that this map is drawn on the globe and that it is not the globe *tout court* but only a portion of it, more precisely the northern quadrant. What follows is the first "method for building a map of the *oikoumene*", the simple conic projection on which the eye moves progressively from one meridian to another. The first projection on a plane presupposes, therefore, that the eye and the globe move, a movement which depends on the particular rule that has been chosen for the drawing of a map on a globe.[11] The second projection is different: here the eye and the globe are fixed in a position and at a reciprocal distance that allows the whole ecumene to be seen in a single glance. The third projection that according to Berggren and Jones is the only one that makes use of a linear perspective (Berggren and Jones 2000: 39) consists in the drawing of an image of the terrestrial globe surmounted by rings, that is to the main circles of the celestial sphere. In this projection, it is difficult to keep globe and table apart, even though this is, from a technical point of view, the only example of linear perspective included in *Geography*. Ptolemy uses the latter to obtain an extraordinary hybrid. On this he writes: "It would not be out of place, however, to add a method of drawing the hemisphere containing the *oikoumene* on a plane as it is seen [on a globe] and surrounded by a ringed globe." (Berggren and Jones 2000: 112).[12] Therefore, representation on a plane must function for a globe made up of rings.

But is there really any difference at all between a table and a map drawn upon a part of a sphere big enough to contain a clear and convincing drawing of the ecumene? Maybe there is no difference. If this is the case, it would not matter much whether the world is represented on a flat surface or has the form of a portion of a globe because in both cases the grid allows a "bit by bit" representation. It is this form of representation, together with the superficial opposition of the two methods, that represents the most convincing sign of their ontological indifference and of the absolute primacy of the grid, whatever its form. The grid is the key term in the globalization process. In other words, the globe and the map are symmetrical and equivalent signifiers. Nancy says that it is only thanks to the grid that it is possible to "to pin one signifier together with another signifier" which are placed among flows of random signifiers "and see then what happens. Something new [...] is obtained that is to say a new signification" (Nancy 1981: 56). This new signification is the globe, while the world that is the meaning lies elsewhere, beyond

---

11   As is well known, the grid of conic projections shows the meridians as straight lines and the parallels as circles. This distorted image is thought to move on meridians as if they were straight lines. Ptolomy at the end of the first Book recommends both projections: the first is simpler because it is possible to inscribe each locality by moving the ruler and with only one of the parallels drawn and divided in grades, the second is more complex because the curvature of the meridians does not allow the use of the ruler in order to find the exact position of a locality. The difference is merely the difficulty of execution.

12   It should be added that Ptolemy suggests another system to flatten the image of the ecumene that should be put inside the rings. On this see Berggren and Jones (2000: 117).

or within the flows. Should one want to find it, one should allow it to resurface from nowhere, "without preambles or models, without beginning or end" (Nancy 2003: 41). Ptolemy, on the other hand, carries the globe into the grid of deceit.

Sphere and map also presuppose two other aspects: the subjective consciousness of the player, and the subject of the projection respectively. Jacques Lacan should be mentioned, as according to him the subject in the symbolic order is born after the image and practices a world that is understood as being symbolic.[13] Lacan, as it is well known, recognizes the identity between the subject of psychoanalysis and the Cartesian subject of modernity, the same one that systematically describes projective geographic knowledge. The subjects of the unconscious and of geography speak the language of certainty but not of the truth: both subjects are interested in what is real in the world, and not in what is true. At the same time, Lacan observes, objective reason contains the necessary principles for transforming what is real into what is true. However, between the two stands there is a difference which is defined as the subject of certainty who refutes any prior knowledge – which is also a realistic definition of geographic production in modernity (see Farinelli 1992: 55–70). We know that "the unconscious subject, prior to entering the realm of certainty, is a thinking subject and that somewhere [the unconscious] manifests itself through a representation marked by deceit" (Lacan 2003: 36–37). This process of revealing which occurs prior to the subject's entering the realm of certainty, is central in Ptolemy's *Geography* because it is precisely here that the unconscious of geography is made apparent through representations of deceit that are responsible for the removal of the subject and of the metaphysical nature of the cartographic image (either as a globe or as a map) and finally of geographic knowledge itself.[14] The mechanisms of this process are very similar to those of deceit and of the game.[15]

The place of displacement is the alienating *vel*, the point that presides over "the circular process" essential for the constitution of the subject in the "sphere of the other". In order to "make us perceive what it is like in the *vel* of alienation" Lacan uses the logic of the *reunion*, which is the same as the mathematical relation of two sets: if in the first and in the second sets there are five elements, then their sum is ten. However, if two elements are common to both sets, their reunion will

---

13   On the relationship between psychoanalysis and geography and the "loss of the body" of the subject, see Blum and Nast (2000).

14   Galison uses the expression "metaphysical images" with reference to those Tables in the scientific Atlas whose goal it is to demonstrate in a universal and objective way a stable truth beyond the world of appearance. This metaphysical function of scientific tables is, according to him, in line with an idea of objectivity popular in the 17th and 18th century. See Galison (1998: 327–359).

15   Lies, according to Tagliapietra, are "the invention of that *ex nhilo*, or *tabula rasa*, where the originary word is blown. To begin means to stay silent, to ignore [...]. The tradition, the subject the states of things ... making such an extreme fiction converge in the game of a symmetrical invention" (Tagliapietra 2001: 48).

give a set of eight elements. What is obtained is a complementary set formed by those elements of the second which do not belong to the first. Therefore the *vel* necessitates a choice whose nature is determined by a lethal factor, for which, at the moment of the reunion "such choice has consequence for either sets". Such a choice concerns the possibility of keeping one of the two parts. Exemplary of this is the formula "your money or your life" (Lacan 2003: 208). This alienating mechanism has two fields related by a *béance* process, that is to say a fissure that allows something from the "field of the subject" to pass into the "field of the Other". In the latter, the subject reveals itself. It is precisely in the "field of the Other" that the signifier can show the subject of its signification. However, as soon as it is asked to function, and to speak, the subject is reduced to function as a signifier and indeed be nothing more than a mere signifier. It is at this moment that the subject reveals itself in the form of an absence. In other words, the *vel* of alienation, the point of balance and of attraction, as lethal factor, condemns the subject to appear "on the one hand as sense produced by the signifier" deprived however of what is his essential being, or "on the other [hand] as disappearance", that is to say disappearance of the subject himself. Lacan continues by saying that, between the signifier (as unitary signifier) and the subject (binary signifier) "there is a question of life and death". This problem revolves around a *beance* relationship (Lacan 2003: 214–16).

"The subject cannot arise at the level of but through its disappearance" (Lacan 2003: 217). The lethal factor forces the subject to be reduced to mere signifier. At the beginning of *Geography* this is described as the question of "life or death" between the unitary signifier and the binary subject. If being is chosen, the subject shuns representation, if sense is chosen the subject can be represented, but it is devoid of his essential signification.

Ptolemy, the subject of the geographic unconscious which "manifests itself somewhere and thinks", is not a subject but the point H: technically this is the last of a series of points that form the ruler or the rule EH, functionally the intersection of all meridians and the centre of all parallels in the map (Berggren and Jones 2000: 86). In the second and third projection – the latter is the most significant as it is the only one that is built through perspective rules and that declares the equivalence between globe and map – this point is imagined as an eye and hence presupposes a real subject (binary signifier). What is true for the second and the third projection must also be true for the first projection, which, according to Ptolemy, is the simplest version of all, but not the least reliable (Berggren and Jones 2000: 93, 112).[16] The principle is in fact always the same: literally an external point. As

---

16   In the first method the idea of the eye is not mentioned. In the second, Ptolemy refers to a central point that is obtained through the point of intersection of the parallel of Soene with the Central meridian; this point is at the same time the centre of the map and of the globe (Berggren and Jones 2000: 88). This is also the point closest to the eye. In the third projection the external eye is a supposition, which however, becomes a certainty (Berggren and Jones 2000: 112).

soon as this enters the realm of certainty with the proper name of a subject, it will be forgotten that originally it was merely a ruler or a rule,[17] a point placed by the alienating *vel*, the lethal factor of geography, where things pass and are always removed, a point H that includes and represents whoever handles the globe.

In the unconscious there is a point, whose signifier is on the level of the representation of a unitary signifier whose intentions are aimed at the world; this is the same signifier that plays with the sphere and aims at the realm of life. The representation characterized by deceit throws a grid onto the planet which "wastes what appeared as a globe which is reduced to its double – a *glomus*" (Nancy 2003: 4), or a map. After a point has been substituted with the eye and the subject, one can proceed to the removal of the metaphysics of geographical knowledge, and Ptolemy's *Geography* becomes a mere "guide to draw the chart of the world" (Berggren and Jones 2000: 4). The modern removal of the original meaning of *Geography* consists in a simple but lethal action: eliminate the italics from the title, and in this way transform the book into the geographical discipline *tout court*.

We can therefore say that Nicola Cusano has provided the warning against a model of the globe which does not differ from a paper model except by some degrees of curvature. Once the removal has happened, this warning remains unheard, or better, it is confused with something else. It is telling that Angus Cameron and Ronen Palan note that several observers of the global system think that this system is a curious phenomenon, that always remains *absent*. The system acts in a space that is beyond States and corporations and has no boundaries. It moves through a temporal horizon that does not belong to the present, because globalism can be defined as a relentless process and moves towards a *telos*, that is to say a global future. Notwithstanding ambiguities and contradictions that characterize this unfulfilled promise, its systematic acceptance produces an institutional structure that serves a global logic. This is possible through techniques and policies which have the purpose of adapting our lives to an *ineluctable imperative* – a sort of teleological ineluctability (Cameron and Palan 2003: 165–184). This curious absence, or deferral, of the globe is the latest result of an original removal which causes the symmetry between globe and map to be forgotten. But what is the shape and the meaning of the sphere that is used in today's game?

The globe in its contemporary image makes its appearance in an American Express advertisement. Here the globe refers to a credit network, and is a metaphor of the economic relation that connects all places in world. It heralds a network in which all parts of the globe can be accessed by money, and for this reason they

---

17   The difference between the map of the world drawn by Ptolemy and the world itself is evident, if only for the fact that this map intends to show the ecumene only. However what is established is a formal and conceptual structure that is shared. As Jean-Jacques Wunenburger notes "with Descartes the difference is postulated as a technique able to endow the image with a suppletive function: the image does not imply the mimetic truth of the thing, it merely serves as substratum so that an intellectual judgement can recognize the referent" (Wunenburger 1999: 175).

are characterized by isotropy, continuity and homogeneity (Harvey 2002:13). It can therefore be concluded that the transformation of the earth into a global space in the last years of the twentieth century is set out by a credit network. There is another version of the globe: the by now notorious "photograph 22 727" together with the picture entitled *Earthrise*, also referred to by Armand Mattelart as the emblems of a new universalism (Matterlart 2003: 344). *Earthrise*, in particular, shows the earth in the image of a globe that fluctuates freely in space; here the natural element prevails over and challenges the artificiality of all cartographic images produced so far.[18] What is inferred is the idea of a globe, freed from a grid of parallels and meridians, decentered and devoid of political boundaries and proper names which could be the beginning for a new project for the world and humanity. The geographical representation in *Earthrise* – which ideologically is seen as natural and devoid of political influence – provides a *tabula rasa* for a political, cultural and economic re-definition of the earth. As Mattelart has noted, *Earthrise* became a public image on 27 December 1968, and the commentaries on this image are reminescent of the Book of Genesis; the captions that appeared on television are literal quotations from the latter. These quotations suggest an underlying promise that concerns globalization: a promise of unity and harmony for earth and humanity.

The two narratives of the origin of the contemporary globe have a double register: one that is economic and whose medium is advertisement,[19] another which is endowed with a universal vocation whose medium is the seductive satellite image and whose truth value "legitimates itself".[20] These narratives are also acting as characters in the plot concerning the contemporary global narrative. Symbolically, the so-called *marketing imagination* is the mechanism through which it functions. As Mattelart notes, it is as if communication made the event real. Managerial imagination forms a global order that makes incomplete realities coherent and consistent the one with the other. It is in other words a "prophecy that creates herself", whose possibility to be realized depends on a sole strategic instrument: security, or the "keystone of the techno-global model for the re-organization of society" (Mattelart 2003: 388–9, 396–7). For geographers it is virtually impossible not to note in these expressions the traces of the projective syntax of modernity: projection/prophesy, image/imaginary, certainty/security. Two models of the world that are commonly defined as opposed are in fact

---

18   According to Denis Cosgrove photograph 22727 taken by the NASA constitutes a novelty in modern geographical imaginary from the Renaissance onwards. See Cosgrove (1994: 280).

19   As Gilles Deleuze and Felix Guattari have noted, "the depth of the shame was reached when computer studies, marketing, design, advertisement [...] took possession of the word, the "concept" [of globe] and said: it is our business, we are the creative ones, we are the thinkers" (Deleuze and Guattari 1994: xix).

20   J. Nancy (2002: 17). This self-referential value provides, according to Nancy, the link between violence and image.

harmonious. Their unchanged nature is guaranteed by the supremacy of the grid, whether this is constituted by parallels and meridians or by credit and information flow, a net that reproduces on the globe the same qualities that characterize the Euclidean geometry of the map. Such a map, as Ptolemy has declared, guarantees a drawing of the world that proceeds point by point no matter whether this is in the form of a table or a globe. To use Lacan's words, what is at stake here is the function of the so-called algorithmic function, which is equivalent to the logical and symbolical function of cartographic discourse. This function critically challenges the Saussurean idea of sign and the impervious nature of the ruler (Farinelli 1991: 9–20). Lacan applies the algorithm to sign and in this way he overthrows its logic and doubles its signifier in two signifiers, one in opposition to the other. However, if in Saussure such a juxtaposition leads to a reinforcement of the difference between the two signifiers, in Lacan it causes the complementarity of their value. As it happens in the case of maps, also in this case the signified is substituted by a function: the symbol of a law. Lacan challenges the Saussurean logic in which the constitution of signifier and signified is subordinated to the line that cuts the sign. On the contrary, according to Lacan, the signifier is always independent and pre-exists the signified, which is destined to precipitate forever unless a sort of theoretical anchorage, called stitching or grid, intervenes.

To return to my argument, two complementary signifiers – the table and the globe – attempt through the grid to block the continuous flux of the signified, which is already substituted by a function that is equivalent to a process of symbolization. The world, the signified, remains in any case absent, despite the *point-de-capiton* (see Lacan 1974: 488–523; Nancy 1981: 36–56). The terms of contemporary geographic discourse on the globe refer to the game of the sphere in two of its main elements: a simplified shape of the world, distance, a space of the game of the capital and, at last, a human being endowed with a body (see Nast and Pile 1998) whose ontological consideration is still universal and transcendental because abstracted from its material, economic, cultural and social relations (see Harvey 2002: 97–101). The point of the game, which is to direct the sphere towards *the realm of life*, presupposes the control of the movement of the world and a violent *coup de main* – precisely the kind of lesson that cartographic reason teaches.

Nowadays the game escapes criticism and, as it is played on a semantic and conceptual level not dissimilar from a teleological one, and makes critical discourse obsolete and ineffective. The fideistic, prophetic and ineluctable vision that characterizes any kind of discourse on the construction of a new global system makes the highest human values similar to the values of capital, realizing in this way the double statute that is at the base of the contemporary globalization, whatever its version. From a logical point of view, on the other hand, the mechanism coincides with the function of the algorithm. Hence, transcendence is characterized by the slipping of an economic signifier which is opposed to a theological one. This annihilates the effectiveness of any discourse that attempts to draw attention to its *essential metaphorical nature*. Such a slipping and removal is literally a question of life or death because it is precisely what dictates the impossibility to control the globe for those who are asked to bear its *l'in-lusio,* that is to say the victims of economic war. What results is a new signification which, however, has nothing to do with the world or life itself.

Let us now take into consideration a fundamental aspect of the game. Towards the end of his dialogue Cusano mentions an "omnipotent man who minted coins". It should be remembered that ancient coins were made from metallic globules, and despite the fact that these were pressed, they still retained their original spherical form. Unlike modern coins, whose two sides are separated by an edging, ancient coins have an even surface (Florenskij 2001: 120). The minter of coins was known to make any coins and to have the power over "the value of all coins". However, what exists is a unique truth from which the value and truths of all coins depend. "Hence," Cusana says, "what follows is that any coin can get along with any other coin, as coins have the same head" (Cusano 1986: 926–928). It is an unexpected ending, but useful, at least to Cusano himself for summing up the whole dialogue. It is also related to the nature of our lives and to the understanding of the game of the sphere, as it explains the beginning of the dialogue, but not only that: it especially suggests that the sphere has a centre. For us, the centre of our globe is somewhere in the United States: from there battles are fought for defending the power of one minter of coins, one who guarantees the agreement with all other coins in order to guarantee techno-global security and the incolumity of the "Davos man" who plays the game (Roberts, Secor and Sparke 2003: 886–897). The divine reference, as Nancy has noted, is marked on the coin itself: the game is played in the name of absolutism and an enigmatic uniqueness of the One and in favour of an extreme monetary capitalization, which is the absolute value, *the realm of life* – a point of projection and a point of view.

We started by mentioning the deceit and the removal produced by the metaphor of the game – the homologation between the globe and the map and the betrayal of the world – which in the meantime has forgotten to be a metaphor. The amusing game of globalization, discovered not long ago and the game that everybody now wants to play, carries on quietly and smoothly, leaving the world in the midst of fluxes, always and already elsewhere, dissipated in the "ugly infinite of a spiral globalization" (Nancy 2003: 28). This is in fact the globalization that Riemann described as a destructive Manichean nightmare dreamt on a complex plane where a point is placed on the sphere, between the poles of zero and infinity.

Bibliography

Berggren, L. and Jones, A. (eds) 2000. *Ptolemy's Geography. An Annotated Translation of the Theoretical Chapters*. Princeton: Princeton University Press.

Blum, V. and Nast, H. 2000. "Jacques Lacan's two-dimensional subjectivity," in *Thinking Space*, edited by M. Crang and N. Thrift. London: Routledge, 183–204.

Blumenberg, H. 1992. *La legittimità dell'età moderna*. Genova: Marietti.

Blumenberg, H. 1987. *Le realtà in cui viviamo*. Milano: Feltrinelli.

Caillois, R. 2000. *I giochi e gli uomini. La maschera e la vertigine*. Milano: Bompiani.

Cameron, A. and Palan, R. 2003. "The Imagined Economy: Mapping Transformations in the Contemporary State," in *State/Space. A Reader*, edited by N. Brenner, B. Jessop, M. Jones and G. Macleod. Oxford: Blackwell, 165–184.

Cassirer, E. 1992. "Il concetto di forma simbolica nella costruzione delle scienze dello spirito," in Cassirer *Mito e concetto*. Firenze: La Nuova Italia, 95–135.

Cosgrove, D. 1994. "Contested Global Visions: One-World, Whole-Earth, and the Apollo Space Photographs," *Annals of the Association of American Geographers*, 84, 270–294.

Cusano, N. 1986. [1463]. *The Game of Spheres*. New York: Abaris Book.

Deleuze, G. and Guattari, F. 1994. *What is Philosophy?* New York: Columbia University Press.

Durand, D. B. 1952. *The Vienna-Klosterneuburg Map Corpus of the Fiftheenth Century. A Study in the Transition from Medieval to Modern Science*. Leiden: Brill.

Farinelli, F. 1991. "In-traduzione: dal bar di de Saussure alla balera di Girard, introduzione a G. Olsson, *Linee senza ombre. La tragedia della pianificazione*. Roma-Napoli: Theoria, 9–20.

Farinelli, F. 1992. *I segni del mondo. Immagine cartografica e discorso geografico in età moderna*. Firenze: La Nuova Italia.

Farinelli, F. 2003. *Geografia. Un'introduzione ai modelli del mondo*. Torino: Einaudi.

Fischer, J. 1930. *Die Karte des Nicolaus von Cusa (1490); die ältestate Karte von Mitteleuropa*. Prag: Kartographische Denkmaler der Sudetenlader.

Florenskij, P. 1989. *Attualità della parola. La lingua tra scienza e mito*. Milano: Guerini e associati.

Florenskij, P. 2001. *Lo spazio e il tempo nell'arte*. Milano: Adelphi.

Florenskij, P. 2003. *La prospettiva rovesciata e altri scritti*. Roma: Gangemi.

Galison, P. 1998. "Judgment against Objectivity," in *Picturing Science Producing Art*, edited by C. A. Jones and P. Galison. New York: Routledge, 327–359.

Harvey, D. 1990. *The Condition of Postmodernity*. Oxford: Blackwell.

Harvey, D. 2002. *Spaces of Hope*. Edinburgh: Edinburgh University Press.

Jameson, F. 1991. *Postmodernism, or the Cultural Logic of Late Capitalism*. Durham: Duke University Press.

Jaynes, J. 1976. *The Origin of Consciousness in the Breakdown of the Bicameral Mind*. Boston: Houghton Mifflin.

Lacan, J. 1974. "L'istanza della lettera dell'inconscio o la ragione dopo Freud," in Lacan *Scritti*. Torino: Einaudi, 488–523.

Lacan, J. 2003. *Il seminario. Libro XI. I quattro concetti fondamentali della psicoanalisi*. Torino: Einaudi.

Mattelart, A. 2003. *Sto ia dell'utopia palnetaria. Dalla città profetica alla società globale*. Torino: Einaudi.

Nancy J. 2002. *Tre saggi sull'immagine*. Napoli, Cronopio.

Nancy, J. 2003. *La creazione del mondo o la mondializzazione*. Torino: Einaudi.

Nancy, J. and Lacaoue-Labarthe, P. 1981. *Il titolo della lettera. Una lettura di Lacan*. Roma: Astrolabio.

Nast, H.J. and Pile, S. (eds) 1998. *Places through the Body*. London-New York: Routledge.

Olsson, G. 2007. *Abysmal: A Critique of Cartographic Reason*. Chicago: University of Chicago Press.

Reclus, É. 1898. "A Great Globe,"*The Geographical Journal*, 12, 401–406.

Roberts, S., Secor, A. and Sparke, M. 2003. "Neoliberal Geopolitics," *Antipode*, 886–897.

Seife, C. 2000. *Zero. The Biography of a Dangerous Idea*. New York: Viking Penguin.

Tagliagambe, S. 1997. *Epistemologia del confine*. Milano: il Saggiatore.

Tagliapietra A. 2001. *Filosofia della bugia. Figure della menzogna nella storia del pensiero occidentale*. Milano: Bruno Mondadori.

Wunenburger J.J., 1999. *Filosofia delle immagini*. Torino: Einaudi.

Chapter 22

To be human (the Secret of the Pyramid)

Ole Michael Jensen

What does it mean to be human? And, why are we so obedient and predictable?

Gunnar Olsson

Like Gunnar Olsson, so also Plato was absorbed by the question of what it means to be human. Part of Plato's answer is built into the allegory of the cave presented in the Republic. In this dialogue Socrates explains to his student Glaukon how "there are men living in a sort of cavernous chamber underground, with an entrance open to the light and a long passage all down the cave". Here, chained to the floor, the prisoners cannot move and cannot see anything but the shadows that are thrown on the limestone-wall in front of them. "A strange picture and a strange sort of prisoners", Glaukon broke in. "Like ourselves," Socrates rhetorically replied. To Plato the point is that the humans throughout their lives have seen nothing of themselves or of one another, merely the shadows cast. It follows that "such prisoners would recognize as reality nothing but the shadows on the wall." [1] What an impressive way of saying that to be human is to be trapped by one's own imagination.

It is easy to understand that this allegory impressed the coupled brains of Gunnael Jensson, when they invented the sculpture *Mappa Mundi Universalis*. In this construction, Plato's stone cave was turned into a crystal palace, more precisely into a glass tetrahedron with its base sunk into a square-shaped rock of granite. As indicated by its name, this imagination is to us nothing less than a universal world-map in which the mappa of Plato's limestone-wall has been replaced by three walls of mirroring glass, projection screens onto which alternative conceptions of what it means to be human are cast; at the same time a rendering of the linguistic prison-house of the taken-for-granted and a descendance into the abysmal gap between categories, the magic theatre of ontological transformations, the taboo-ridden place of power, the cultural melting pot par excellence.

A major distinction between conventional mapping and the cartography of the *Mappa Mundi* is that whereas the former presents a visible landscape on a single sheet of blank paper, the latter shows an invisible mindscape on a triple of panels ready to capture particular standpoints. Grasping the true nature of this construction is exceptionally difficult, perhaps impossible, for even though the observer seems to be looking into the crystal palace from the outside, we are in

---

1 *Plato:* Republic, *514a–515c.*

reality always inside it, the semiotic animal by its nature unable to escape. As in Plato's allegory we too are tied to our own imaginations, the main difference being that in his case there was a cave with a single fire casting shadows of human existence, while to us there is a wide-screen movie theater with three projectors one in each corner of the tetrahedron's bottom plane, each of them projecting its specific conception of human existence onto its respective screen.

This is consequently the idea of *Mappa Mundi Universalis*: a figurative mode of representation; a multifaceted allegory; a triangulation of human life; a blueprint of mapping; a display of cartographical reason; a Global Positioning System abstract enough to tell us not only where we are in the material world of sticks and stones but also where we lead our lives in the social world of knowledge, power and meaning: a castle with a Sleeping Beauty waiting for a prince to enter and break the spell.

Driven by a desire beyond control, our imagination eventually took the form of a slab of Kalmar granite; three equilateral pieces of German Weissglass; three spoke-like wires of pure gold; one gem of red ruby – all of it visible to our outer eyes. To our inner eyes, however, these same materials have miraculously changed so that the granite now stands for the foundation of existence; the glass for the walls of depiction; the wires for the projection of alternative perspectives; the ruby for the blood from the first sacrifice, the cultural Big Bang through which a primordial difference was killed and turned first into three forms of identity, eventually into three modes of understanding.

In September 2000, the year when the second millennium flipped into the third, this merger of holy object, scientific instrument and work of art was presented at the Uppsala International Contemporary Art Biennial, the piece itself exhibited in the nave of the city's cathedral, the major church in the Lutheran Kingdom of Sweden. Immediately following the opening show, the crowd moved across the street to the Anatomical Theater, where the contraption was expertly dissected by an art historian and a theologian.

Ten years later the Jensson couple is continuing their expeditions into the realm of the pyramid,[2] asking particularly where its remarkable spell comes from, why it seems impossible to treat it merely as the thing it is, why we cannot stop asking what it actually means. And the most remarkable of everything remarkable is that whenever we ask, it always responds, often with answers never heard before. The trick, we have learned, is to become one with the physical structure itself, to experience with our own bodies the differences between the richness of the crystal palace and the poverty of the cave.

This said it should be stressed that the pyramid came to life neither through observational induction nor logical deduction but rather through a process of abduction, that rare moment which in religion is called "revelation", in art "inspiration". There is no doubt that in his subsequent visits to the world

---

2   To the Jensson couple, "The Pyramid" early became the nickname of the tetrahedron, in its physical performance named "The Crystal Palace" (Jensson 2000).

of cartographic reason Gunnar has been constantly guided by the maps of the pyramid; in his own words, "the present volume may be read as a record of the silent conversations I have subsequently had with this material expression of desires non-suppressed" (Olsson 2007: ix).

For myself, the lessons have been equally profound, especially as they have taught me more than I know about daily life itself. It is in that light that the present article should be read, an updated report from a decade's struggles with the question of understanding how I understand.

But before I proceed into that presentation – and with the double purpose of making sure that both I and the reader will know where we are – I must briefly return to the fix-point of fix-points, the well that is located at the center of the tetrahedron's base, the place where one of our ancestors for the first time became the bodily symbol of a difference different enough to make a difference. Others gathered around the expelled and through their curiosity effectively turned the original difference into a set of identities. To the architects of the crystal palace this first sacrifice is nothing less than the birth-place of the semiotic animal, the moment when it learned how to live with the tension between identity and difference without going crazy.

Rephrased, it is the memory of this cultural Big Bang that is etched into the foundation of the tetrahedron, the red ruby at its center symbolizing the blood from the murder. In the legend of the *Mappa Mundi* this point is marked with an **a**, the trace of the primordial distinction which as three variants of identity moved to the corners of the tetrahedron's base: 1/ the tautological a=a, 2/ the informative a=b, and 3/ the shadowy a. Translated into a picture:

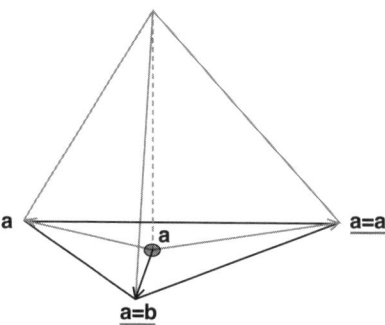

In my frequent walks through the palace – always afraid of getting lost – this base map has often helped me find the way. The reader might wish to keep it ready at hand.

Outside the insight

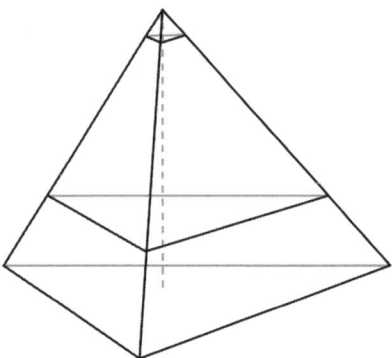

Viewed from the outside the pyramid looks like an intelligible piece of architecture. The real challenge, though, is to connect the reasoning from the outside with the experiences on the inside. "Mapping" is the name of this Kantian endeavor; on the one hand the drawing of a picture – on the other the telling of a story; all in all an invitation to new explorations.

To enter the three-dimensional pyramid is to enter a horizontal structure made of three forms of life and a vertical structure made of three levels of abstraction. On the ground floor, fixed by the sunken base, the various life-forms are found in three fields, easy to imagine because they are close to what we experience in daily life. On the top floor there is an exclusive penthouse in which everything is condensed into the abstract conception of a Sovereign power. In-between these two levels there is yet another floor, in my imagination a splendid hall occupied by three types of prophets, all of them on their way from the different corners in the base to the screening walls of the tetrahedron.

An astonishing discovery is that deep in the corners of the ground floor are not only the a, a=b and a=a of the basic reformulations, but also the pronouns of 'you', 'it' and 'I'. From the outside these fix points may be only faintly sensed, but inside the palace they appear in a different light; as already noted, the interior walls function as projection screens, each one prepared to receive its own very specific image of the world, the most persuasive pictures projected onto the corresponding screen at a right angle. Whenever the projections are not right-angled, the imperfect representations are merged into a blend of misunderstanding and creativity.

Breaking the spell of the pyramid is to detect its deep secret, a secret which lies in its form rather than in its content. It is therefore crucial to realize that the forces that have given form to the pyramid belong to the birth of the construction, the genesis itself; the result of the primordial distinction through which the creation and murder of a scapegoat established the emptiness in which power can hide. This point – **a** – is marked by the ruby, in Jensson's mind the lid of the blood-filled spring that welled up from the first sacrifice.

The Commoners' Floor

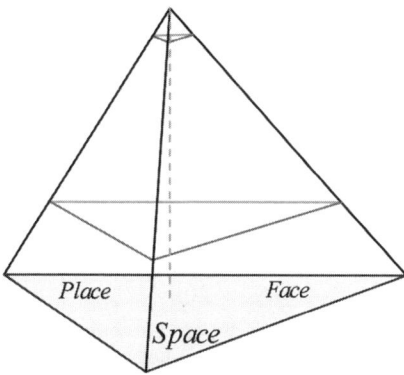

When the explorer gets into the pyramid from the outside, he encounters a myriad of existences. On closer inspection, though, the seeming chaos turns into three modes of being and three modes of understanding. Rephrased, this means that we are living in three different worlds, each with its own logic, each with its own language, each with its own conception of knowledge. In this manner every person – you and I included – is living his own life on the commoners' floor. It follows that our varying cultural compositions are direct consequences of our positions on this ground floor of the pyramid; like the music that fills the cathedral, so also the resonances in the crystal palace depend on exactly where you are. Sensible to your own vibrations, you experience your actual mode of life.

When the laughing HA-HA shakes the body, then the cultural sounding board is one with the taken for granted. Since laughter often makes fun of the story of creation, this sound tends to be disliked by all religions; indeed it stands for those paralyzing critiques of power that only the jester may create. As a consequence, the HA-HA is released only when a person's childhood is stunned. It was this sounding board that Martin Heidegger investigated in his studies of the *Dasein*, the place where being and time show themselves to be inseparable twins.

When the surprising AH-HA leaves the body, then the sounding board is that of the Saussurean Bar, the place where new understandings burst out of the old. In that moment of clarity the "overstanding" signifier oscillates with the "understanding" signified, a feeling of relief that the sage creates whenever he uses the principles of pure reason; as René Descartes put it: "Cogito ergo sum: I think, therefore I am". Immanuel Kant later examined the limits of this room, extracting from it the essence of pure reason, making it ready for the disciplines of science in the process.

Finally, when the relieving AAHH thrills the body, then the resonance of the chamber and the sounding board of the body are vibrating together. Such experiences are what the artist creates in the room of aesthetics, the space that Søren Kierkegaard's Don Juan explored and the room in which Friedrich Nietzsche later declared that God is dead and that the idea of logical reason is nothing but a human invention.

On the commoners' floor the alternative modes of being and understanding have found their way also into concrete artefacts, the layout of the city an outstanding example. To be precise, any city, no matter whether it is ancient or modern, capital or provincial, has its defining places, streets and squares. We do not usually distinguish between these elements and in fact notice them only when they are seemingly absent, i.e. when the city shows a striking lack of places, when the streets carry no traffic, and when the squares offer no room for contemplation.

Notable for the *place* is that it exists by virtue of itself, i.e. by virtue of its origin. Every place therefore comes with its own history, just as any religion comes with its own genesis. This is where you can laugh and cry, because this is the territory of the narrative. Surveying the place is nevertheless impossible, for being in a place is by definition to be part of it; you fall in love *with* the place just as you fall in love *at* the place. Time is out of the question, because place is itself one with the time-dimension. That gives to place the status of a one-dimensional floor of the city. In the actual city place takes the form of pubs, workshops, offices, homes and other meaningful venues. The home is the place par excellence, the blind alley of all movements, the furniture, the renovations and the redecorations the very signs of a good life.

What is typical of the *streets* is that they can be seen only from above. Although they are meant for transport, it is only the map that can tell you about new destinations, just as the researcher can make new discoveries only in relation to what he already knows. To find one's way among the streets is therefore to travel over the roof-tops. Yet, without a map with the street-net drawn in, the city is a labyrinth without exit. In turn this means that a topographical street-map is not a map of places but a map of the connections between places. And just as the experience of the street requires distance, so distance requires simplification. The desired overview emerges only if the cartographer manages to translate the city into a geometry with names. And by no coincidence this phrase – geometry with names – is Gunnar's homemade definition of geography.

In real cities the streets take the form of walkways, roads, boulevards, subways and motorways. In all these instances the associated signs are aimed at the traveller and his attempts to reach a given goal; the independently acting subject is consequently transformed into a set of functional movements. Perhaps this explains why the street often creates mixed feelings of euphoria and angst, euphoria for being a part of the city's organism, angst for getting lost in the street spaghetti.

The main characteristic of *squares* is that they exist only through their own design, the great urban squares originally laid out as drill grounds but inadvertently functioning as ideal settings for revolutionary performances as well. Like art, no square can exist without an audience, for a square is essentially an installation, a public event, a watchful eye. Just as a stage is not a stage unless delimited, so a square is not a square unless draped by the city itself. To put it differently, a square is a multidimensional expression of the world, a symbol which unfolds at the moment when the spectator approaches it as if it were a piece of art. In the concrete city, the square is located where many streets come together. Hence, there are small squares and large squares, the former grown by accident, the latter constructed at the pleasure of the prince. Like their conceptual cousin, the auditorium, the squares are intended for temporary stays, intermodal transport nodes which offer alternative destinations refreshing pauses and pointers to the previously unknown. While cathedrals and statues indicate that the church and the monarchy look upon themselves as eternal guides, secular institutions like banks, universities and city halls try in vain to do the same. Indeed it is only street performers, vendors and cafés that dare to enter the square; the reason is that since its beginning the square was designed for gathering an audience, the ultimate stage for the theatre of the moment.

The city's places, streets and squares are inextricably connected with the phenomenon of the city walk, i.e. with our attempts to break loose from one place in search for another. It is in that struggle for liberation that the place becomes visible and the street comes to serve as a means of connection. The square, in contrast, arises only when someone stops to ask about the shortest way to a given destination. It is in this moment of hesitation, in this unstable location, that competing routes are generated and the choice between them becomes unavoidable. The fork in the road is the origin of the square (Stjernfeldt 1996).

As we have just seen, the city walk alternates between contemplation, transfer and hesitation, a movement that has not taken place in a real town of rocks upon rocks but in the imaginary space of the crystal palace. Just like the city itself, that edifice has shown itself to be paved with three kinds of stone, one to dwell on, one to move on, and one to reflect on: place, space and face.

Walls of imagination

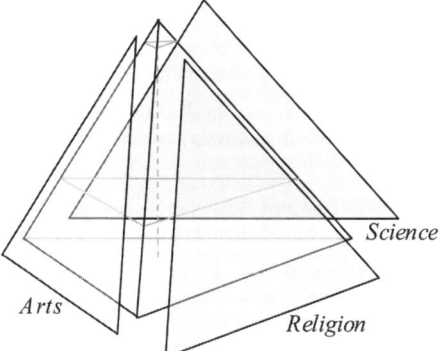

No walk would ever be possible were it not for the human faculty of imagination: imaginations stuck to the remembrance of place; to the map of space; and to the self-reference of face. Exactly as the floors of the crystal palace are paved with three kinds of stone, so the three walls of tetrahedron are prepared to reflect of three modes of imagination: belief; knowledge; and creativity. These imaginations, actually three modes of mapping, are as old as the world itself. More commonly they are known as religion, science and art. What all these institutions share in common is the question of how we are believed when we tell the truth, i.e. the question of how competing visions of the world are transferred into alternative pictures and stories of the taken for granted. To help us perform this magic trick of geometry, narrative and signing is the key function of cartography.

In *religion*, the strength of the given map is in its capacity to present the world as it looks from the elevated perspective of the divine. While this is the view from the Olympus, Heaven, Paradise, Nirvana and Infinity, the geometry that makes it possible comes from Euclid – a depiction of the world *without* background, i.e. a world in which mortal subjects cannot interfere. The accompanying story is by necessity one-dimensional, creation myths, dictatorships and mathematical proofs being typical examples. To reach a deeper understanding of these maps, one should recall that the reflecting wall has been prepared to capture the projections that come from the corner of a, and to send them back dressed up as Peircean icons: "Let there be light, and there is light". As in the Genesis, the utterance and what the utterance is about are one and the same, a perfect sign in which signifier and signified have been woven into a text of perfect communication, indeed into the hosts of a communion. Out of this combination of geometry, storytelling and signs comes a representation in which the picture looks at the viewer, the map telling you where you are. To art historians this is the reverse or Byzantine perspective of Orthodox icons. In other contexts it is known as the mental map of the taken for granted, by definition the context of contexts.

In *science*, the strength of the given map is in its capacity to capture the world as it seems to the individual eye. This enables the third persons of he and she to share the world and to identify it as their common viewpoint. While the invented position belongs to the subject, the geometry comes from Girard Desargues, who depicted the world *on* a background. To identify with this type of representation is straight-forward, because it is the impression of the object itself. These maps are by necessity two-dimensional and the stories that make them believable are preached at the university; it follows that according to this myth all knowledge must be universal and reproducible. To reach a deeper understanding of these maps, one must know that the reflecting wall has been made to capture the impressions that come from the corner of a=b, and to send them back addressed as Peircean indexes. Since in this perspective whatever can be pointed to can also be represented, the corresponding signs function as bridges between the map and the territory. Out of this combination of geometry, storytelling and signs comes a representation in which the picture is one with the shared reality. To art historians this is the linear or renaissance perspective, in common parlance knowledge par excellence, the only mode of understanding that is deemed acceptable by the academy. In daily life it is called "common sense".

In *the arts*, the strength of the map is in its capacity to move with the world through a process of becoming. For this purpose the fine arts establish a viewpoint in which the interplay between creator and created, i.e. between the individual I and the a=a, is captured in one and the same representation. From this meta-viewpoint "nothing" can be perceived before the existence of a "no thing" has been imagined. The position invented is consequently that of the subject of the subject of the subject, of the subject …. The relevant geometry stems from Eugenio Beltrami, the Italian mathematician who taught us how to depict the world *with* a background, that background itself the object of yet another background. To identify with this perspective the artifacts that come with it must be at least three-dimensional. Although this universe of desiring desire can never be fully grasped, the reflecting wall has been prepared to capture the reformulations that issue from the a=a corner and to send them back in the Peircean suit of symbols. In this way only that which can be imagined leaves its traces, symbols that bridge the gap between map and territory. Out of this combination of geometry, storytelling and signs comes a perspective in which the picture appears not as a picture of reality but as the reality of reality.

In summary there would be no imagination, no belief, no uttering, no exchange of signs without the screens of imagination just as in the case of Plato's cave there would be no shadows without the limestone wall. If not expressed in the modes of religion, science or arts, there is no message. It follows that there is no mapping outside the *Mappa Mundi Universalis*, not even the *Mappa* itself.

The Prophets' Hall

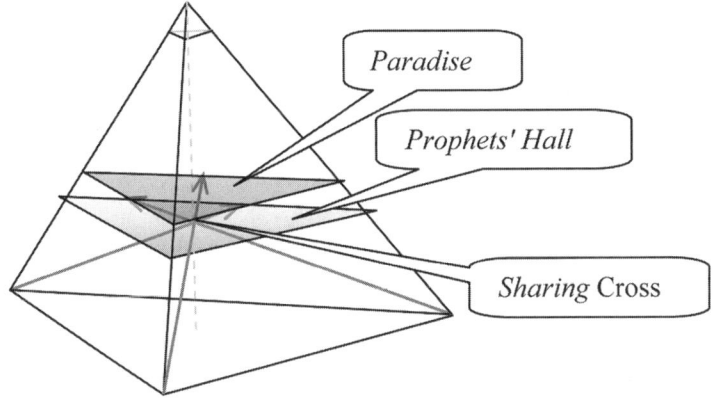

As I climb up the interior of the pyramid I realize that the crystal palace reaches to heaven. The air gets thin and the task of forming cohesive imaginations grows increasingly demanding. Half way to the top I find myself in the Prophets' Hall, home of some remarkable mediators between heaven and earth, all of them known for their prophecies and clear-sightedness: Jesus, Mohammed, Buddha, Marx, Keynes, Smith, Michelangelo, Wagner, Joyce and with them a number of other luminaries reverently worshipped by their disciples. Unlike the gods, however, these persons are not persecuted and sacrificed scapegoats but expert performers of the game of identity and difference. The victors of that game are rewarded with a ticket that takes them all the way from the alternative corners of a, a=b and a=a to the final truths on the walls of imagination: the icons of religion, the indexes of science, and the symbols of the arts. With my inner eye I see three cable cars, each with its own trajectory, each with its own passengers of saints, scientists and artists.

The most exciting station of these trajectories is the intersection where the three lines are crossing each other. There, from the platforms of the Sharing Cross Station, the attentive observer can catch a glimpse of the three end-stations, because it is from this intersection that walls can be hit exactly at a right angle; three times Janus' eyes and three times perfect representation: the dream point of faultless triangulation, the ideal position for intentional planning; everything in the common name of God, Man and Ego.

The Sharing Cross – the point where the trajectory from one corner intersects with that from the other two – is the point where each prophet meets his likes from alternative congregations. To the not yet recognized, this encounter is of course what changes the chosen from being an emissary of the community to becoming a crusader of his own conviction. Given this situation he is faced with the choice of either remaining in the station's waiting room or of continuing to the final destination on the wall. To those who are satisfied with life on the platform, the station pub provides space for eternal contemplation.

Despite its splendors, the Prophets' Hall must not be mistaken for Eden. Paradise is instead the landscape that lies between the points of perfect representation. Paradise is therefore one with the ceiling of the Prophets' Hall, a heaven that can be seen from the Sharing Cross. For those rare individuals who are ambitious enough to fulfill their mission by continuing to the wall of representation, there is still a distance to go, a final journey that the chosen performs alone or in the company of a trusted Sancho Panza. Finding the way is nevertheless easy, for the guiding star shines right above the cross. The prophet, who is strong enough to go on to the end will eventually reach the point on the wall where he can "light a star" that tells his congregation on the commoners' floor that an institution can be built; the Vatican with its Pope, the Academy with its professors, the Museum with its curators. While Jesus needed the help of Saint Paul, and Marx the help of Lenin, their colleagues Mohammed and Immanuel Kant did it all alone.

Given these circumstances it comes as no surprise that the Prophet's Hall with its Sharing Cross is crowded with potential truth-sayers. And no surprise either that their disciples, who originally took off from the corners of the commoners' floor, believe that their leader charts the way to heaven.

The Sovereign's Penthouse

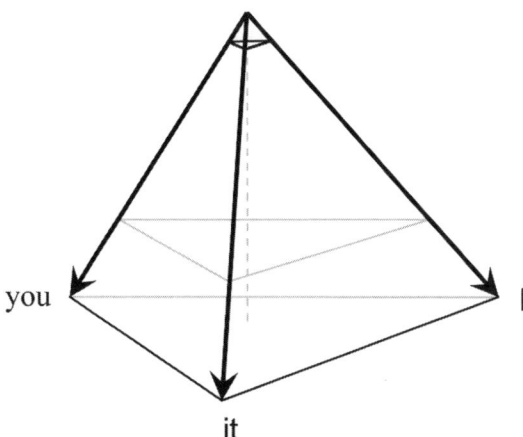

In a penthouse located at the pyramid's top lives the commander-in-chief, a mind-filled force without body. JAHVHE is the Hebraic name of this entity sometimes also known as the tautological "I am who I am". The power radiating from this point is a power beyond representation, a force generated by the self-referential Sovereign and his identical twin, the scheming Lucifer. Although registered on the same address they are continuously moving around, always ready to exchange seats on their travel between heaven and earth, always afraid of being caught in the either-or net of logical reason.

The penthouse is a part of the pyramidal construction, its residents dust from the Big Bang of culture. Like other ghosts they are haunting the castle, never leaving a trace, never making themselves known. To see their faces is strictly forbidden. The Almighty and his brother are well aware that were it not for the pyramid, they would not be at all. For that reason, and to that end, life in the crystal palace is governed by a constitutional law, a set of commandments which are designed to keep the ruling power in power and to censor all critique before it is uttered. In particular, no one is allowed to imitate the Lord, seriously or for fun, because imitating emanates from desire and desire comes out of a lust for what the Sovereign himself stands for. The paradigm is in the constitutional law of Moses' first stone tablet.

These basic commandments all address the positions of the a, a=b and a=b, where 'you', 'it' and 'I' occupy the bottom-corners of the pyramid. It is not for nothing that the most powerful paragraph is addressing the 'you', the occupants of the position a which faces the projection plane of religion, itself in reality a form of politics: "You shall attend to all party meetings, never enter into the no-man's land between clean and unclean, never bite the hand that feeds you" (Jensson 2000: 9). In the original version, this commandment bears the words: "Remember the Sabbath day to keep it holy".[3] In short this is a social control-paragraph, because every power-wielder knows that the biggest threat comes from people who are shut out of the community.

Two closely related paragraphs are addressing the "it", the occupants of the position a=b that faces the projection plane of science: "You shall not make for yourself a graven image or any likeness of the Almighty, for no power-wielder allows his secret to be revealed. No statue, no picture" and "You shall not misuse the name of your Leader, never tie his proper name to a definite description" (Jensson 2000: 9). In the original version, these decrees bear the words: "You shall not make for yourself a graven image or any likeness of anything that is in heaven above, or that is in the earth beneath, or that is in the water under the earth," and "You shall not take the name of the LORD your God in vain."[4] In short the two commandments state that the Sovereign must never be addressed by the pronoun "it". Taken together these words function as a censorship paragraph and an anti-caricature paragraph. For every Führer knows first that all matter-of-fact signification undermines his power and then that irony is the most effective way to break through the barriers behind whatever he may be hiding.

The key paragraph of the law addresses the "I", i.e. the occupant of the position a=a which faces the projection plane of the arts: "You shall have no other ruler before me" (Jensson 2000: 8). In the original version, "You shall have no other gods before [or beside] me."[5] The strategy is obviously to threaten, for every Sovereign knows that his best defense lies in the self-referential proclamation of an untouchable dictatorship.

In total an exclusion-paragraph directed at the individual "you", two censorship-paragraphs meant to prohibit the creative use of "it", and a threatening-paragraph, designed to protect the self-referential "I" – these are the forces that permeate the tetrahedron of power, these are the pillars which keep our construction from collapsing. Three injunctions pointing directly to the reformulations of the **a;** a, a=b and a=a in the bottom-corners of the pyramid, exactly where the transcended pronouns of you, It and I are rooted. From there, they are sent on to the fields of religion, science and the arts, where they are seen as the signs of Peircean symbols, indexes and icons. Engraved in the walls of the palace is the blue-print of the construction, a double story of birth and death a secret hidden since the beginning of time.

---

3   Exod. 20:8, Deut. 5:12.
4   Exod. 20:4-10, Deut. 5:8-1.
5   Exod. 20:3, Deut. 6:7.

Inside the insight

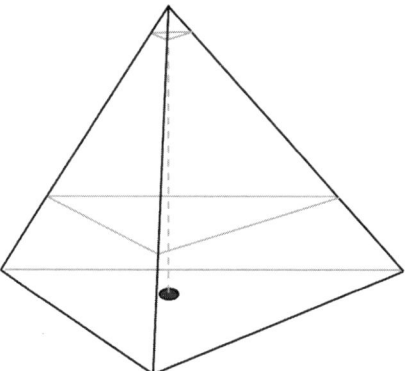

When the police enter the penthouse to search for the pyramid's secret, they will not find what they are looking for. The reason is that the key is not with the Sovereign on the top floor but with the commoners in the basement, more precisely in the murder story of Cain and Abel, the violent act through which the original difference **a** is turned into a set of alternative identities – a, a=a and a=b. It is through the Big Bang of this violent act that all meaning is created, a meaning so crucial that its understanding is forbidden.

And yet there are also the symbolizing events of René Girard's scapegoat and Gunnar's references to Aristotle's excluded middle, the latter neither the either-or of conventional logic nor the both/and of dialectics, but something entirely outside the realm of naming. So deep are the imprints of these exclusions that they structure the entire pyramid.

A crisis, an excluded third, a murder, and a ban on representation, such are the signs of the scapegoat, at the same time victimized and sacralised. The constitutional law is itself the scar of the mimetic crisis which played a pivotal role not only for the psychoanalyses of Sigmund Freud and Jacques Lacan but also for the socioanalysis of René Girard; by first being killed and then worshipped, the scapegoat became the creator of the community.

This explanation, however, does in my mind not present the whole story. Instead there is much to suggest that what actually happened in the first second of culture lay not in the sacrificing of an excluded but in the exclusionary practice itself; "thingification is the price we pay for not accepting that there is a beyond beyond the beyond of expression, an absence inscribed without trace in every discourse" (Olsson 1991: 176).

It is the memory of this absence that fills the Sovereign's penthouse with the ghosts of the cultural Big Bang. The realization of this emptiness is the price we pay for the language we speak. It follows that the idea of original sin is not a Christian invention, but a conception deeply engraved into every culture and into every language. As it turns out, both Genesis and Nietzsche were wrong, for the truth is that it was neither the Lord who created man nor man who created the Lord, but language that created the Lord in the name of man. The gospel is right: "In the beginning was the Word, and the Word was with God, and the Word was God."[6]

Finally: we humans are so obedient and so predictable because like Plato's prisoners we too are children of our own imaginations. Unlike them, however, we are not chained to the floor with our gaze fixed on the dancing shadows, but we are free to move between three rooms of resonance, each framed by its own wall of representation and its own form of geometry. Yet, no one – the two twins of the Lord and Lucifer included – can liberate himself from this collective imagination. No way out.

And so it is that even the many-faceted pyramid can be summarized in a diagram:

	a	you	place	home	religion	icon
a	a=b	it	space	street	science	index
	a=a	I	face	square	Arts	symbol

Bibliography

Badcock, C.R. 1980. *The Psychoanalysis of Culture*. Oxford: Basil Blackwell.

Bible, the Holy. Revised Standard Version. 1971. Dallas: Melton Book Company.

Carse, J.P. 1986. *Finite and Infinite Games*. New York: Free Press.

Girard, R. 1987. *Things Hidden Since the Foundation of the World*. London: The Athlone Press.

Girard, R. 1996. Mimesis and Violence, in *The Girard Reader*, edited by James G. Williams. New York: Crossroad.

Jensen, O.M. 1993. Red River Valley: Geo-graphical studies in the landscape of language. Environment and Planning D. *Society and Space*, 11, 295–301.

Jensen, O.M. 1995. *Vægge/Walls, Kundskabens projektioner/Projections of Knowledge*. Oslo, Copenhagen, Stockholm, Boston: Scandinavian University Press.

Jensen, O.M. 2001. Rum/Space, in *Boase, Fremtidens bolig/The Future Home*, edited by C. Bech-Danielsen. Copenhagen: SBi and The Royal Danish Academy of Fine Arts.

Jensson, G. 2000. *Mappa Mundi Universalis*. Uppsala: Uppsala International Contemporary Art Biennial. (Jensson alias Ole Michael Jensen and Gunnar Olsson, reproduced in Olsson 2007: 410–437).

Koestler, A. 1975. *The Act of Creation*. London: Picador.

---

6   John 1:1.

Kvasz, L. 2004. History of Geometry and the Development of the Form of its Language. *Synthese*, 116 (2), 141–186.

Olsson, G. 1980. *Birds in Eggs/Eggs in Birds*. London: Pion.

Olsson, G. 1991. *Lines of Power/Limits of Language*. Minneapolis: University of Minnesota Press.

Olsson, G. 2007. *Abysmal: A Critique of Cartographic Reason*. Chicago: University of Chicago Press.

Peirce, C. S. 1955. *Philosophical Writings of Peirce*. Edited by J. Buchler. New York: Dover.

Plato. 1945. *The Republic of Plato*. Transl. by F. MacDonald Cornford. Oxford: Oxford University Press.

Plato. 1961. Timaeus in *The Collected Dialogues of Plato*, edited by E. Hamilton and C. Huntington. Princeton: Princeton University Press.

Reichert, D. 1996. Räumliches Denken. Zürich: Hochschulverlag AG an der ETH Zürich.

Stjernfeldt, F. 1996. Sted, gade, plads – en naiv teori om byen, in *Byens pladser*, edited by M. Zerlag. Copenhagen: Borgen, 11–30.

Chapter 23
Double Crossed

Marcus A. Doel

> Thought, which knows it will fail in any case, is duty-bound to fail.
>
> Jean Baudrillard, *The Perfect Crime*

At the outset of the collection of essays gathered together in *Abysmal*, Olsson suggests that "the best way to approach [it] is to read it as a minimalist guide to the landscape of western culture," although he cautions the would-be reader that "anyone who enters these lands travels on a tourist visa, the specialized expert no exception" (Olsson 2007: xi). For while the landscape that Olsson wishes to show us is the all-too-familiar—and, in his hands, the strangely unfamiliar—Human landscape splayed out according to the vicissitudes of cartographical reason, it is a landscape that is fundamentally alien for reasons that will become apparent.

On the one hand, Olsson wishes to show how the world has been made and remade in the ever-shifting image of what it means to be Human in continuous struggle and negotiation with the Gods and the Earth. On the other hand, Olsson wishes to show that in so doing the Humans are ineluctably estranged from their world. The world is not so much the home of Man—or the homeland of Man—as it is the refuge of Man—or, as we shall see, His Prison House. To enroll the reader in the guise of a tourist is therefore to grant the reader a privilege ordinarily denied to those Humans who have come to occupy the unknowable and inhospitable landscape of Western culture. Indeed, no sooner has Olsson tempted the would-be reader with the counterfeit promise of a tourist visa than he revokes it. At the threshold of *Abysmal* proper—in an essay entitled 'Border-Man,' which asks rather innocuously: "what does it mean to be human?"—Olsson informs the reader about the true nature of the escapade to come:

> **A mission no agent can refuse**: How do I find my way in the power-filled world of thoughts-and-actions, things-and-relations? These were the cryptic orders handed down to me by the Secret Service Headquarters, a set of commands as impossible to fulfill as not to accept:
> - lay bare the familiar of the unknown;
> - find the principles of imagination and specify the rules of ontological transformation;
> - draw a map of the Territory of the Humans, (re)trace its fluctuating boundaries and find its stable centre;
> - produce an atlas of what it means to be human;
> - initiate a critique of cartographical reason. (Olsson 2007: 3.)

The order has been given, the order shall be executed! What follows is therefore a report of what the agent discovered during his clandestine operation. (Olsson 2007: 9.)

Hereinafter, the reader of *Abysmal* is no longer a disinterested tourist on an edifying Grand Tour of the landscape of Western culture, but a participant in espionage, cunningly disguised as a bourgeois tourist. In fact, Olsson tells us that the dossier bearing the title *Abysmal* is divided into six sections: an account of how the agent's bespoke surveying instruments were assembled and calibrated; a set of findings from a series of reconnaissance raids and incursions into lands unknown; a report of a debriefing and a pep talk in readiness for battle; the text of a Cabinet meeting prior to the production of an atlas containing maps never before seen; a Requiem Mass for the souls of the dead and the kicking corpse of modernity; and a few scattered memorials to those who have fallen. Yet at this stage it remains unclear whether the reader is an innocent bystander unwittingly given access to this dossier or whether the reader is in receipt of this dossier precisely because s/he is part of the Secret Service: the agent's handler, perhaps; or another clandestine agent, perchance; or the mission's controller, conceivably; or a double agent, even. What is seemingly not in doubt is that Olsson is working for the Humans, who have found themselves pinned down beyond enemy lines and splayed out in enemy territory. The **pinning down** of the Humans beyond enemy lines is most clearly articulated in the final installment of *Abysmal*, in an essay entitled *Mappa Mundi Universalis*. This essay sketches out "a beautiful crystal palace," "a linguistic prison-house" (Olsson 2007: 412), within which the Humans are caged and domesticated behind a plethora of invisible and untouchable bars. The **splaying out** of the Humans in enemy territory is most clearly articulated in the first installment of *Abysmal*, in the essay entitled 'Border-Man':

Homo erectus caught in its own net of:

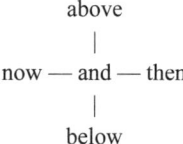

above

|

now — and — then

|

below

Such is the image that will guide us throughout [*Abysmal*]. Two perpendicular lines, map and compass turned into one. Disguised as a geometry with names this cross begins to live, in the process transforming the system of spatial coordinates into a Vitruvian Man. (Olsson 2007: 12.)

The key text within the *Abysmal* dossier is entitled 'Mission Impossible,' which opens thus:

From the beginning ... I have pursued the question of what it means to be human, more exactly what it means to live in the oikumene [the inhabited Human world]. .. What I have found is that the limits of the oikumene are one with the limits of my world, these limits themselves one with the limits of the languages in which the semiotic animal thinks-and-acts. In my self-appointed role as a (post)modern land-surveyor I dream of mapping the outer reaches of that universe, a world which fills the void between the five senses of the body and the sixth sense of culture. (Olsson 2007: 239.)

The Humans, as semiotic animals, are condemned forever to make sense of their world within the confines of the Prison House of Language. For their world is a world of sense. They know nothing but their wor(l)d. Whence the fact that the secret agent's "various raids into the unknown have all taken off from the legendary Bar de Saussure" (Olsson 2007: 239). And since there is literally no-thing, no-body, and no-sense beyond the Bar de Saussaure—"*Il n'y pas de hors-texte* [There is nothing outside the text; there is no outside-text]," as Derrida (1978: 158) once put it—, the machinations of the Bar trace out both the limits, the form, and the content of their world. "The boundary between the oikumene and the anoikumene can be reached only from the oikumene side"—from the side of sense, from the side of the wor(l)d—; it can be reached "only from the interior of the Saussurean Bar" (Olsson 2007: 240).

For Olsson, then, the wor(l)d is not 'out there'—awaiting Human discovery and meaningful expression—but 'in here.' It takes place *within* the Saussurean Bar.

[T]he Bar-in-Between is nothing less than the trace of the real, the true home of the semiotic / rhetorical animal, the consecrated place where we live and die, the stage on which we perform our dual roles as actors and spectators. (Olsson 2007: 86.)

What, then, is the Saussurean Bar? It is the line that articulates the two key components of that which makes sense: the sign—that which signifies (denoted by the letter '*S*') and that which is signified (denoted by the letter '*s*'). It is the line that holds together, whilst keeping apart, that which *makes* sense, and that which makes *sense*:

$$\frac{S}{s} \quad \text{or} \quad \frac{s}{S}$$

Five aspects of this articulation of sense need to be emphasized. First, signifiers and signifieds are mutually constituted and never travel alone—there is "no S without an s, no s without an S" (Olsson 2007: 106). One cannot grasp sense (s) without making sense (S). Second, signifiers and signifieds necessarily fail to coincide, not only because each is defined through negative difference, but also because $S \neq s$. This is best illustrated by the Lacanian algorithm, S/s, such that sense slips beneath—and is occluded by—that which makes sense. Third, precisely because of this non-coincidence and slippage, signifiers and signifieds are enchained in a series of displacements without origin or end. This is best illustrated by the Derridean notion of *différance* (i.e. the perpetual differing and deferring of meaning, reference, and sense, the trace—or serial erasure (*seriasure*)—of which testifies to the fact that there is truly nothing outside the text). Fourth, since signifieds (including referents) slip beneath the interminable play of signifiers, they are placed under erasure. This is best illustrated by the Saussurean Bar: '—.' Fifth, and finally, the Bar '—' is not only that which *separates* signifiers and signifies (/)—perpetually holding their incommensurability and dissemination both together and apart—, but that which *simulates* their exfoliation (– –). "The essence of power is thus in the slanted /, its appearance in the repetitive – –" (Olsson 1982: 29). Signifiers and signifieds are simulacral effects of the solicitation and vacillation of the Bar. This is best illustrated by the double crossing of signifiers and signifieds—'× ×'—or, what amounts to the same thing, the barring of the Bar ('–/–'): "How can I simultaneously anchor – – in / and / in – –?" (Olsson 1982: 32). Hence the fact that Olsson (2007: 240) insists that:

> The limits of the oikumene can be studied through experimentation *with*, *on* and *in* the Saussurean Bar, invaluable insights gained in the process. The trick is to take the fraction line of the sign:
>
> $$\frac{s}{S}$$
>
> literally, in other words to treat it as the line [that] it really is:

In other words, the Bar goes all the way across (—), all the way down (|), and all the way around (O). "Everything interesting happens in that thin penumbra that simultaneously keeps signifier and signified together and apart" (Olsson 1991: 152). There is, in short, nothing outside the Bar. Each 'S' and each 's' is merely a localized twisting of the Bar—a twist in the fabric of space and time in which sense becomes loopy, and in so doing sense always undoes itself as non-sense in the round. Figured thus, sense—and the wor(l)d that makes sense—flickers in and out of non-sense, into which it ineluctable plunges, like an asymptotic singularity, a vanishing point or the degree-zero of dematerialized abstractness:

This—*and this alone*—is what it means to be Human in the Prison House of Language.

Elsewhere, Olsson (2007: 107) will speak of the "braided chain" of signifiers and signifieds (× × ×), which continuously transposes sense and the making of sense, thereby confounding the Cartesian distinction between matter and meaning, between the sensible and the intelligible, and between the body and the soul: "Flesh turned to word, matter to meaning" (Olsson 2007: 138); "To explore possible worlds is to be a geographer with a mind that matters and a matter that minds" (Olsson 1991: 81). However the Bar is conceived ('/,' '—' or ' × '), Olsson is effectively saying that there is nothing outside the Bar. The Human wor(l)d only makes sense from *within* the machinations of the Bar: "we have no choice but to live in and of language, much as the fish lives in and of water. The Bar is our home, the fraction line the bed in which our signifying descendants are being conceived" (Olsson 2007: 84). And yet the Bar *itself* is by definition bereft of sense.

Like the tain of the mirror (the reflective tinplate or tinfoil surface that is affixed behind a sheet of transparent glass), *which renders visible without itself being visible*, the Bar lets sense take place, without itself taking a place amongst the sense that it sustains. Hence the fact that the title of the secret-service dossier is *Abysmal*. The Bar is the asymptotic abyss of the excluded middle, "the abysmal topos of ontological transformation" (Olsson 2007: 123). It is the vanishing point into which everything plunges, where sense and non-sense become indiscernible and undecidable. For although the Bar is that which enables the sign to signify (intermittently, falteringly, and faultily), the Bar itself is asignifying. The Bar is neither a signifier nor a signifed. It is senseless and abyssal. And yet it is what enables "the void" between "the five senses of the body" (i.e. '*S*') and "the sixth sense of culture" (i.e. '*s*') to be "filled" with a world (Olsson 2007: 239). The Bar comes *between* and *before* sense and orientation. This is why Olsson places 'and' at the center of the Human wor(l)d, whilst insisting that this 'and' is not only con-junctive (and therefore dis-junctive) but also pre-positional (and therefore de-positional):

Or, what amounts to the same unhinged thing:

Consequently, the Bar is not simply the centre of the sign through which sense and senslessness will come to be played out (–/–), it is also a force of spatio-temporal displacement through which place is splayed out:

$$\begin{array}{c} | \\ \text{— and} \rightarrow \textit{Metonymy} \text{ (syntagmatic)} \\ \downarrow \\ \textit{Metaphor} \text{ (paradigmatic)} \end{array}$$

This is why "The sign is a map" (Olsson 2007: 115). "Two perpendicular lines, map and compass turned into one" (Olsson 2007: 12). When all is said and done, then, we must appreciate that *Abysmal* is entirely devoted to "life in the Saussurean Bar," the Bar in which "we learn what it is to be human" (Olsson 2007: 240).

Now, drawing upon his long-standing penchant for spatial science, Olsson takes the Bar ('/,' '—,' '×' or '–/–'), this abyssal "dividing/unifying divisor between Signifier and signified" (Olsson 2007: 240), and treats it "as if it were a mathematical function" (Olsson 1991: 61). He forces it to approach, asymptotically, the limit conditions of zero and infinity, becoming 'minimally thin' (———————————) and 'maximally thick' (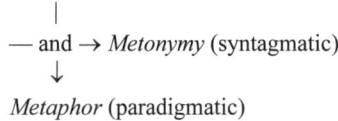), respectively. In the minimal case, the Bar that differentiates and exfoliates signifiers and signifieds appears to erase itself in the congruence of the performative speech act, such that the '*S*' appears to meld seamlessly with the '*s*' ('saying as doing' and 'doing as saying'—"The Saussurean Bar is here as thin as it can ever become. Not only are word and object the same; so are object and meaning" (Olsson 1991: 61). As Humpty Dumpty once said: "When *I* use a word, it means just what I choose it to mean—neither more nor less.") In the maximal case, the Bar that differentiates and exfoliates signifiers and signifieds appears to "grow" and "solidify," and in so doing its insistent presence reveals *two* limits, each of which is *internal* to the Bar. It eludes both sense ('*s*') and that which makes sense ('*S*'). The boundary of the Human Territory qua meaningful wor(l)d-view (i.e. the Prison House of language; the limits of language, representation, and thought-and-action; and the oikumene) is therefore akin to an impenetrable cavity wall. On one side, the wall appears to hold back the "muteness of pure spirituality" (the "non-sign"—or asignifying sign—*s/s*), beyond which one might imagine a "*Mindscape of Pure Meaning,*" home to the "noumena of the signifieds," a no-man's land of non-sense invariably reserved for the Gods. On the other side, the wall appears to hold back the "stuttering noise of pure physicality" (the "non-sign"—or asignifying sign—*S/S*), beyond which one might imagine a "*Rockscape of Pure Materiality,*" home to the "phenomena of the Signifiers" (Olsson 2007: 241), a no-man's land of non-sense invariably reserved for the base material of the Earth.

As semiotic animals, the Humans and their wor(l)ds exist and make sense **entirely within** the unhinged limits of the Saussurean Bar '—,' **between** its asignifying cavity wall '=,' and **around** its central abyss '□.' Beyond this cavity wall "lies the vastness of what can be neither said nor shown, by definition unreachable on its own terms. This is the land of the anoikumene, the realm of pure imagination. The boundary between the oikumene and the anoikumene can be reached only from the oikumene side, only from the interior of the Saussurean Bar" (Olsson 2007: 240). The wor(l)d that "fills the void" with a semblance of sense is determined by the structure of the sign as a divisor '/' and a Bar '—.' Consequently, the Human Territory is bisected horizontally (by the Bar '—' that separates meaning and matter) and vertically (by the divisor '/' that ensures the non-coincidence and slippage of what we talk *about* and the language that we talk *in*). This twofold bisection produces four sectors within the Territory of the Humans that Olsson maps onto Kant: the Intelligible and the Sensible, which yield objects of cognition (ontology); and Knowledge and Opinion, which yield kinds of cognition (epistemology). Where the Bar and the divisor intersect ('–/–,' '+,' '×' or 'χ'), there is the void, the empty square, the dematerialized point, and the vanishing point. This is the abyss whence everything came as a simulated effect (e.g. the braided chains of *S*'s and *s*'s) and into which everything plunges as indiscernible and undecidable non-sense (*S̶/̶s̶*). *Abysmal* is Olsson's attempt to map and narrate this singularity as it perpetually vanishes.

Accordingly, Olsson (2007: 146) constructs a "metaphysical base-map" based on Plato's *Republic* to determine the "limits of language." Its rectangular form marks off the boundary between the Human Territory and four "Terræ Incognitæ" beyond the reach of language that Olsson dubs Mindscape, Rockscape, Blindland, and Deafland. "Mindscape contains nothing but pure and non-expressible meaning (a signified without a Signifier, an *s* in search of an *S*), while Rockscape is permeated by pure and non-observable matter (a Signifier without a signified, an *S* with its *s* aborted)" (Olsson 2007: 147). Blindland is something akin to sensation without sense (an impossible chain of 'pure' signifiers devoid of signifieds: *S'S*), while Deafland is akin to sense without sensation (an impossible chain of 'pure' signifieds bereft of signifiers: *s's*). "Bound between these non-penetrable limits of language lies the realm of the oikumene itself, by definition homeland of the semiotic animal" (Olsson 2007: 241). And yet, despite being constrained to signify and make sense (*S/s*), the semiotic animal may nevertheless asymptotically approach these four Terræ Incognitæ beyond the limits of language through the faltering and *seriasure* of signification (*S̶/̶s̶*):

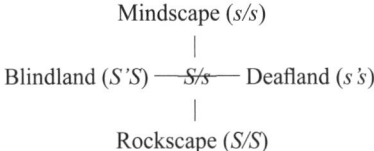

Although such an abstract, formalist, and (post)structuralist mapping of what it means to be Human may appear eternal, Olsson is at pains to show how the limits of the Bar, its cavity wall, and its twisted loops have shifted back and forth over time and place: "in general the humans have come to occupy more and more of what once belonged to the aliens" (Olsson 2007: 243). Apparently, so much more now seems to make sense, and so much less seems to evade sense. This "process of gerrymandering" is illustrated in the second half of *Abysmal* by four maps of "decisive battle-sites in the perpetual war of what it means to be human" (Olsson

2007: 243): the Babylonian epic of Gilgamesh and Enkidu; the Jewish account of Jacob's struggle with the Lord; the Greek tragedy of Oedipus; and the fate of Jesus Christ at Nicaea. On each occasion, the Humans were able to push back the frontiers of sense. And yet, on each occasion, the newly expanded Territory of the Humans continued to exist entirely within the abysmal limits of language, between its cavity wall, and around its central abyss. However, since the dominion of sense is secured by the unhinged machinations of the Bar, it is vulnerable to the latter's ineluctable slippage and *seriasure*. As always, our wor(l)d is slipping away. "Sign off," says Olsson (2007: 437). The end of the wor(l)d is nigh, says *I*. This—*and this alone*—is what it means for the Humans to be double-crossed by the Bar. In the not too distant future, one can only hope that a day will come when the wor(l)d will once again make no sense at all (Baudrillard 1996).

> As he watched, the stars began to slide about, to realign themselves upon the black canvas of the sky as though to spell out some message for him. A warning maybe. But it was all just a sluggish scramble, like the shuffling of dominoes, nothing he could make sense of, and he grasped thereby some small portion of his fate: that anything the universe might have to say would remain forever incomprehensible to him. … Whut do they say, oletimer? he asked. Whut do the stars say? … After a long silent time, the Indian said: They say the universe is mute. Only men speak. Though there is nothing to say. (Coover 1998: 83.)

Meanwhile, the double agents operating in the Bar de Saussure will have to content themselves with the draining away of sense.

> So leastward on. So long as dim still. Dim undimmed. Or dimmed to dimmer still. To dimmost dim. Leastmost in dimmost dim. Utmost dim. Leastmost in utmost dim. Unworsenable worst. (Beckett 2009: 95.)

When all is said and done, the secret-service dossier that bears the title *Abysmal* recounts Olsson's leastmost unworsenable forays into the oikumene's utmost dim. In the social space of silence: *He must sign on. He can't sign on. He'll sign on.*

Bibliography

Baudrillard, J. 1996. *The Perfect Crime* (trans. C. Turner). London: Verso.

Beckett, S. 2009. Worstward Ho, in *Company / Ill Seen Ill Said / Worstward Ho / Stirrings Still*, edited by D. Van Hulle. London: Faber and Faber, 79–103.

Coover, R. 1998. *Ghost Town*. New York: Henry Holt.

Derrida, J. 1976. *Of Grammatology* (trans. G. Spivak). Baltimore: Johns Hopkins University Press.

Olsson, G. 1982. – / –. *Sub-Stance*, 35, 24–33.

Olsson, G. 1991. *Lines of Power/Limits of Language.* Minneapolis: Minnesota University Press.

Olsson, G. 2007. *Abysmal: A Critique of Cartographical Reason.* Chicago: University of Chicago Press.

Chapter 24
De Nobis Ipsis Silemus

Franco Farinelli

I am attached to nothing and from time to time I turn upside-down the whole edifice of my thoughts with profound indifference towards my own and other people's opinions. Then I observe things from all possible perspectives in the hope to find the right and truthful one from which to begin to reconstruct.

Immanuel Kant (Excerpt from a letter from Kant to Herder dated May 1768.)

Kant taught physical geography from the summer semester of 1756 until his retirement in 1797 when his course was cancelled (Gedan 1923: 509–10).[1] In 1797, the *Ornithorynchus paradoxus* made its first appearance in what was then known as New Holland. A four page description of this creature appears in the Italian edition of Kant's *Physische Geographie* and is conflated with the *Ornithorynchus aculeatus* (see Eckerlin 1807: 56–60). This detail is significant as it demonstrates how non-Kantian the Italian version of Kant's geography is.[2] This is even more true if we consider that subsequent editions, (Gedan 1923: 513), have added more misunderstandings. The fact is that, from the beginning of the nineteenth century to the present time, all those who have come into contact with Kant's *Geographie* have added a new piece to its body, in the same way that Kant himself did throughout his life.[3] Despite this, Kant's *Geographie* cannot be likened to an *Ornithorynchus*, that is to an animal which seems to contain in itself all other animals. Despite all its mutations, the structure of Kant's *Geographie* in fact remains unaltered. It is important to specify that the word "structure" here is not

---

1   P. Gedan (Hrsg). *Immanuel Kants Physische Geographie*, in *Kants Gesammelte Schriften* (hereafter quoted as *GS*). Berlin-Leipzig: Königlichen Preussischen Akademie der Wissenschaften zu Berlin, 1923.

2   On the fact that this animal was unknown to Kant, see Eco (1997: 71–80).

3   As is well known, Kant always refused to leave his native town and invitations from Universities in Halle and Berlin were turned down. According to J.A. May this decision was related to a geographical interest, and in particular the ideal location of Königsberg especially for those who wanted to explore the world without travelling; one could suppose that Königsberg was for Kant what in 1628 and 1649 Amsterdam was for Descartes. When Kant lived in Königsberg it was a vibrant and very cosmopolitan port town with communities of people from Scandinavia, Holland, England, Poland, Russia and other Slavic countries as well as a German speaking population. On this see, *GS*, VII: 120–121, and May (1970: 5).

used to signify simply the order of discourse or the order of things.[4] The question at stake is this: what does Kant imply with the term "physical geography"?

During Kant's times, earth, sky and sea ceased definitely to be what they had been in ancient times. In particular, at the end of the fifteenth century, the earth ceased to be the smallest sphere in the cosmos, which, according to Aristotle, was surrounded by the spheres of water, air and fire. In mediaeval times – when it was believed that the ratio between earth and sea (that is the proportion between the volume of the elements) – was one to ten, people believed that the earth was a small isle which had sprung out of the Ocean waters thanks to a providential (and local) divine power. Throughout the entire sixteenth century, after the great oceanic enterprises, according to European culture man can live anywhere on the surface of the earth, and that earth is a terraqueous globe, in which land and water are not two different spherical bodies but together constitute a unique ball, whose centre is the centre of the world. As a consequence, the ecumene, that is to say the populated earth, which until that time had been thought to be a small flat place lost in the immensity of the sea is seen to extend throughout the whole surface of the globe; water expanses are no longer external limits, but internal lakes (Randles 1986). At the beginning of the seventeenth century Galileo erased what Bertolt Brecht later called the Aristotelian sky. For the abyss to be erased, one must wait till the beginning of the eighteenth century, when Ferdinando Marsigli for the first time talked about sea depth, which, until that time was far from being taken for granted. His *Histoire Physique de la Mer* (Marsigli 1725), published when Kant was learning to walk, described the bottom of the sea as homogeneous and continuous.

In short, Kant belongs to the first generation for whom the terrestrial globe ceased to be what it had been and instead became something continuous, homogeneous and isotropic. For Heinrich von Kleist one of the consequences of this change is the fact that the only possibility that is now left is to move around the globe "in order to see whether by any chance this will offer a way in on its rear side" (von Kleist no date: 76). The question then was how one could put order into the immeasurable universe of forms; more precisely, how is it possible to grasp and systematize an increasing flux of information, the results of the experience that seamen and travellers continuously poured in? If one wants to reduce Kant's studies into one single thought, one could state that, according to

---

4   Before Kant's death in Germany two editions of his geography were published: *Immanuel Kants physische Geographie*, Mainz und Hamburg 1801, by Gottfried Vollmer and *Immanuel Kants physische Geographie. Auf Verlangen des Verfassers aus seiner handschrift herausgegeben und zum Theil bearbeitet von D. Friedrich Theodor Rink*, Königsberg 1801, by Göbbels und Unzer. A comparative analysis of these two editions is not the purpose of this study. Suffice here to say that generally both editions follow the same sequence: mathematical elements-water-earth-atmosphere. However, while the first edition is followed by references to the alteration of the terrestrial mantel, different kinds of animals, plants, metals, rocks and peculiar aspects of single regions of earth, divided in different sections, in the second edition all these aspects are included in the initial part.

himself, experience depends on the application of a scheme. The question that follows is this: which scheme does Kant refer to?

Forty years ago, Michel Foucault illustrated with subtle irony what happens when a classification is deprived of its "operating table", which according to him is "what enables thought to operate upon the entities of our world, to put them in order to divide them into classes, to group them according to names that designate their similarities and differences – the table upon which since the beginning of time language has intersected space" (Foucault 1967: xvii). By way of clarification, Foucault quotes a famous text by Borges about a certain Chinese Encyclopaedia in which animals are divided into 13 categories, ranging from those that belong to the Emperor to those that from a long way off look like flies, frenzy animals, those drawn with a very fine camelhair brush, those that make love and so forth – a heteroclite, or, in Foucault's words, an impossible atlas because this classification lacks the mute ground upon which it is possible for entities to be juxtaposed (Foucault 1967: xvii–xix). In Kant's Geography, as transcribed by Vollmer, this place or table is earth itself, and the criterion for order is the degree to which we experience its parts:

> Earth is classified according to what we know, though this method does not lead to a geophysical classification such as 1. lands of which we know the circumference and the interior completely, EUROPE: 2. lands of which we know the circumference and the majority of the interior, ASIA: 3. lands of which only the circumference is known and the interior is completely unknown, AFRICA: 4. lands of which the circumference is only partly known and of the interior we know less than of the circumference, AMERICA and NEW HOLLAND, 5. lands that have been seen but never found again 6. lands well known by the ancients, but still lost 7. lands that can only be hypothesized.[5]

It might be that Borges, who declared that he had read Ritter's *Erkunde* while looking for a certain place, knew of this classification. Nowadays, this kind of classification might well cause unease, producing the same effects that Borges' classification did on Foucault. However, what causes unease nowadays is not the lack of a common ground as it was for Foucault – indeed in this case the classification concerns the whole planet – but the fact that such a classification does not conform to the schemata we have become accustomed to after Kant, and precisely *because of* Kant. That schemata classifies all objects endowed of natural or historical existence through the principle either of resemblance, that

---

5   Rink's version did not differ greatly, *GS*, IX, 228: "1. Lands of which we know the circumference and the interior; 2. Lands that we know only in part; 3. Lands of which we know only the coasts; 4. Lands that cannot be found any longer; 5. Lands known by the ancients and that are now lost; 6. Lands whose existence can only be hypothesized.

is metaphor, or of genealogy, that is metonymy and synecdoche (Tort 1989)[6] – a principle which is referred to and explained at the beginning and at the end of the Vollmer edition.[7]

This is precisely the opposition between a sort of Linnaean logical classification, which produces natural systems based on semblance or similarity, and a Kantian physical classification, founded on proximity or things, that has lately been at the centre of attention of those geographers who wish to explore a criterion which, similarly to the Kantian one, avoids the logic of the "intemporal rectangle" (i.e. the table) inside which there is no space for Kant's geography (Gregory 1994: 25–6).

In actual fact, things are not so simple, for as testified by Kant's Geography itself, the classification of places on the earth is not geophysical. As Kant writes in 1757, at the beginning of his programme (GS, II: 3):

> Physical geography only considers earth's natural conditions and what it contains: seas, continents, mountains, rivers, the atmosphere, mankind, animals, plants and minerals. All this however is not considered with the thoroughness and philosophical exactitude typical of physics and natural history, but rather with the reasonable curiosity of a traveller who is searching everywhere for anything that is worth noting, peculiar or marvellous and then compares his observations following a certain plan.

The plan comes before observation in the same way in which, as said above, schemata comes before experience. Such an order of priority can also be noted in the above- mentioned Kantian classification of the lands above sea-level. Let us take the last case – the one that concerns lands that are only hypothesized. These are divided into two sub-categories: those whose existence can be hypothesized for historical reasons and those whose existence can be hypothesized for physical reasons. Among the latter is the mythical land of Agisymba, the Land of the South imagined by the ancients and mentioned by Ptolemy (see Berggren and Jones 2000: 145–47). This is positioned at the lower limit of the ecumene, and, in the modern period, was believed to interrupt the Indian Ocean; as a consequence, this area became an immense closed sea, a sort of Antarctic cap whose only end is – as it was later discovered – Australia. Kant was the first to challenge the reasons that attested to its existence (Vollmer edition, III: 592–94.) Even though Kant does not actually specify it, such a presence could be theorized because of the supposed analogy between the Southern and the Northern hemisphere, which extended up to 70° of latitude. What was then the reason for this analogy, the origin of the matrix?

---

6   Pages 12 and 17 contain some reflections by Kant on different human races (1775–1777).

7   The principle of genealogy or of kinship in Kantian anthropology became in his Geography a principle of proximity or vicinity (Vollmer edition I, 23–4 and VI, 409–10): for this reason it is correct, as Tort does, to refer to metonymy which includes both cases.

Let us take now for example the case of a real land, Greenland. Kant hypothesized its insular nature which, at that point, could not be proven. Kant's hypothesis was based on the symmetry of the two parts of the globe positioned eastwards and westwards of a vertical axis of the new and the old continent. For the same reason (that is to say a logic which nowadays appears analogical and symmetrical) perhaps also America reached further west, according to Kant, than what the maps showed. Also, let us think of the insistence with which Kant posits the reciprocal correspondence of coasts facing each other (for example the coast of Sicily and the African coast, or the western protuberance of the African continent and the gulf of Mexico), anticipating and inferring in this way genetic hypotheses on telluric motions that were formulated only a century later with the theory of continental drift by Wegener. All this is possible thanks to the "Plan", a plan that means "map", that is geographical representation. In Kant's geography, such a map or schemata (i.e. the table or the "intemporal rectangle"), contrary to what happens with the Chinese Encyclopaedia, comes before experience.

In other words, if we posit a schemata that lies at the source of all Kantian architecture (Pierobon 1990: 65), this scheme must be cartographic. This is a delicate and complex issue that here can merely be hinted at. However, it is significant that in the introduction to his Geography Kant has provided "architectonics" with one of its most perspicuous definitions (Vollmer edition, I: 23):

> The world cognition should be organized as a system; otherwise we could not be sure either to have grasped it wholly, or to be able to remember it, because our gaze cannot dominate what we know. In the system the whole comes before its parts, in the aggregate parts come first. It [the system] is the architectonic idea, without which science cannot build – let us a say – a home for itself. Whoever wants to build a home for himself must first have an idea of the whole, from which all its parts can be deduced. The description of the world and the earth as a system must start with the description of the globe, the idea of the whole, and always go back to it.

Even clearer and more faithful to the section on "architectonics" in the Critique is the variation in Rink's transcription: "The idea is architectural and it creates sciences". After the example of the home, Rink adds that "an architectonical concept is a concept in which multiplicity derives from totality, that is to say from globality (GS, IX: 158; Kant 1998: 691 ff; also Hohnegger 2004: 71–158). Kant himself explains the objectivity (validity) of knowledge as a function of transcendental subjectivity. He re-interprets the "venerable Aristotelian couple" of form and matter and identifies such re-interpretation with a philosophical Copernican revolution (Landucci 1993: 26). However, Kant's transcendental philosophy is nothing but the architectonics of the passage from the *a-priori* to the *a-posteriori* (Pierobon 1990: 33–4). Differently from Aristotle, for Kant, forms are always pure, because they are functions, or structures, applied to a subject that transcends materiality, and is therefore autonomous from experience. Therefore "formal" for

Kant means "a-priori", pertinent not to the sensible world but to the intelligible, and the two are clearly distinguished the one from the other. This equivalence of meaning marks what is new and radical in Kant's thought, even though at a close analysis what Kant discovered was actually part of Ptolemy's geography which dates back to the second century A.D. Here the sphere is translated into a Plan and the art of projection, that is to say the technique through which the complexity and the totality of the globe is broken down into a multiplicity of cartographic representations, is illustrated. The *Critique of Pure Reason* can be defined as nothing but the protocolar, systematic, radical and conscious realization of Ptolemy's assumptions.

Confronted with the Ptolemaic way of flattening the curves of the world into tables, we can suppose that Kant came to consider the problem of the "way of knowledge" or of the "cognition of the world", and made the transition from empirical geography to the "geography of reason", or, as he himself put it, to the geography of "the dark space of our intellect" (Kant quoted in Cassirer 1977: 173–4). Ancient metaphysics moved along the Ptolemaic precept and attempted to posit being, understood here as the point of projection, as the source of complex and precise determinations. Despite one being antithetical to the other, both Empiricism and Rationalism believed in the existence of such a being, which corresponded to the "actual reality of things which mind must absorb and then reproduce as a copy" (Kant quoted in Cassirer 1977: 174). According to the Kantian analytics of pure intellect, metaphysical being is not an original data but a problem or a postulate. This in turn poses questions about the nature of the projection point and, in particular, about the transition from its "objectivity" to the subjective form of representation, a movement that in the context of Kantian thought seems to represent the metamorphic character of the projection mechanism. This is called "the design" according to which "reason has insight only into what it itself produces", as it is written in the preface of the second edition of the Critique (Kant 1998: 109), a critique which in short is the illustration of a sort of mental map of the projective drawing and its analysis that begins not from the objects that are produced, but from the recognition of its function as producer of a particular mode of knowledge.

At the end of the Analytic, and in order to introduce the difference between phenomena and noumena, Kant confesses the cartographic nature of his thought (Farinelli 1995: 145; Farinelli 1996: 267–302). He writes (Kant 1998: B 295): "We have now not only travelled through the land of pure understanding, and carefully inspected each part of it, but we have also surveyed it, and determined the place for each thing in it." However, he also adds, "it will be useful first to cast yet another glance at the map of the land that we would now leave". It is in the appendix of his transcendental dialectics that he describes how this model works and explains the necessary regulative use of transcendental ideas. Such a use consists in (Kant 1998: B 672):

directing the understanding to a certain goal respecting which the lines of direction of all its rules converge at one point, which, although it is only an idea (*focus imaginarius*) – i.e. a point from which the concepts of understanding do not really proceed, since it lies entirely outside the bounds of possible experience – nonetheless still serves to obtain for these concepts the greatest unity alongside the greatest extension

And he goes on saying "Now of course it is from this that there arises the deception, as if these lines of direction where shot out from an object lying outside the field of possible empirical cognition (just as objects are seen behind the surface of a mirror)". With "deception" here Kant refers to ancient metaphysics against which he warns his readers, but which, at the same time, he considers necessary: "if besides the objects before our eyes we want to see those that lie far in the background, i.e., when, in our case, the understanding wants to go beyond every given experience (beyond this part of the whole of possible experience), and hence wants to take the measure of its greatest possible and uttermost extension" (Kant 1998: B 673). If it might appear excessive to identify the *focus* with the Ptolemaic point of projection and the "lines of direction" with the axis that descend from such a point of projection (Farinelli 1995:145; Farinelli 1996), we need only consider what is written further on (Kant 1998: B 675, emphases in original):

The hypothetical use of reason is therefore directed at the systematic unity of the understanding's cognitions, which, however, is the *touchstone of truth* for its rules. Conversely [Umgekehrt in a reverse movement, not bottom up but from the top down] systematic unity (as mere idea) is only a *projected* [projektierte] unity, which one must regard not as given in itself, but only as a problem.

It is significant to point out that the modern term for projection comes from alchemy and concerns precisely the effect produced by the Philosophical Stone, that is to say the transformation of vile metals into gold (Eco 1990: 76). This implies an ontological transformation that concerns not merely form (as naive geographers believe) but rather the very nature of things, and of the world. Referring back to the above quotation, the reference to "touchstone" is reminiscent of the very process through which metals are transformed into gold; the reference to projection shows that Kantian reflection resides in the problematic acceptance of the systematic unity of this process; in other words the Ptolemaic projection point is Pure Reason, and its Critique is the cartographic description of projection.

The famous motto on silence that Kant borrows from Bacon and includes in the foreword to his Critique marks the intention to say no more concerning himself as geographer and the cartographic nature of his thought. As a matter of fact, as Strabo also noted, the first ones to study geography where philosophers (Strabo, I :1). Kant's example is not therefore a novelty, just as the term "globalization" is not a novelty. Globalization indicates an intention to consider earth in its real form, that is to say as a globe, and this is precisely the problem from which

Kant's reflections began. What is new, however, is the fact that while Kant relies completely on the mediation of the cartographic scheme, and consequently on the reduction of the globe into a map, and of knowledge into a geography of the mind, this type of reasoning is no longer possible.

According to Hegel, Minerva's noctule, that is philosophy, flies up in the twilight. But if the earth is a globe, where, anthropologists nowadays wonder, is twilight? Provided we can find a place for the twilight, this cannot be the same for everyone (Clifford 1997: 20–21). Precisely because there is no answer to this question, it is urgent that we return to Kant, and more particularly to his dense geographic lessons, the originary form of his thought and raw material of his philosophy. If it is true that our world is different from Kant's world, it is also true that we continue to regard it in the way that Kant taught us. Fortunately, his geography also leaves open another possibility, the possibility of a physical geography based on the principle of physical classification, a geography that remains yet to be constructed, the geography we need if we want to understand globalization. A geography which was the first, and still remains, the only dream left to Western culture.

Bibliography

Berggren, J.L. and Jones, A. 2000. *Ptolemy's Geography. An Annotated Translation of the Theoretical Chapters*. Princeton: Princeton University Press.

Cassirer, E. 1977. *Vita e dottrina di Kant*. Firenze: La Nuova Italia.

Clifford, J. 1997. *Routes. Travel and Translation in the Late Twentieth Century*. Cambridge, MA: Harvard University Press.

Eckerlin, A. (ed.) 1807. *Geografia Fisica di Emanuele Kant*. Vol. 3. Milano: Tipografia di Giovanni Silvestri.

Eco, U. 1990. *I limiti dell'interpretazione*. Milano: Bompiani.

Eco, U. 1997. *Kant e l'ornitorinco*. Milano: Bompiani.

Farinelli, F.1995. "L'arte della geografia," *Geotema*, 1.

Farinelli, F. 1996. "Von der Natur der Moderne: eine Kritik der kartografischen Vernunft," in D. Reicher (Hrsg), *Räumliches Denken*. Zürich: ETH.

Farinelli, F. 2004. "Experimentum mundi," introduction to I. Kant, *Geografia fisica*. Begamo: Leading Edizione, i–xxix.

Foucault, M. 1967. *The Order of Things. An Archaeology of the Human Sciences*. London, Routledge.

Gedan, P. (Hrsg). 1923. *Immanuel Kants physische Geographie*, in *Gesammelte Schriften*. Berlin-Leipzig: Königlichen Preussischen Akademie der Wissenschaften zu Berlin.

Gregory, D. 1994. *Geographical Imaginations*. Cambridge, MA: Blackwell.

Kant, I. 1998. *Critique of Pure Reason*. Cambridge: Cambridge University Press.

Von Kleist, H (n.d.): "Über das Marionettentheater," in von Kleist *Werke*, V. Berlin-Leipzig: Bong.

Landucci S. 1993. *La "Critica della Ragion Pratica" di Kant. Introduzione alla lettura*. Roma: La Nuova Italia Scientifica.

Marsigli, L.F. 1725. *Histoire Physique de la Mer*. Amsterdam: Aux Depens de la Compagnie.

May, J.G. 1970. *Kant's Concept of Geography and its Relation to Recent Geographical Thought*. Toronto: University of Toronto Press.

Pierobon, F. 1990. *Kant et la fondation architectonique de la métaphysique*. Grenoble: Millon.

Randles, W.G.L. 1986. *De la Terre plate au globe terrestre (Une mutation épistémologique rapide 1480–1520)*. Paris: Colin.

Strabo. 1917 32: *Geography*. Translated and edited by Horace L. Jones. London: Heinemann.

Tort, P. 1989. *La Raison classificatoire*. Paris: Aubier.

Publications by Gunnar Olsson

a. Monographs

Distans och mänsklig interaktion. Unpublished licantiate thesis, Uppsala Universitet, Kulturgeografiska Institutionen, 1964.

Distance and Human Interaction: A Review and Bibliography. Philadelphia, PA: Regional Science Research Institute, 1965 (reprinted 1968).

Distance, Human Interaction and Stochastic Processes: Essays on Geographic Model Building. Ann Arbor, MI: Department of Geography, University of Michigan, 1968 (PhD Thesis, University of Uppsala, Sweden).

Spatial Sampling: A Technique for Acquisition of Geographic Data from Aerial Photographs and Maps. Ann Arbor, MI: Regional Research Associates, 1971.

Birds in Egg. Ann Arbor, MI: Michigan Geographical Publications, No. 15, 1975.

Birds in Egg/Eggs in Bird. London: Pion, 1980.

Ucelli nell'uova/Uova nell'ucello. Roma: Edizione Theoria, 1987.

Krisens tegn/Tegnets krise. Nimtofte: Forlaget Indtryk, 1989.

Antipasti. Göteborg: Korpen, 1990.

Linee senza ombre: La tragedia della pianificazione. Roma: Edizione Theoria, 1991.

Lines of Power/Limits of Language. Minneapolis: University of Minnesota Press, 1991.

Abysmal: A Critique of Cartographic Reason. Chicago: University of Chicago Press, 2007.

b. Collections

Meddelande från ett symposium i teoretisk samhällsgeografi. Uppsala Universitet: Forskningsrapporter från Kulturgeografiska Institutionen, No. 1, 1965 (co-edited with Olof Wärneryd).

Meddelande från ett simuleringssymposium. Uppsala Universitet: Forskningsrapporter från Kulturgeografiska Institutionen, No. 6, 1966.

Special issue on "Trends in Spatial Model Building," *Geographical Analysis*, Vol. 1, No. 3, 1969.

Special issue on "Geography, Epistemology and Social Engineering," *Antipode*, Vol. 4, No. 1, 1972.

Geographic Humanism, Analysis and Social Action. Ann Arbor, MI: Michigan Geographical Publications, No. 17, 1976 (co-edited with George Kish and John D. Nystuen).

Philosophy in Geography. Dordrecht: D. Reidel 1977 (co-edited with Stephen Gale).

A Search for Common Ground. London: Pion, 1982 (co-edited with Peter Gould).
Limits of Representation. München: Accedo, 1994 (co-edited with Franco Farinelli
and Dagmar Reichert).
Chimärerna. Porträtt från en forskarutbildning. Stockholm: Nordplan, 1996.
Poste Restante. En avslutningsbok. Stockholm: Nordplan, 1996.
Att famna en ton. En kulturgeografisk minnesbok. Uppsala: Acta Universitatis
Upsaliensis, Ser. C, No. 63, 1998.

c. Articles and Chapters

"Utflyttningarna från centrala Värmland under 1880-talet", *Meddelande från
Uppsala Universitets Geografiska Institutioner*, Ser. A, No. 178, 1962, pp. 1–20.
"The Spacing of Central Places in Sweden", *Papers and Proceedings of the
Regional Science Association*, Vol. 12, 1964, pp. 87–93 (co-authored with
Åke Persson). Reprinted in R.G. Putman (ed.): *A Geography of Urban Places*.
Toronto: Methuen, 1970.
"Distance and Human Interaction: A Migration Study", *Geografiska Annaler*, Ser.
B, Vol. 47, 1965, pp. 3–43.
"Deterministiska och stokastiska interaktionsmodeller", in Gunnar Olsson
(ed.): *Meddelande från ett symposium i teoretisk samhällsgeografi.* Uppsala
Universitet, 1965, pp. 6–14.
"Gravitations- och potentialformuleringar samt distansfunktionen i rumslig
simulering. En preliminär bibliografi", in Gunnar Olsson (ed.): *Meddelande
från ett symposium i teoretisk samhällsgeografi.* Uppsala Universitet, 1965,
pp. 63–91.
"Värmland och Dalsland", in H.W. Ahlman (ed.): *Sverige: Land och Folk.*
Stockholm: Natur och Kultur, 1966, pp. 681–695.
"Deductive and Inductive Approaches to Model Formulation", *Komitet
Przestrzennego Zagospodarwania Kraju Polskiej Nauk, Studia*, Tom 17, 1967,
pp. 173–193.
"Approaches to Simulations of Urban Growth", *Geografiska Annaler*, Ser. B,
Vol. 48, 1966, pp. 9–22 (co-authored with Roger Malm and Olof Wärneryd).
Reprinted in *Quantity and Quality*, Vol. 3, 1969 and in H.M. Blalock (ed.):
Quantitative Sociology. New York: Academic Press, 1975.
"Beskrivning och testning av rumsliga simuleringsresultat", in Gunnar Olsson
(ed.): *Meddelande från ett simuleringssymposium.* Uppsala Universtet, 1966,
pp. 28–67.
"Central Place Systems, Spatial Interaction, and Stochastic Processes", *Papers
and Proceedings of the Regional Science Association*, Vol. 18, 1967, pp.
13–45. Reprinted in W.H. Leahy et al. (eds): *Urban Economics.* New York:
Free Press, 1970; and as "Zentralörtlische Systeme, Raumlische Interaktion,
und Stochastische Prozesse", in D. Bartels (Hrsg): *Wirtschafts- und
Sozialgeographie.* Köln: Kiepenhauer und Witsch, 1970.

"Lokaliseringsteori och stokastiska processer", in T.F. Rasmussen (ed.): *Forelesninger i Regionale Analysemetoder*. Oslo: Norsk Institutt for By- og Regionsforskning, 1967, pp. 55–86.

"Kulturgeografi och planering", *Svenska Dagbladet*, 16 mars, 1967.

"Teori, modell och planering", *Choros*, No. 3, 1967, pp. 1–13.

"Geography 1984", Department of Geography, University of Bristol, *Seminar Paper*, Ser. A, No. 7, 1967, pp. 1–28.

"Spatial Theory and Human Behavior", *Papers and Proceedings of the Regional Science Association*, Vol. 21, 1968, pp. 229–242 (co-authored with Stephen Gale). Reprinted in W.K.D. Davis (ed.): *The Conceptual Revolution in Geography: Selected Essays in Method*. London: London University Press, 1972.

"Urbanisering och industrialisering", *Ymer*, Vol. 88, 1968, pp. 178–187.

"Complementary Models: A Study of Colonization Maps", *Geografiska Annaler*, Ser. B, Vol. 50, 1968, pp. 115–132. Reprinted in R.L. and K.N. Singh (eds): *Readings in Rural Settlement Geography*. Varanasi: N.G.S.I., 1975.

"Ekshärads befolknings- och bebyggelseutveckling", in Signe Borlind (ed.): *Ekshärad socken*. Karlstad: Värmlandstryck, 1969, pp. 54–67.

"Inference Problems in Locational Analysis", in K.R. Cox and R.G. Golledge (eds): *Behavioral Models in Geography: A Symposium*. Evanston, IL: Northwestern University Press, 1969, pp. 14–34. Reprinted in K.R. Cox and R.G. Golledge (eds): *Behavioral Problems in Geography Revisited*. New York: Methuen, 1981 (translated into Japanese, 1987).

"Trends in Spatial Model Building", *Geographical Analysis*, Vol. 1, 1969, pp. 219–224.

"Explanation, Prediction and Meaning Variance: An Assessment of Distance Interaction Models", *Economic Geography*, Vol. 46, 1970, pp. 223–233. Reprinted in L.A. Brown (ed.): *Population and Migration in an Urban Context*. Columbus, OH: Ohio State University, 1970.

"Logics and Social Engineering", *Geographical Analysis*, Vol. 2, 1970, pp. 361–375.

"Correspondence Rules and Social Engineering", *Economic Geography*, Vol. 47, 1971, pp. 545–555.

"Analogs, Theories and Decision Making", *Papers and Proceedings of the Regional Science Association*, Vol. 27, 1971, pp. 39–43.

"Migration and Resettlement: Some Comments on Action Suggested by Scientific Models", Paper prepared for the *South East Asian Development Advisory Group*, Washington, DC, 1971, 37 pp.

"Geography and Social Engineering", *Papers from the IVth International Congress for Logic, Methodology and Philosophy*, Bucharest, 1971.

"Some Notes on Geography and Social Engineering", *Antipode*, Vol. 4, 1972, pp. 1–22.

"On Reason and Reasoning, On Problems as Solutions and Solutions as Problems, but Mostly on the Silver-Tongued Devil and I", *Antipode*, Vol. 4, No. 2, 1972, pp. 26–31.

"Servitude an Inequality in Spatial Planning: Methodology and Ideology in Conflict", *Antipode*, Vol. 6, No. 1, 1974, pp. 16–21. Reprinted in R. Peet (ed.): *Radical Geography*. Chicago, IL: Maroufa Press, 1977.

"The Dialectics of Spatial Analysis", *Antipode*, Vol. 6, No. 3, 1974, pp. 50–62.

"On Words and Worlds", *Papers and Proceedings of the Regional Science Association*, Vol. 35, 1975, pp. 45–49.

"Societal Efficiency and Individual Freedom: The Ecological Fallacy Revisited," Paper prepared for the Population Commission, *International Geographical Union*, Palmerston North, New Zealand, 1974.

"Social Science and Human Action or On Hitting Your Head Against the Ceiling of Language", in Stephen Gale and Gunnar Olsson (eds): *Philosophy in Geography*, 1979, pp. 287–307.

"Of Ambiguity or Far Cries From a Memorializing Mamafesta", in David Ley and Marvin Samuels (eds): *Humanistic Orientations in Geography*. Chicago: Maroufa Press, 1978, pp. 109–120.

"Om samhällsplaneringens mytologi", *Nordplans Årsrapport*, 1977, pp. 62–69.

"Identitet och förändring eller Om hemlängtan som ontologiska transformationer", *Acta Universitatis Upsaliensis*, Symposia No. 11, 1978, pp. 101–114.

"Framtiden är poesi", *Futuriblerne*, Nr 4–5, 1978, pp. 45–46.

"Om makt", in Per Andersson (ed.): *Nordplan 10 år*. Stockholm: Nordplan, 1978, pp. 60–64.

"On the Mythology of the Negative Exponential or On Power as a Game of Ontological Transformations", *Geografiska Annaler*, Ser. B, Vol. 60, 1978, pp. 116–123. Reprinted in Michael J. Dear and Steven Flusty (eds): *The Spaces of Postmodernity: Readings in Human Geography*. Oxford: Blackwell, 2002, pp. 85–94.

"The New Social Science: Toward a Mandala of Thought and Action", in I.C. Cullen (ed.): *Analysis and Decision in Regional Policy*. London: Pion, 1979, pp. 7–19.

"Om makt och planering eller En kamp om tankesystem och verklighetsuppfattningar", in Hans-Åke Jansson (ed.): *Nya tider – Ny regionalekonomi*. Sveriges Lantbruksuniversitet, Arbetsgruppen lantbruk och samhälle, 1979, pp. 3–14.

"Planering som ideologi", in Hans-Åke Jansson (ed.): *Nya tider – Ny regionalekonomi. Sveriges Lantbruks-universitet*, Arbetsgruppen lantbruk och samhälle, 1979, pp. 15–29.

"Planlægning som ideologi", *Byplan*, Vol. 31, 1979, pp. 106–109.

"Helhet, sammanhang och förståelse: Frågor om gränser", in Kenneth Abrahamsson (ed.): *Den gränslösa samhällsförvaltningen*. Stockholm: Statskonsult, 1980, pp. 114–123.

"Ord om tankar/Tankar om ord", *Svensk Geografisk Årsbok*, Årg. 56, 1980, pp. 89–94.

"Att tänka i säregna öglor", *Svenska Dagbladet*, 23 August, 1980.

"On Yearning for Home: An Epistemological View of Ontological Transformations", in D.C.D. Pocock (ed.): *Humanistic Geography and Literature*. London: Croom Helm, 1981, pp. 121–129.

"Om snedstreck och snedsprång", in Anssi Paasi (ed.): *Maantiede 1980-luvulla. University of Joensuu*, Publications of Social and Regional Science, No. 26, 1981, pp. 9–22.

"Thunderbolt on Herons Shore", in Allan Pred (ed.): *Space and Time in Geography: Essays Dedicated to Torsten Hägerstrand*. Lund: Gleerups, 1981, pp. 122–126.

"Den nya samhällsvetenskapen: Frågor ur statskapitalismens gränsland", *Futuriblerne*, Nr. 3–6, 1981, pp. 60–63.

"Nysvenska tankar i gammeldansk tappning: Frågor runt tecknet / i individ/ samhälle", *Nordisk psykologi*, Vol. 34, 1982, 79–85.

"Planleggingens øye", in Noralv Veggeland (ed.): *Planleggingens muligheter, 1. Teorier for handling*. Oslo: Universitetsforlaget, 1982, pp. 212–232.

"Från Epimenides till modern geografi eller Från postmodernismen tillbaka till Epimenides", in Sverre Strand (ed.): *Geografi som samfunnsvitenskap*. Oslo: Universitetsforlaget, 1982, pp. 11–16; 190–191.

"–/–", in Peter Gould and Gunnar Olsson (eds): *A Search for Common Ground*. London: Pion, 1982, pp. 223–231. Reprinted in *Sub-Stance*, Vol. 35, 1982, pp. 24–33.

"Epilogue: A Ground for Common Search", in Peter Gould and Gunnar Olsson (eds): *A Search for Common Ground*. London: Pion, 1982, pp. 261–264.

"Toward a Sermon of Modernity", in Mark Billinge et al. (eds): *Geography as Spatial Science: Recollections of a Revolution*. London: Macmillan, 1983, pp. 73–85.

"Expressed Impressions of Impressed Expressions", *Geographical Analysis*, Vol. 15, 1983, pp. 60–64.

"Mot en svensk ideologi", *Tvärsnitt*, Årg. 5, 1983, pp. 38–44.

"Ajan teksti: politiikan ja suunnittelun sirpaleita", *Yhteiskuntasuunnittellu*, Vol. 21, 1983. pp. 1–5.

"I begynnelsen var ordet", *Expressen*, 10 February, 1983.

"Jag lovar att", *Expressen*, 9 April, 1983.

"Marxismen som en modernism", *Svenska Dagbladet*, 15 April, 1983.

"Roland Barthes, den ensamme", *Svenska Dagbladet*, 8 August, 1983.

"Research and Research Training: A Futuristic and Global Perspective", in Bengt Sigurd and Olof Wärneryd (reds): *Forskning och forskarutbildning*. Stockholm: Universitets- och Högskoleämbetet, 1983, pp. 79–87.

"Of Creativity and Socialization", *Archivio di Studi Urbani e Regionale*, No. 18, 1983, pp. 143–154.

"Kreativitet och kreativitetens villkor", in Gunnar Arpi and Frank Petrini (eds): *Regional förnyelse – hur?* Stockholm: ERU-rapport No. 31, 1984, pp. 7–15.

"Vem är mäktigast?", in Jan-Evert Nilsson (ed.): *Et alternativt Norden*. Oslo: Gruppen för Resursstudier, 1984, pp. 279–292.

"Sociala experiment: Sandwich à la Russe", *Futuriblerne*, No. 4–6, 1984, pp. 13–18. Reprinted in an English version as "Of Experiments: Sandwich à la Russe", in Erling Olsen (ed.): *The Rules of the New Game*. København: Det danske selskab, 1984, pp. 35–45.

"Tidens text: Skärvor av politik och planering", *Artes*, No. 6, 1984, pp. 75–81.

"Marxism as a Modernism: Review Essay", *Environment and Planning D, Society and Space*, Vol. 2, 1984, pp. 241–244.

"Svenska visdomständer", *Expressen*, 21 February, 1985.

"Om planeringens paradoxer", *Nordisk Samhällsgeografisk Tidskrift*, No. 2, 1985, pp. 3–9. Reprinted in Johannes Møllgaard and Gunnar Olsson: *Tidmaskinen – två föreläsningar om planläggaren som totalisator*. Stockholm: Nordplan, 1985, pp. 15–28.

"Blodets rytm, strändernas svall", *Expressen*, 6 July, 1985.

"Redan-men-inte-ännu", in Jan F. Bernt et al.: *Ramlagstiftning*. Stockholm: Nordplan, 1985, pp. 3–14.

"The Eye and the Index Finger: Bodily Means to Cultural Meaning", *Geotopiques*, 1985, pp. 61–69.

"Makten är en ful gubbe", *Expressen*, 31 December, 1985.

"Creativity and Socialization", in Raymond Dandel and Nicole Lemaire D'Agaggio (eds): *Life Sciences and Society*. Amsterdam: Elsevier, 1986, pp. 219–227; 229–232.

"The Social Space of Silence", *Poetica et Analytica*, Vol. 3, 1986, pp. 6–30. Reprinted in *Environment and Planning D, Society and Space*, Vol. 5, 1987, pp. 249–262.

"Maktens lapptäcke", in Mikael Löfgren and Anders Molander (eds): *Postmoderna tider*. Stockholm: Norstedts, 1986, pp. 413–419. Reprinted in *Svensk linje*, 1988.

"Ekonomisk Geografi och Samhällsplanering", *Nordrefo*, Årg. 17, 1987, pp. 96–102.

"Sui possibili centri del cerchio", *Montedison Progetto Cultura*, Anno 2, No. 5, 1987.

"Human Action as a Magician's Trick", in Pasquale Alferji (ed.): *Frontiere della Scienza a della Tecnologia*. Milano: Montedison, 1987.

"The Eye and the Index Finger: Bodily Means to Cultural Meaning", in Reginald G. Golledge et al. (eds): *A Ground for Common Search*. Santa Barbara, CA. Santa Barbara Geographical Press, 1988, pp. 126–137.

"Ögat och pekfingret", in Gunnar Broberg et al. (eds): *Kunskapens trädgårdar. Om institutioner och institutionaliseringar i vetenskapen och livet*. Stockholm: Atlantis, 1988, pp. 234–252.

"On Doughnutting: Discusssion of Jaakko Hintikka's paper *Exploring Possible Worlds*," in Sture Allén (ed.): *Possible Worlds in Humanities, Arts and Science*. Berlin: Walter de Gruyter, 1989, pp. 74–81.

"Om tragedins struktur: En läsning av Oidipus Rex", in *Tal över blandade ämnen, 1984–86. Collegium Curisorum Novum, Årsbok* 1985–86, pp. 21–37. Reprinted in Veijo Meri et al.: *Att se in i en själ*. Stockholm: Nordplan, 1988, pp. 17–30.

"The Language of Geography and the Geography of Language", in Gabriele Zanetto (ed.): *Langages de représentation géographiques*. Venezia: EST, 1989, pp. 135–156.

"Bumerangens återkomst", in Kirsten Simonsen (ed.): *Planlægning under postmoderne vilkår*. Roskilde: Roskilde Universitetsforlag, 1988, pp. 121–145.

"Bjälken i ögat", *Expressen*, 26 August, 1988.

"Mödom mod och morske män", in Yvonne Hirdman (ed.): *Maktens former*. Stockholm: Carlssons, 1989, pp. 114–148.

"Braids of Justification", in Pekka Kosonen and Anneli Levo-Kivirikko (e): *Vapaus, veljeys ja vallankäyttö*. Helsinki: Hakapaino, 1989, pp. 209–223.

"Pärla för svin", *Wermlandus*, Årg. 30, No. 4, 1990.

"Lines of Power", in Trevor J. Barnes and James S. Duncan (eds): *Writing Worlds*. London: Routledge, 1991, pp. 86–96. Reprinted in *Nordisk Samhällsgeografisk Tidskrift*, No. 11, 1990, pp. 106–116; and in Trevor Barnes and Derek Gregory (eds): *Reading Human Geography: The Poetics and Politics of Inquiry*. London: Arnold: 1997, pp. 145–155.

"Don Quijote ser katten springa till skogs", *Expressen*, 6 October, 1990.

"Individ/samhälle", *Kungliga Musikaliska Akademiens Årsskrift*, 1990, pp. 53–60.

"Inivisible Maps: A Prospectus", *Geografiska Annaler*, Ser. B, Vol. 73, 1991, pp. 85–91.

"Malivic sfigurato", *Slam*, No. 3, April 1991, pp. 7–10.

"The Eye and the Ear", in Bruno de Lescure (ed.): *La Qualité Sonore des Espaces Habités*. Grenoble: CRESSON, 1991, pp. 157–166.

"Transformación e identidad del Centro Histórico", in *Primeres Jornades ciutat vella: Revitalització urbana, econòmica i social*. Barcelona: Ajuntament de Barcelona, 1991, pp. 159–162.

"Hur viljans makt blev maktens vilja", *Expressen*, 22 August, 1992.

"Människans gränser", *Rockhammarbygdens Jultidning*, 1992.

"Ethics in Planning", *Nordisk Samhällsgeografisk Tidskrift*, No. 15, 1992, pp. 62–65. Reprinted in *Giornale del dottorato in pianifacazione territorial*, No. 5, 1993.

"Chiasm of Thought-and-Action", *Environment and Planning D, Society and Space*, Vol. 11, 1993, pp. 279–284. Reprinted in Franco Farinelli et al. (eds): *Limits of Representation*, 1994, pp. 29–53.

"Smaterilizzato", *Slam*, No. 5, Giugno, 1993, pp. 1–4.

"Tankens estetik", in Lise Bek et al. (eds): *Syn for rum: Om byens og landskabets æstetik*. Aarhus: Aarhus Universitetsforlag, 1993, pp. 188–197.

"Landet Tanke-och-Handling", in *Tal över blandade ämnen, 1993–94, Collegium Curiosorum Novum, Årsbok*, 1994.

"Heretic Cartography", *Ecumene*, Vol. 1, 1994, pp. 215–234. Reprinted in *Fennia*, Vol. 172, 1994, pp. 115–130.

"Job and the Case of the Herbarium", *Environment and Planning D, Society and Space*, Vol. 12, 1994, pp. 221–225.

"Vasallerna", in *Programblad* till Heinrich von Kleists *Schroffenstein*. Malmö: Malmö Stadsteater, 1994.

"Makt och mening", in *Programblad* till August Strindbergs *Gustav Vasa*. Stockholm: Stockholms Stadsteater, 1994.

"Detta eviga sommarlov", *Wermlandus*, Nr. 5, 1994, pp. 12–13.

"Signs of Persuasion", in Wolfgang Natter et al. (eds): *Objectivity and Its Other*. New York: Guilford Press, 1995, pp. 21–32.

"La granata di Malevic", *Geotema*, Vol. 1, 1995, pp. 63–64.

"Du skall icke göra dig ett beläte", *Johan Gottlieb Gahn-akademiens Årsbok*, 1995.

"Landscape of Desire: Review Essay", *Economic Geography*, Vol. 71, 1995, pp. 435–439.

"Alumbrando el muro", in Fernando Castro Flórez: *Nacho Criado. La idea y su puesta en escena*. Sevilla: Sibilina, 1996, pp. 11–12.

"Tanke-och-handling", in Gunnar Olsson (ed.): *Chimärerna: Porträtt från en forskarutbilning*, Stockholm: Nordplan, 1996, pp. 15–28.

"Semikolon", in Gunnar Olsson (ed.): *Chimärerna: Porträtt från en forskarutbildning*. Stockholm: Nordplan, 1996, pp. 277–282.

"Die Projektion des Begehrens/Das Begehren der Projektion", in Dagmar Reichert: *Räumliches Denken*. Zürich: Vdf Hochschulverlag an der ETH, 1996, pp. 225–249; 251–266.

"Placenta", in Gunnar Olsson (ed.): Gunnar Olsson: *Poste Restante: En avslutningsbok*. Stockholm: Nordplan, 1996, pp. 13–16.

"Aningars aning: Glimtar av en humanvetenskaplig handlingsteori", in Gunnar Olsson (ed.): *Poste Restante: En avslutningsbok*. Stockholm: Nordplan, 1996, pp. 54–70.

"Places of Desire", in Pauli Tapani Karjalainen and Pauline von Bonsdorf (eds): *Place and Embodiment*. Helsinki: University of Helsinki, 1997, pp. 37–43. Translated into Slovenian as "Mesta zelje," *Anthropos*, Letnik 28, 1996, pp. 62–69.

"Misión imposíble", *Anales de Geografía de la Universidad Complutense*, No. 17, 1997, pp. 39–51.

"Horror vacui", in Martin Gren and P.O. Hallin (eds): *Svensk kulturgeografi. En exkursion inför 2000-talet*. Lund: Studentlitteratur, 1998, pp. 117–132.

"Towards a Critique of Cartographical Reason", *Ethics, Place and Environment*, Vol. 1, 1998, pp. 145–155.

"Stadier på livets väg", in Gunnar Olsson (ed.): *Att famna en ton*. Uppsala: Acta Universitatis Upsaliensis, 1998, pp. 103–119.

"Den dematerialiserade punkten", in Lise Bek et al. (eds): *Perspektiv på rum*. Hørsholm: Statens Byggeforskningsinstitut, 1999.

"Malevich Torpedoed", chapter from *Lines of Power/Limits of Language* reprinted in Jed Rasula and Steve McCaffery (eds): *Imagining Language: An Anthology*. Cambridge, MA. MIT Press, 1998.

"Heretic Cartography: Report from a Mapping Expedition", in Jale Sanart (ed.): *Art and Environment*. Ankara: SANART, 1998.

"Landscape – Border Station Between Stonescape and Landscape", in Girolmano Cusimano (ed.): *La Construzione del Paesaggio Siciliano: Geografi e scrittori a confronto*. Palermo: Alloro, 1999, pp. 135–145.

"Glimpses", in Peter Gould and Forrest Pitts (eds): *Geographical Voices*. Syracuse: Syracuse University Press, 2002. Freench version as "Aperçus," in Peter Gould and Antoine Bailly (eds): *Mémoires de Géographes*. Paris: Anthropos, 2000. Japanese version 2008.

"From a=b to a=a", *Environment and Planning*, Ser. A, Vol. 32, 2000, pp. 1235–1244.

"Skattkammarön", in Martin Gren, P.O. Hallin and Irene Molina (eds): *Kulturens plats/Maktens rum*. Stehag: Symposion, 2001.

"Washed in a Washing Machine™", in Claudio Minca (ed.): *Postmodern Geography: Theory and Praxis*. Oxford: Blackwell, 2001.

"Costellazioni", *Sistema Terra – Rivista internazionale di telerilevamento*, Vol. 8, 1999, pp. 139–144. English version as "Constellations," *Sistema Terra – Remote Sensing and the Earth*, Vol. 8, 1999.

"Torsdagen den 16 december 1999", in Gert Nilson (ed.): *En text för var dag*. Göteborg: Korpen, 1999.

"Detta är detta", *Nordisk Estetisk Tidskrift*, Vol. 19, 2000, pp. 39–54.

"*Mappa Mundi Universalis*", exhibition catalogue Uppsala: Eventa 5, 2000. Author: Gunnael Jensson alias Ole Michael Jensen and Gunnar Olsson.

"The Uppsala Bull," in the exhibition catalogue *Per aspera aspera ad astra*. Uppsala:Eventa 5, 2000.

"Mappa Mundi Universalis: Kartografik Aklin Gücü Üzerine bir Yorum", *Mimarlik Kültüri Dergesi*, Vol. XXI, 2001, p. 156.

"Eggs in Bird", in Michael J. Dear and Steven Flusty (eds): *The Spaces of Postmodernity: Readings in Human Geography*. Oxford: Blackwell, 2002.

"Orto Botanico", in Marco Picone (ed.): *Bodies and Space: Gunnar's Travels*. Palermo: Università di Palermo – Laboratorio Geografico, 2002.

"Glimpses", in Peter Gould and Forrest Pitts (eds): *Geographical Voices: Fourteen Autobiographical Essays*. Syracuse: Syracuse University Press, 2002.

"Situations of Madness and Ethics in Planning", in Organizing Team (eds): *Suburbia*. Copenhagen: Planet, 2003.

"Nordiskt dunkel", in Birgitta Östlund (ed.): *Nordiska självbilder*. Holstebro: Nordisk kulturfond, 2003.

"Landscape of Landscapes", in Kirsten Simonsen and Jan Öhman (eds): *New Trends in Nordic Human Geography*. London: Ashgate, 2003.

"Koden till Sanningens ö", *Biblis*, No. 23, September 2003.

"Charta Magna", in the exhibition catalogue Stefan Lundgren: *The Measures of Abraham*. Lund: Lunds konsthall, 2003.

"Placing the Holy", in Tom Mels (ed.): *Reanimating Places: A Geography of Rhythms*. London: Ashgate, 2004.

"Tracciando la strada", *Equilibri. Rivista per lo sviluppo sostensibile*, Vol. 9, 2005.

"A Conversation", in Kajfes, Ariana: *Occular Witness*. Stockholm: Glänta Produktion, 2005.

"Kränksmideriets ABC", in Programblad till *3xNU*, Stockholm: Stockholms Stadsteater, 2006.

"När Himmelbjerget kom till Muhammed", in Kjeld Buciek et al. (eds): *Rumslig Praxis*. Roskilde: Roskilde Universitetsforlag, 2006.

"————", *Tidskrift för litteraturvetenskap*, No. 3–4, 2006.

"The Gone is not Gone", *Progress in Human Geography*, Vol. 31, 2007.

"A Local Habitation and a Name", in Niklas Forsberg and Susanne Jansson (eds): *Making a Difference: Rethinking Humanism and the Humanities*. Stockholm: Thales, 2011, pp. 325–352.

"Regional Development: NordPlan and NordRegio", *International Encyclopedia of Human Geography*, Vol. 7. Oxford: Elsevier, 2009, pp. 469–472.

"Tanritanimaz Haritclik: Bir Haritacilik Kesif Seferinin Raporu", in Jale N. Erzen (ed.): *Sant ve Cevre*, pp. 51–59. Ankara: Yalcin Matbaacilik, 2007.

"Min Kant han är tre kanter, en trekant är min Kant", in Christian Abrahamsson, Fredrik Palm and Sverre Wide (eds): *Sociologik. Essäer om socialitet och tänkande*. Stockholm: Santérus, 2011.

"Untitled", *Environment and Planning D, Society and Space*, Vol. 26, 2008, pp. 752–757.

"Mappae Mundi Medievalis", *Pavilion*, No. 12, 2008, pp. 200–235 (in the English and Romanian Reader of Bucharest Biennale 3). Bucharest: Artphoto, 2008.

"Palimpsest", *Cartographica*, Vol. 44, 2009, pp. 101–109.

"Mapping the Forbidden", *Fennia*, Vol. 188, 2010, pp. 3–10.

"The man Who Was One With His Body", *Progress in Human Geography*, Vol. 34, 2010.

"Huset där allting tar sin början", in Olle Matsson et al. (eds): *Värmlands nation 350 år – och lite mer därtill*. Uppsala: Värmlands nations skriftserie, 2010.

"Matthew 5:37", *Parallax*, Vol. 56, 2010.

"Mappare il pribito", *Bolletino della Società Italiana Roma*, Ser. XIII, 2010, pp. 441–451.

"Mapping the Taboo", in Stephen Daniels et al. (eds): *Envisioning Landscapes, Making Worlds: Geography and the Humanities*. London: Routledge, 2011.

"Un manual del terror institucionalizado", *Victoria de Durango*, 31 March, 2011.

"Book Review Forum of Elden's Terror and Territory" (guest edited by Christian Abrahamsson), *Dialogues in Human Geography*, Vol. 1, 2011.

"Limits to the taken-for-granted" (in a review section on "Fictional Worlds" edited by Mary Thomas and Christian Abrahamsson), *Environment and Planning D, Society and Space*, Vol. 29, 2011.

"Space and Spatiality in Theory" (together with Peter Merriman, Martin Jones, Eric Sheppard, Nigel Thrift and Yi-Fu Tuan), *Dialogues in Human Geography*, Vol. 2, 2012.

Index of Names